Policy for American Agriculture

CHOICES AND CONSEQUENCES

POLICY FOR

American
Agriculture

CHOICES AND CONSEQUENCES

M. C. HALLBERG

IOWA STATE UNIVERSITY PRESS / AMES

M. C. Hallberg is a professor of agricultural economics at Pennsylvania State University.

© 1992 Iowa State University Press, Ames, Iowa 50010

♾ Printed on acid-free paper in the United States of America

First edition, 1992

Library of Congress Cataloging-in-Publication Data

Hallberg, M. C. (Milton C.)
    Policy for American agriculture: choices and consequences / M. C. Hallberg. —1st ed.
        p.   cm.
    Includes bibliographical references (p.   ) and index.
    ISBN 0-8138-1368-9 (alk. paper)
        1. Agriculture and state—United States.   2. Agriculture subsidies—United States.   3. Family farms—Government policy—United States.   I. Title
    HD1761.H353   1992
    338.1'873—dc20   91-36670

CONTENTS

PREFACE

THE MAJOR THRUST of farm policy over the course of the past sixty years has been targeted for enhancing farm incomes. By almost any measure conceivable, this support has been quite generous. As might be expected, this support has elicited polar reactions. At one pole are those who argue that farmers are an indispensable element of society producing the most basic of human needs. As a consequence, this group argues, farming and the farmers' way of life must be preserved even at the expense of significant taxpayer cost. At the other pole are those who argue that while all this is true, farmers should be, and can be, guided by an economy that is permitted to dictate adjustments in resource use unfettered by government intervention, as is the case for many other entrepreneurs.

It is not my aim to advocate adoption of either of these polar positions. In fact, as will become evident as this study unfolds, an absolutely free market solution for agriculture is not likely to be "workable." At the same time, however, many of the intervention strategies we have pursued to date may well be too expensive to sustain, may be out of touch with the needs of an agricultural economy that is part of a highly integrated national and international economy, may do too little to ensure the pursuit of socially desirable environmental goals, and may even be counterproductive in terms of meeting farm sector income distribution goals.

The trick is to find a reasonable middle ground—a compromise position consistent with the social values to which individuals at either end of the spectrum subscribe. Unfortunately this is never easy. Policy-making is anything but an exact science. Lauren Soth, nationally acclaimed former editor of the *Des Moines Register,* said in *Farm Trouble* (1957, 217–18): "In a democracy social intervention has to proceed slowly. Quick panaceas are viewed with suspicion, and rightly so. . . . So we will continue to muddle along by compromise, experimentation, and half-measures. This is democracy, and it is the best way." Soth may well have been correct. If so, one may hope that we are well informed as we muddle through. A clear vision of the nature of the agricultural sector itself is needed. The impact of policy decisions on many different people, resources, and nonagricultural interests must be assessed. The proper role that government can and should play in directing this sector must be carefully thought out.

Economics can provide much of the information needed to help guide us in this endeavor. Standard tools of economic analysis are well suited to such an inquiry. Some of the newer quantitative techniques are invaluable aids to a more thorough and precise understanding. But neither reliance on theoretical models nor measurement alone will get the job done. The tools of the economist cannot provide unambiguous answers and solutions. Economics is not well equipped to solve distributional or equity problems. Social valuations and political realities must also be reckoned with. This is a lesson that students of policy must learn and learn well.

This book is intended for all who are interested in the welfare of farmers and farm families and in the future of agriculture. It is aimed at providing the basic tools and information needed for future policy analysis and development. As such, this book should be useful to farm leaders and agricultural policymakers, to urban residents, to industry leaders, and to others interested in expanding their knowledge about the agricultural sector and about policy for this sector. Since the book has resulted directly from my efforts to teach both undergraduate and graduate students something about the subject, an important audience is the college student. In particular, it is targeted for the undergraduate student who has a reasonable background in basic economic principles and thought. The book should also serve as an introductory text in graduate-level courses on agricultural policy. For the latter, however, this book would need to be supplemented by additional readings, some of which are suggested at the end of each chapter. A graduate-level course would delve much deeper into the various subjects covered, and particularly into the more sophisticated theoretical and empirical methods used to analyze policy choices.

The book is not intended to be a history book. I believe quite strongly, however, that if one is to thoroughly comprehend agricultural policy, one must have at least a minimal familiarity with the historical background. A start in this direction is provided here by an examination of trends both in the economic evolution of the sector and in policy for the sector since 1950. To gain a fuller appreciation of agriculture's past, however, it is necessary to immerse oneself in a much broader and deeper study. The references listed at the end of the book should be of help here. I suggest the following as particularly useful historical texts for the beginning student: T. W. Schultz, *Agriculture in an Unstable Economy;* Murray Benedict, *Farm Policies of the United States: 1790–1950;* Gilbert C. Fite, *American Farmers: The New Minority;* and John T. Schlebecker, *Whereby We Thrive: A History of American Farming, 1607–1972.* The last is particularly useful as a reference to the technologies and institutions of agriculture. Finally the Appendix to this book provides a convenient chronology of legislation and executive orders affecting agriculture since 1862. This listing will be useful not only in helping the reader follow the path of legislative and executive actions affecting agriculture, but also in suggesting the issues agricultural policymakers faced over the course of nearly 130 years.

The often heard expression "policy is whatever governments choose to do or not to do" is partly true, but at the same time it is a gross oversimplification. How do governments choose and what is the basis for their choices? Who participates in the policy process? Is the process chaotic and irrational or does it follow a consistent path and a logical set of rules? Is there a place for the scientist in the policy arena and, if so, what is the scientist's role? Why do we need policy and what issues are amenable to policy treatment? Is agriculture somehow unique so that it needs special policy treatment?

Discussion of these questions gives real content to a study of policy. In the introductory section of this book, I examine such questions in some detail. More specifically, the first section will provide (1) a brief exposure to the policy process, (2) a summary of past policy for the agricultural sector, and (3) a short description of the characteristics of the agricultural sector of relevance to the formation of agricultural policy. This section, then, lays the groundwork for the subsequent sections of the book, which evaluate policy for agriculture.

The second section examines the benefits and costs of price and income support for agriculture. Assessing the benefits of farm programs is complicated by the fact that there are both primary and secondary beneficiaries. Here the primary focus is on measurable benefits to farmers and the farm input sectors. Also we must not ignore benefits to consumers; unfortunately it is more difficult to assign quantitative measures to some of the alleged consumer benefits. Assessing the costs of farm programs is equally complex. The direct budgetary outlays are fairly readily available. There are also indirect costs, however, for which account must be taken. These are expressed in such terms as resource misallocations, resource misuse, higher food costs, as well as opportunities forgone by those encouraged to stay in agriculture and related industries when they could have adjusted to other activities.

In spite of the inherent difficulties and imprecision in assessing benefits and costs, this is of vital importance. It provides the framework for realistic analysis of policy options—the primary aim of the third section. More specifically, this section treats in detail the merits and demerits of agricultural policies focused on expanding domestic demand, providing price support to farmers, controlling farm output, providing various forms of market control, restricting or expanding international trade, conserving farm resources, and stabilizing farm prices and/or quantities.

The final section is intended as a guide to policy choices for the agricultural sector in the years ahead. It reviews some of the issues encountered in conceptualizing and analyzing policy objectives and choices; it reviews agriculture's capacity to adjust to economic incentives and changes and, from this perspective, contemplates the likely need for future policy; and it outlines and discusses some additional key issues that must be decided in choosing future policy objectives for agriculture.

Many individuals have contributed to shaping my approach to the study of policy for the agricultural sector and thus to this book. The students in my classes and those I have had the pleasure to work with on thesis and dissertation research have been a continual source of inspiration as well as prime motivators for directing my own studies. The faculty of which I have had the pleasure of being a part for the past twenty-five years has been helpful and supportive in innumerable ways. In particular, I single out George Brandow, who shared much of his policy lore and experience, and the late A. P. Stemberger, who was always willing to discuss policy issues and in the process taught me a good deal more about the economics of agriculture than I imagined I needed to know. To those who critically reviewed parts of this book and made valuable suggestions (Ted Alter, Tom Brewer, Jim Herendeen, and Jim Shortle), I am most grateful. Thanks also to the book reviewers, who not only offered many helpful suggestions but also asked some very penetrating questions, all of which led to substantial improvements. I am most grateful to Nancy and Brian Hallberg for their efforts in preparing the graphics for the entire manuscript. Finally, a special thanks to my wife for her patience and understanding during the long gestation period of this book.

I

The Policy
Environment

1 Agricultural Policy and Its Environment

This book is concerned with policy for or directed to the agricultural sector. This is an extremely wide and diverse topic. The agricultural sector is itself very broad and complex, encompassing many different activities and involving many different people and occupations. It certainly includes farming. But it also includes the farm input industries, marketing and processing, transportation of farm inputs and outputs, and wholesale and retail trade in agricultural inputs and food products. A variety of issues are of concern to this sector: income of farm families, stability of farm and food prices, food safety, a food reserve, international trade, standardization and grading of farm and food products, resource conservation and environmental protection, demand expansion, farm credit, farm labor, and the vitality of rural communities, to name a few.

To limit the scope of the study somewhat, I direct primary attention to farming and farm families. While many agricultural sector activities and issues are impossible to keep apart, the main focus of this book will be on programs aimed at providing support to farm incomes rather than on programs relating to market development, performance, and control or to the food industries. Admittedly this is rather restrictive. It is dictated not so much by the biases of the author as by the fact that this is the area about which there has been the most unrest among farm people over the past six decades.

When I speak of policy, I mean *public policy* as opposed to household or firm policy. Many individuals are prone to equate *policy* with government regulation of private activity. While it is true that implementation of public policy does most frequently involve some type of government regulation, this need not always be the case, so we must have a broader interpretation of policy.

This chapter begins by providing an operational definition of public

3

policy that will guide our study. It also briefly examines three questions: Why is agricultural policy sought or needed, what are the important determinants of agricultural policy, and who participates in the agricultural policy-making process? The chapter ends with a very brief sketch of the more important policy thrusts for American agriculture over the past two centuries.

Policy Defined

It is useful at the outset to have in mind an operational definition of *public policy*. Anderson (1975) offers several suggestions as to what this concept encompasses. First of all, public policy is purposive or goal directed rather than a result of chance. Second, public policy is implemented by a governmental unit on behalf of the constituents this unit serves. Hence, in a democracy (if not in any political system), public policy must sooner or later meet with the approval of the body politic. Third, public policy consists of a plan of action or of inaction—what will be done and how and when, or what will *not* be done and why—rather than being a simple statement of plans with no action specified or intended. That is, public policy is what the action agency will do or not do, not what it suggests be done or not be done. Note that a specified policy requiring no action on the part of the governmental unit does not necessarily mean that no policy is implemented. A hands-off policy (or, in the jargon of economics, a policy of *laissez-faire*) is a viable policy alternative. This emphasizes that policy need not necessarily involve regulation of private activity. Finally it should be clear that several possible options are available to the governmental unit, but this unit consciously chooses only one.

Accordingly I define *public policy* to be *a course of action consciously chosen from among the available alternatives and designed to fulfill a specified need or set of needs of the body politic; it includes specification of the means by which this course of action is to be implemented.* Policy statements are merely formal expressions of the policy adopted. *Policy decisions* are choices made by public officials that authorize or give direction and content to the policy ultimately adopted. *Policy outcomes* are the societal consequences (intended as well as unanticipated) that flow from implementation of the specified policy.

Why Policy Is Needed

As the definition establishes, public policy is formulated and implemented for the purpose of achieving some particular objective or set of objectives. Why is policy necessary? The question suggests its own answer: When left to its own devices, the social system to which the policy is directed does not work well enough to meet the specified objective or objectives. In the economic sphere, this might happen for one of three reasons: (1) The free market does not cause the wealth (or, more generally, the benefits) generated by the economy to be distributed in a way that is deemed to be equitable or fair or acceptable by society; (2) the free market, when left to its own devices, fails to allocate resources so as to lead to the most economically efficient output level, and consequently, policy is needed to correct for or to prevent such failure; or (3) the free market does not lead to growth and development of the economy consistent with society's needs and/or aspirations. The first and third reasons are fairly clear. The second requires elaboration.

This is not the text in which to explain the workings of the perfectly competitive model. Suffice it to say that perfect competition is consistent with economic behavior that results in the most economically efficient use of the economy's scarce resources. It must be emphasized here that perfect competition does not ensure that an optimal or equitable or desirable distribution of income will occur—only that the most efficient use of resources will occur given the existing distribution of wealth among individuals.

When market imperfections exist, aggregate private (marginal) economic benefits diverge from social (marginal) economic benefits. In such instances, competitive firms acting in their own self-interest and free from any government regulation will not produce the socially optimal (i.e., most efficient) output level. Such market imperfections include the existence of monopoly or monopsony; collusion among firms in the industry; entry barriers that result from economies of scale or other factors; lack of perfect information about prices, quantities, or product grades; and externalities.

The concept *externality* and the special problems associated with the existence of externalities are very important to agriculture. These issues have received a good deal of attention in recent years and will likely continue to confound policy choices for the industry in the years ahead. Thus it is useful to have a good grasp of the basic concept before proceeding further. A formal definition of *externality* is given in the Glossary. A simple example will best illustrate the concept. Consider milk production in southeastern Pennsylvania. Assume there is no government interference in the dairy industry and that milk producers in this region are profit maximizers who individually produce so little milk relative to the total U.S. milk supply that they cannot influence market price. The supply function for milk, which is the aggregate of producer marginal costs for producing various quantities of milk, is represented by the curve labeled FF' in Figure

1.1. The (derived) market demand function for milk is represented by the curve labeled DD'. Equilibrium price and quantity resulting from rational private behavior in the dairy sector are P_e and Q_e, respectively.

Unfortunately pollution in the form of nitrate seepage in underground water supplies and in the Susquehanna River, which ultimately finds its way into Chesapeake Bay, also accompanies milk production in southeast Pennsylvania (not to mention some air pollution as well). This pollution is for the most part not of major concern to individual producers seeking to maximize profits from their dairy operation, but it is a major issue for the authorities responsible for providing safe drinking water and for minimizing Chesapeake Bay pollution. The costs society must bear for cleaning up this pollution are not negligible and must somehow be added to the costs of producing milk.

Combined costs of production and pollution abatement are reflected in the "social" marginal cost function represented by the curve labeled SS' in Figure 1.1. Now we see that the "socially optimal" equilibrium price and quantity are given by P_e' and Q_e', respectively. Since milk producers acting out of self-interest have no incentive to produce at the socially optimal output level, some type of regulation is most likely necessary to force them to do so. This regulation might take the form of production quotas designed to restrict aggregate milk production to Q_e'. Alternatively it might take the form of taxes on milk output sufficient to cover the costs of cleaning up the pollution and hence causing milk producers' aggregate supply curve to shift up to the position now occupied by SS'. I return to an analysis of these options in Chapter 10. The point to keep in mind here is that externalities such as this constitute an instance of *market failure* and generally necessitate some government action to correct or ameliorate.

The Basis for Policy Decisions

Public policy, then, is to be viewed as a conscious choice from among alternatives currently available. But how is the choice made? What criteria are used to assist in the choice?

The need to develop the heartland of America so as to foster the nation's desire for sustained growth led directly to much *developmental policy* that had a marked effect on the structure of and support for American agriculture. The Great Depression, which led to reduced demand for food and fiber, to low farm prices, and to low farm incomes at a time when a full 25 percent of the population lived on farms and nearly 44 percent of the people were rural residents, prompted the *compensation policies* that evolved out of the 1930s. Federal budget pressures in recent years are alleged by some to have been the cause of efforts to seek a

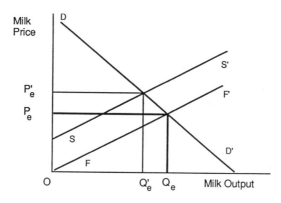

Figure 1.1. Equilibrium prices and output in milk production, with externalities

reduction in federal expenditures on the farm programs, particularly on the dairy program (see Petit 1985). High interest rates, leading not only to high producer costs but also to an overvalued U.S. dollar, can have a marked effect on U.S. agricultural exports. Clearly, economic conditions can influence policy choices.

Societal values, beliefs, and attitudes concerning what government should do and how it should operate also condition policy choices. Widespread adoption of agrarian fundamentalism as a moral philosophy has had a decided impact on agricultural policy in America. This philosophy, generally attributed to Thomas Jefferson, stresses the value of an atomistic, family-oriented, independent-ownership structure for agriculture with limited government interference. One of the most remarkable attributes of modern society is the tenacity with which its people (both rural and nonrural) have valued and fostered agrarian fundamentalism—even in the face of a farm population that has by now dwindled to less than 2.5 percent of the total population in America. Through the years, this basic value has been given expression largely through pleas to maintain the "family farm" even though this concept is so difficult to define unambiguously that most people have given up trying.

Brewster (1961) identifies four creeds or basic values held by both rural and urban people in America that underlie this agrarian philosophy. These creeds are the *work ethic*, the *democratic creed*, the *enterprise creed*, and the *creed of self-integrity*. The fact that these values are still quite widely held today even though conflicts among them have intensified appears to explain the persistence of the agrarian philosophy.

The first of these creeds encompasses the general notion that the more proficient person is worthy of respect, higher status, and emulation. The work ethic is to be sanctioned in general because it leads to a hardworking labor force, which makes greater contributions to economic growth and progress than would be the case if this ethic did not exist. Also, according to the work ethic, society owes an individual according to his or her contributions and should ensure that each individual has access to the means with which to develop his or her creative and earning potential. All this leads to support for perfect competition as opposed to other forms of economic organization where factors of production (including labor and management) are paid according to the value of their marginal product.

The *democratic creed* encompasses the idea that all persons are of equal worth and dignity and that no one individual or group (certainly not government) is wise enough to have dictatorial power over others. At the very least, all should have an equal voice in policy decisions.

The *enterprise creed* holds that individuals are responsible for their own economic security and should be free to manage their own affairs without government interference or restrictions.

According to the *creed of self-integrity,* innovators in the development of new knowledge and methods are to be prized by the community and must not be discouraged from dissenting from the group norm to chart new paths.

Collectively these beliefs lead to support for market mechanisms and opposition to government interference, the welfare state, and big business. There are, though, conflicts among these basic values and beliefs. Hard work, innovativeness, and adoption of new methods and new technology lead to surplus production, which in turn leads to low producer returns. The result is usually a call for government assistance to maintain farm incomes at levels consistent with the work expended, but this is undesirable if one strictly adheres to the enterprise creed. Pollution from animal production and the unrestricted use of chemicals in crop production leads to calls for regulation, but this also is at odds with the enterprise creed. Furthermore surplus production leads to the migration of people out of agriculture, which is at odds with a basic tenet of agrarian fundamentalism.

Such conflicts have intensified since World War II with the result that policy debates today are much more acrimonious and prolonged and policy development is much more difficult. Nevertheless this system of beliefs has had a significant impact on the type and amount of federal support for agriculture via support for a family-oriented farm structure as opposed to a corporate agricultural structure, protection from foreign competition, income parity for farmers, support for the right to bargain collectively with those having superior market power, and subsidized credit.

Participants in the Policy Process

Official Participants. The official participants in the policy-making process at the national level are the administration and its agencies, Congress, and the courts. Today the Executive Office is expected to provide leadership in developing policy proposals for agriculture. Congressional leadership is currently so fragmented by the committee system and by lack of strong party leadership that initializing proposals in the House or Senate is nearly impossible. Thus today we typically see new farm legislation being proposed by the administration and sent to Congress where the House and Senate agricultural and appropriations committees rework it into a bill that can pass both houses. Key members of Congress may participate informally in shaping the administration's proposal, but the secretary of agriculture, as the president's chief agricultural advisor, can generally be expected to play the more important role. The administrative agency of relevance here is the U.S. Department of Agriculture (USDA), which serves as an advisor to the secretary of agriculture and through him to the president. It plays a vital role in this capacity, primarily in analyzing the consequences of policy alternatives.

The path a new farm bill takes from its inception to final presidential signature is long and arduous.[1] The bill is formally introduced in the respective chambers of Congress by a legislator of the president's party. Once it has been introduced, it is numbered by order of introduction, referred by the chamber's parliamentarian to an appropriate committee (or committees), and then assigned to a subcommittee of that full committee. In the agricultural sphere, the appropriate committees are the House Committee on Agriculture and the Senate Committee on Agriculture, Nutrition, and Forestry.

The subcommittee holds hearings and invites testimony from government officials, outside experts, and interest groups. The purpose of these hearings is to obtain expert opinions, to test public opinion, to build support for the bill, and/or to delay action on the bill. Sometimes these hearings serve primarily as forums for establishing positions.

Once these hearings are completed, and if the subcommittee intends to see the bill through, it will meet in a closed session for "markup"—that is, to redraft the bill in appropriate legislative language and agree to any amendments and/or substantive changes in the original provisions of the bill. After markup, the bill is voted upon by the subcommittee and, if approved, moves on to the full committee. The full committee can table the bill and thus kill it or "order it reported out" to the full House or Senate with either a favorable or an unfavorable recommendation. A bill will generally be approved by the full committee; otherwise, it is not likely to be sent up by the subcommittee or indeed even marked up. *Thus the first point at which a bill can be killed is by the subcommittee failing to mark it up.*

When the bill is sent to the House or Senate floor, it is accompanied by a report describing its purpose and scope, amendments that the committee added, existing law that the proposed legislation changes, estimates of the expected costs to the government of the proposed legislative changes that will be incurred, and perhaps the views of the administrative departments and agencies on the legislation. The language of this report can be very important in building or failing to build support for the bill.

In the House, consideration of the bill is governed by a "rule," that is a set of resolutions passed by the House Rules Committee governing the time limit for general debate and whether or not amendments from the floor are possible at all or limited to specific items filed by a specific deadline. The rule must be accepted by vote of the full House before actual debate on the bill begins. *Here, then, is a second point at which a bill can be killed—by defeating the Rules Committee resolutions before the bill itself can come up for debate on the floor.* In the Senate, procedures are less formal because the Senate is a much smaller body (100 senators versus 435 representatives). Thus any senator can bring up a bill in that chamber at almost any time with few restrictions. In practice, though, scheduling of legislation is usually done in both chambers by the majority leader in consultation with the minority leader.

After floor debate and amendments made by members not on the committee that handled the bill are added, the bill is voted upon. If approved, it is transmitted to the other chamber. Once both chambers have approved the bill with their own amendments, it will be sent to a joint chamber conference committee to reconcile any differences in the versions the two chambers passed. The conference committee is made up of ranking members of the House and Senate committees that sent the bill to the floor. Their job is to re-draft the bill in such a way that both chambers will find it acceptable. This is a crucial stage in policy-making because bills passed by the two chambers often contain very different provisions on policy, programs, and spending. Thus the members from each chamber will find occasion to consult with the leaders of their chamber to make sure that the compromises worked out will be accepted by their respective chambers. The operation of conference committees can be very time consuming and take months to complete. Most bills, though, are passed by both the House and the Senate once they get through the conference committee.

After the bill has been to conference and subsequently passed by both chambers, it is sent to the president for signature. If the president does not approve of the amended version of the original bill, he or she may veto it.

A presidential veto can be overridden by a two-thirds majority in each chamber.

The actual process of policy formation at this level is thus slow, tedious, and political, but by no means irrational. It is not necessarily the most efficient way to pass legislation, but by design it is a system that preserves the separation of powers as required by the Constitution, and it permits the widest possible input. The rules of the game are somewhat unclear and change as procedures and membership of the respective chambers change. Petit (1985) describes the game as for the most part "political bargaining" in which the actors include interest groups as well as legislators. Political bargaining is necessary for two reasons. On the one hand, not everyone agrees on the need for, direction of, and appropriate implementation of policy, so compromises must be made and deals on voting on different policy or legislative packages must be consummated. On the other hand, legislators must satisfy their constituents in order to get reelected, and constituents themselves have different priorities.

Downs (1957) and Peltzman (1976) have attempted to formalize political activity in a democracy based on the premise that the politician's objective is to maximize votes and thus gain election or get reelected. Downs emphasizes the role of imperfect knowledge in this process and likens a politician's behavior to that of an oligopolist in persuading people to buy more of a product via brand preferences, product differentiation, and collusion. Peltzman's approach is similar except that he emphasizes the dynamics of different voting groups along with the role of imperfect information and transaction costs.

Krueger (1974) and Buchanan, Tollison, and Tullock (1980) emphasize the importance of "rent seeking" as a motivating factor. The idea here is that resources are spent for the sole purpose of gaining a right to the economic rent or economic payoff generated by the policies or legislative packages under consideration. The action of lobby groups then is seen as rent seeking.

More recently, various authors have attempted to develop theories of altruistic behavior with which to explain political behavior and policy choices (see, e.g., Collard 1978). Altruism has not, however, been as widely accepted as have some of the other arguments or theories used to explain political decisions.

Unofficial Participants. The unofficial participants in the policy-making process consist of various interest groups, the most important of which are farm organizations. Other unofficial participants include the consumer groups, the environmentalists, the conservationists, and the agribusiness groups. The consumer groups are of course interested in low and stable food prices, stable food supplies, nutritious and safe food, food aid to the needy, and appropriate information about food via labeling, grading, and standardization. There was a spate of political activity by consumer groups in the 1970s which was probably responsible for the fact that in 1977 food got first billing in the title of a farm bill and food aid programs were incorporated into the same bill that detailed the farm programs. However, this is about as far as it went. Consumer groups have yet to become well enough organized to speak effectively on price and income policy for agriculture.

One might expect consumers to vigorously oppose agricultural policy that is a boon to farmers but costs consumers millions of dollars in the form of higher food prices. As I indicated before, agrarian fundamentalism is fairly strongly adhered to even among consumers, so they might just be willing to pay the price to further this basic philosophy. Clearly this is the case in Western Europe and one of the reasons why the European Economic Community (EEC) protects its farmers to a higher degree than does the United States.

Some students of policy, however, see another reason. Consider U.S. sugar policy. Using estimates provided by Babcock and Schmitz (1986), this policy cost each and every person in the United States $11.50 in 1983. This is hardly sufficient to cause an individual consumer much alarm. It would clearly cost each consumer considerably more than $11.50 per year to organize or participate in an effective campaign opposing the legislation.

Each of the 12,600 U.S. and Puerto Rican sugar producers in 1983, on the other hand, gained an estimated $107,143 from this policy! Sugar producers can afford to spend considerable time and money ensuring the continuation of sugar policy. Sugar consumers may not be indifferent to U.S. sugar policy, but they have little economic incentive to oppose. This by no means justifies current sugar policy. It does, though, point out one of the economic realities of political life (see Downs 1957). In cases where the benefits of policy are concentrated and the costs widely dispersed, we can expect unequal pressure to be exerted on policymakers by the opposing groups.[2] It is not at all clear that consumers should be chastised for their apathy. Nor does it seem appropriate that consumers collectively should be made to bear the cost of such policies without being heard.

Environmentalists and conservationists have long been active in seeking agricultural legislation that would serve to curb erosion. Congress has generally been responsive when there are surpluses of agricultural commodities but less so when supplies are rather more tight (a subject to which I return in Chapter 3). In recent years, additional pressures have been exerted by these groups to seek legislation dealing with both conservation and pollution.

The large grain dealers and input suppliers are interested in maintaining large volumes. They support demand expansion programs, but otherwise seek a deregulated agriculture. Presumably they feel they stand a better chance to be competitive in the international market if American agriculture is market oriented. There are conflicts here though. The international grain traders who own significant amounts of grain storage space are more than a little interested in government programs that foster levels of production sufficient to fill these storage facilities. In general, the profit margins on grain storage operations can be expected to exceed profit margins on grain sold overseas.

By far the most influential unofficial participants in farm policy formation are the farm organizations. The five principal national farm organizations are the Grange, or the National Order of Patrons of Husbandry, founded by the Masons in 1867; the National Farmers' Union, founded in 1902; the American Farm Bureau Federation, organized in 1920 and consisting of a federation of local farm bureaus set up by county extension agents; the National Farmers Organization, founded in 1960 as a militant organization; and the American Agricultural Movement, organized in 1977.

The Grange was organized during a period when farm prices were low (following the Civil War) and nearly everyone connected with farming attributed the high farm input prices and low farm product prices to the monopolistic corporations. Farmers were also antagonistic toward the railroads who engaged in discriminatory pricing practices, and bankers who charged high interest rates and foreclosed on farmers too quickly. The Grange's most important early activity was seeking public regulation of rail rates via the so-called Granger Laws. The Grange was also active in seeking the tripartite farm legislation of the Lincoln administration,[3] in seeking to raise the USDA to cabinet status, and in pushing for free delivery of mail to rural households. Today the Grange reflects the conservative values of older, small-scale farmers and is more active in the East and West than in the heartland of American agriculture. The Grange has traditionally supported protectionist legislation for agriculture, marketing orders, and two-price plans which call for higher prices on the domestic market and lower prices on the international market.

The National Farmers' Union originated as an organization which emphasized business approaches to solving farmers' problems in buying inputs, selling produce, and obtaining credit. It originated in Texas and is most active today in the Wheat Belt area of the Great Plains. The National Farmers' Union is probably the most liberal of the national farm organizations and is the strongest supporter of production controls, parity returns for farmers, subsidized food consumption programs, subsidized exports, and other government measures with which to raise farm incomes (see Glossary for a definition of *parity*).

The American Farm Bureau Federation leans heavily toward individual freedom and equity, limited government interference, and free-market solutions to farm problems. It is the largest and strongest of the national farm organizations, being particularly strong in the Corn Belt and the Old South. It tends to align itself with big business, whereas the Grange tends to align itself with organized labor. The federation strongly opposed the supply control programs of the Kennedy and Johnson administrations. It also generally opposes labor legislation and welfare programs.

The National Farmers Organization was founded on a platform of bargaining with first-handlers (buyers of farm produce) for higher farm prices, with the threat of withholding actions if necessary. As early as 1959, it called for withholding actions of limited scope and duration. It orchestrated an all-out livestock withholding action in the fall of 1962 and attempted to bargain collectively for milk price increases using the withholding action in 1963. These actions were not very successful, however. Withholding actions involving products of limited storability have never been very effective.

The American Agricultural Movement was founded in an era when farm prices were declining and farm debt was rising (1977–78). Its membership consisted largely of young farmers from the Great Plains, the Corn Belt, and the Old South. The members of this organization were disenchanted with existing policy approaches and organized a tractor caravan with which to descend upon downtown Washington, D.C., and into the Washington, D.C., offices of U.S. legislators. In the short run, American Agricultural Movement members advocated a general farmers' strike. In the longer term, they wanted legislation that would peg farm prices at no less than 100 percent of parity, marketing quotas tied to the land, no government storage except for strategic reserves, export prices identical to domestic prices, and prices of competing imports pegged at 110 percent of parity. The American Agricultural Movement did have some success in raising support levels for wheat, corn, and cotton in 1978. It has not, however, had a sustained impact on agricultural policy.

These national organizations participate actively in policy debates. They all have a permanent lobby in Washington, D.C. From time to time they have attempted to develop a coalition which would speak with one voice. Their biggest problem, however, is that their membership, while large, is quite heterogeneous because of the wide array of commodities represented and regional interests reflected. Thus a consensus on general agricultural policy is difficult to achieve.

In addition to the national farm organizations, there are organizations for almost all of the individual commodity groups. These can be somewhat more effective because their agendas are much more restricted in scope. Three of these groups that should be singled out for special attention are the National Cattlemen's Association, the National Milk Producers' Federation, and the National Association of Wheat Growers. In each case, these groups are by and large quite effective for their members. The cattle raisers go to Washington and implore Congress to maintain beef quotas so as to reduce beef imports, and then they go back home! The National Milk Producers' Federation concentrates on legislation designed to enable them to use marketing orders more effectively so as to restrict the flow of milk from one market area to another, to maintain milk price supports at attractive levels, and to keep dairy product imports low. The National Association of Wheat Growers is interested in high wheat price supports and wheat export subsidies. Each of these groups has been very effective because they are not required to prioritize their demands among more than one commodity group. Cattle raisers want beef import controls and nothing else. Dairy farmers want support for the dairy programs and nothing else. Wheat growers want high price supports and assurance of a market for their wheat and little else. In contrast, the American Farm Bureau Federation goes to Washington with a full agenda, with such concerns as free-market prices for feed grains, cotton, rice, wheat, wool, and milk; no production controls but, if absolutely necessary, voluntary acreage set-asides; export subsidies; and land conservation subsidies. Further the Federation must present this slate on behalf of a membership that is divided on the relative importance of individual items on the agenda and on behalf of farmers that may not even be in complete agreement that all items should be on the agenda (see, e.g., Petit 1985).

Policy for American Agriculture

To attempt a rigorous delineation of different types of policy would take us far afield and would have little utility. One can, however, identify some specific forms that policy for American agriculture has taken since colonial days.

Development Policy. Development policy focuses on the enhancement of one or more productive resources. This might take the form of developing entirely new production regions so that more resources are available for production, or it might take the form of seeking to change production processes so as to increase the aggregate amount of output possible with a given bundle of resources. In sum, development policy is aimed at shifting the supply schedule to the right through technological or managerial processes or by reducing the cost of inputs.

Most of the early legislation affecting American agriculture should rightly be considered development policy—policy directed at expanding food production for a growing nation and for developing the country so it could sustain growth. But this development was to be achieved in a very specific way as far as farming and farm people were concerned. Agrarian fundamentalism, expressed most eloquently by Thomas Jefferson but espoused by others of his day as well, shaped this development process and had a very profound impact on the nature of farming that evolved. Small, family-sized farms, for example, were deemed to be the most appropriate units with which to effect an egalitarian, agrarian society. Various acts of the Continental Congress, the Homestead Act of 1862 and related legislation, and the Land Reclamation Act of 1902 specified how public lands and publicly developed resources (e.g., irrigation water in the case of the Land Reclamation Act of 1902) were to be distributed. Collectively these acts represented formalization of an agrarian policy for America.

Transportation policy in America over the years also facilitated the development of an agrarian society consisting of family-sized farms by providing farmers low-cost access to both input and output markets. The development of some 3,000 miles of waterway canals from 1815 to 1840 was accomplished with state and federal subsidies. The most successful of these was the Erie Canal, running from the Hudson River to Lake Erie. Federal support to railroad companies via generous grants of public lands and government loans between the years 1840 and 1860 resulted in the construction of over 30,000 miles of rail lines in this country. Many historians claim this is one of the most important factors contributing to the rapid and effective development of the heartland and western regions of the United States. The 50-mile long Lancaster Turnpike (running from Philadelphia to Lancaster, Pennsylvania) was completed in 1794, and the 600-mile long National Turnpike (running from Cumberland, Maryland, to Vandalia, Illinois) was completed in 1850, both with the aid of federal support. This and other road-building activity also contributed to the expansion of this nation and to support for the chosen agrarian structure. One might make the case that the 41,000-mile Interstate and Defense Highway System, authorized in 1956, also helped to sustain the rural infrastructure built up to that time.

Public support for agricultural education, research, and extension also contributed directly to the maintenance of an agrarian society founded on small, family-sized farms. The Morrill Land-Grant College Act of 1862 made low-cost education available at the college level. It was intended to foster the application of science to agriculture and to enable the sons and daughters of farmers to further their education in the agricultural sciences. The Morrill Act gave to each state public lands (30,000 acres for each senator and representative the state had in Congress at the time) with which to endow an agricultural or mechanical arts college to be established in the state.

Application of science to agriculture could not have proceeded, however, without provisions for agricultural research. The Hatch Experiment Station Act of 1887 was subsequently passed to provide for an annual grant (initially $15,000) out of public funds to each land-grant college for research in the agricultural sciences. A system of state experiment stations was established to administer these funds and to conduct the requisite research.

To extend the results of this research and knowledge directly to farmers, the Smith-Lever Act of 1914 established the Cooperative Extension Service. This act initially provided each land-grant college a federal allotment of $10,000 out of public funds. To provide federal support for the teaching of vocational agriculture in high schools across the country, the Smith-Hughes Vocational Education Act was passed in 1917.

Additional federal support dealt with the specific needs of farm families in an agrarian society. The Rural Electrification Administration Act of 1936, for example, was passed to provide long-term, low-interest loans to local electric cooperatives which extended lines to farms and farming communities. Rural Free Delivery, begun in 1896, did much to improve the lot of farm families. Boorstein (1973, 133) argues this was a most important communication revolution in American agricultural history for "it lifted the farmer out of the narrow community of those he saw and knew, and put him in continual touch with a larger world of persons and events and things read about but unheard and unseen."

Regulatory Policy. Regulatory policy focuses on making an economic system operate more efficiently, on coping with externalities, and/or on preventing abuses of market power. Such policy might take the form of grades and standards regulation, regulation relating to the provision of market information, prohibitions against specified unfair trade practices, regulation permitting farmers to organize so as to offset the superior market power of firms with which farmers must deal, or regulation against excessive pollution.

The Sapiro movement pointed out the advantages of collective action on the part of farmers. Agricultural cooperatives were found to be an effective means by which farmers could market their products when faced with a limited number of buyers who had much greater market power than did an individual farmer. This movement also emphasized the importance of clarifying the legal status of agricultural cooperatives in light of the antitrust laws existing at the time. The Capper-Volstead Act of 1922 gave farmers the legal right to form farmer cooperatives without fear of prosecution under the antitrust provisions of the earlier Sherman and Clayton Antitrust Acts, thus strengthening the economic power of farmer organizations.

Credit Policy. Credit policy focuses on providing farmers access to sources of credit when the private sector deems agriculture to be too risky to provide credit at reasonable rates. The credit system available to farmers prior to World War I was quite unsatisfactory. Again farmers were at the mercy of an industry with superior market power and not very excited about lending money to entrepreneurs engaged in such a risky business. Congress responded on behalf of farmers with the Federal Farm Loan Act of 1916, which created twelve cooperative Federal Land Bank Associations, whose purpose is to offer credit to farmers and farm businesses on a long-term loan basis (five to forty years). The Farm Credit Act of 1933 created twelve Federal Intermediate Credit Banks, which provide loan funds to a system of countrywide local Production Credit Associations, which in turn are authorized to provide short-term credit (for up to seven years) directly to farmers and farm-related businesses. The Farm Credit Act also created a system of thirteen Banks for Cooperatives, whose purpose is to make loans to agricultural cooperatives.

Conservation and Environmental Policy. Conservation and environmental policy focuses on preserving natural resources for current and future uses and on ensuring that the environment is kept clean and available for the purposes society deems desirable. Over the years, conservation of natural resources (soil and water) has not been a major focus of agricultural policy. As we shall see in the next chapter, conservation policy has from time to time been incorporated into farm policy but has, by and large, been implemented when agricultural surpluses mount and then relegated to a lesser status when supplies become tighter. Environmental policy has certainly come to the fore in recent years but has not as yet been fully integrated into agricultural policy.

Stabilization Policy. Stabilization policy is aimed at

minimizing or reducing fluctuations in prices, quantities, or both. Almost all agricultural price and income policy legislation over the last several decades has had as one of its explicit objectives providing consumers with an adequate and uninterrupted supply of food or at least avoiding prolonged periods of food shortages. In general, production has been sufficient to meet U.S. consumers' needs, or stocks have been adequate to cover production shortfalls so that food shortages have been avoided. Whether or not this has been due to the legislation implemented is debatable.

Most agricultural legislation has also had as an implicit or explicit goal, the stabilization of farm commodity prices. Such a goal is usually fostered on the basis that reducing price uncertainty leads to more rational and less costly production decisions and thus to a more efficient and stable production sector. As we shall see in Chapter 11, however, American farm policy has not been very successful at stabilizing farm prices. One reason is that the type of policy instruments we have used over the years are not well suited for this purpose. A second reason is that the U.S. agricultural sector is not isolated from the world. Hence when conditions in other parts of the world change (as a result of droughts, wars, or political upheavals), these changes impact farm prices in the United States via our trade relations with other nations.

Compensation Policy. Compensation policy focuses on redistributing income to effect society's equity goals. This can be done in a variety of ways: constraining supply for an industry such as agriculture where demand is generally considered to be inelastic; enhancing demand through a rightward shift of the demand curve; supporting product prices at a level higher than would otherwise be the case with government purchases or subsidized consumption and exports; subsidizing the use of such inputs as fertilizers, machinery, or credit; or making direct income payments out of tax revenues.

Compensation policy in agriculture was ushered in with the Great Depression. The first piece of legislation in this area was the Agricultural Marketing Act of 1929, but the real beginnings of compensation policy for agriculture can more properly be assigned to the Agricultural Adjustment Act of 1933. Legislation designed to effect the farm compensation policies of the nation was dominant in the 1930s and is dominant still because prices and farm incomes were and are the subject about which there was and is the most unrest. Consequently price and income policy for agriculture will get most of our attention in subsequent chapters of this book.

The Need for Policy Analysis

It is generally true that policy decisions have a direct bearing on the well-being of every member of society. Thus it can almost always be said that each and every one of us has a personal stake in the outcome of policy decisions. To the extent that policy decisions can have an impact on future generations, each of us also has an obligation to preserve and improve the system of which we are now a part and which we will pass on to future generations.

In a democracy, all citizens have an opportunity to manifest their ideas and preferences concerning policy decisions at the voting place. Some of us may even have an opportunity to participate actively in the policy-making process. We might become responsible for carrying out policy, or we might become advocates for a particular policy position on behalf of a larger group or simply because of a personal commitment.

To perform effectively in any of these roles, it is obviously of importance to become as well informed as possible about the policy choices and policy outcomes. We most certainly must be aware of the content of current policy, how it was formed, and for what purpose it was intended. To decide whether to react positively or negatively to given policy choices (via our voting behavior or in our role as policy participants or advocates), we must also be informed about the societal consequences of public policy. For example, which farm families will benefit from the proposed farm policy and which will not? Will farm product export enhancement programs designed to make U.S. agricultural products more competitive in international trade have serious effects on the conservation of soil and water resources? Will marketing quotas help or hinder the resource adjustment responses of the nation's dairy farmers? Will production controls in feed grains have a debilitating effect on rural communities, which in turn reduces feed grain farmers' ability to survive? What are the social costs associated with specific types of supply control measures? Are price supports the most effective means of achieving our income distribution objectives?

At the professional-scientific level, policy analysis is the dominant concern. Here scientists conduct rigorous research into the causes and consequences of public policy through the use of accepted standards of scientific inquiry. They will attempt to answer some fundamental questions: Why is the policy in question necessary? How is it formed? Who will be affected by it and to what extent? What are the important variables in the response functions needed to evaluate policy outcomes? What are the numerical values of the parameters of these response functions? The analyst proceeds by developing and testing general propositions about the causes and consequences of public policy and accumulates reliable research findings of general significance.

The professional-scientific role just outlined would seem to suggest that scientists are all business and must take a neutral position on policy issues as they go about their work. Some contend that the scientist must operate in this fashion, free from value judgments about what constitutes good or useful or desirable policy or about the costs and benefits of policy choices, in order to be an objective analyst of policy issues and consequences.

Quite frequently, scientists will be cast in the role of advising policy-makers on the basis of their analyses. This advice might take the form of assessments of the costs and benefits of policy choices or of suggesting alternative policy options to be chosen. Maintaining neutrality in this role is not so easy—some would argue it is impossible!

Certainly pure science is concerned only with establishing the facts and the causal relations between these facts. It is difficult, however, to prevent the intrusion of value premises in carrying out scientific work. Indeed it can be argued that the scientist (the chemist, the biologist, the plant scientist, as well as the social scientist) inserts value judgments by the very act of choosing a subject to research. The economic scientist makes a value judgment in asserting that a *Pareto Optimal* state is to be preferred to a *non–Pareto Optimal* state (see Glossary, s.v. *Pareto optimum*). The social scientist also makes a value judgment when choosing a policy option only on the basis of quantitatively measurable net benefits. Weights are rarely assigned to the nonmeasurable benefits!

Faced with the task of advising on policy, value premises *must* be chosen and inserted. Clearly this is extrascientific in that it does not emerge from the scientific inquiry itself. Nevertheless I agree with Myrdal (1958, 1):

> There is no way of studying social reality other than from the viewpoint of human ideals. A "disinterested social science" has never existed and, for logical reasons, cannot exist. The value connotation of our main concepts represents our interest in a matter, gives direction to our thoughts and significance to our inferences. It poses questions without which there are no answers.

Extreme care must be exercised, though, if objective research is to be done and if objective advice is to be given. Value-loaded concepts (such as the *family farm, small-scale agriculture, regenerative agriculture, environmental protection,* or *animal welfare*) should be identified as such, and value premises underlying the analysis must be highlighted. Myrdal (p. 2) goes on to warn:

> A value premise must not be chosen arbitrarily; it must be relevant and significant in relation to the society in which we live. It can, therefore, only be ascertained by an examination of what people actually desire.

People's desires are to some extent regularly founded on erroneous beliefs about facts and causal relations. To that extent a corrected value premise—corresponding to what people would desire if their knowledge about the world around them were more perfect—can be construed and has relevance.

By no econometric trick, however, can a value premise be generated by pure reasoning or inferred from facts other than people's actual valuations.

Notes

1. The description of the legislative process outlined here follows closely that given in World Perspectives (1988).

2. This has long been recognized by policy analysts. As early as 1927 Pareto (1971, 379) observed the following:

> In order to explain how those who champion protection make themselves heard so easily, it is necessary to add a consideration which applies to social movements generally. The intensity of the work of an individual is not proportionate to the benefits which that work may bring him, nor to the harm which it may enable him to avoid. If a certain measure A is the cause of the loss of one franc to each of a thousand persons, and of a thousand franc gain to one individual, the latter will expend a great deal of energy, whereas the former will resist weakly; and it is likely that, in the end, the person who is attempting to secure the thousand francs *via* A will be successful.
>
> A protectionist measure provides large benefits to a small number of people, and causes a very great number of consumers a slight loss. This circumstance makes it easier to put a protection measure into practice.

3. The Homestead Act, the Morrill Land-Grant College Act, and the act that established the U.S. Department of Agriculture. For a chronological listing and brief explanation of agricultural legislation since that time, see the Appendix.

Suggested Readings and References

Anderson, James E. 1975. *Public Policy-Making.* 2d ed. New York: Holt, Rinehart, and Winston.

Babcock, Bruce, and Andrew Schmitz. 1986. "Look for Hidden Costs." *Choices,* fourth quarter, 18–21.

Benedict, Murray. 1953. *Farm Policies of the United States: 1790–1950.* New York: Twentieth Century Fund.

Boorstein, Daniel J. 1973. *The Americans: The Democratic Experience.* New York: Vintage.

Brewster, John. 1961. "Society Values and Goals in Respect to Agriculture." In Iowa State University Center for Agricultural and Economic Adjustment. *Goals and Values in Agricultural Policy*. Ames: Iowa State University Press.

Myrdal, Gunnar. 1958. *Value in Social Policy*. Edited by Paul Streeten. New York: Harper Bros.

Petit, Michel. 1985. "Determinants of Agricultural Policies in the United States and the European Community." International Food Policy Research Institute, Research Report 51. Washington, D.C. November.

Robinson, Kenneth L. 1989. *Farm and Food Policies and Their Consequences*. Englewood Cliffs, N.J.: Prentice-Hall.

Tweeten, Luther. 1979. *Foundations of Farm Policy*. 2d ed. rev. Lincoln: University of Nebraska Press.

World Perspectives. 1988. *U.S. Agricultural Policy Guide*. World Perspectives Policy Guides. Washington, D.C.

Advanced Readings

Buchanan, James M., Robert D. Tollison, and Gordon Tullock, eds. 1980. *Toward a Theory of the Rent-Seeking Society*. College Station: Texas A & M University Press.

Collard, David. 1978. *Altruism and Economy: A Study in Non-Selfish Economics*. New York: Oxford University Press.

Downs, Anthony. 1957. "An Economic Theory of Political Action in a Democracy." *Journal of Political Economy* 65 (2): 135–50.

Krueger, Anne O. 1974. "The Political Economy of the Rent-Seeking Economy." *American Economic Review* 64 (3): 291–303.

Pareto, Vilfredo. [1927] 1971. *Manual of Political Economy*. Translated by Ann S. Schwier. Edited by Ann S. Schwier and Alfred N. Page. New York: A. M. Kelley.

Peltzman, Samuel. 1976. "Toward a More General Theory of Regulation." *Journal of Law and Economics* 19 (2): 211–40.

2 Price and Income Policy for Program Commodities

This chapter provides a summary of the various policy instruments used to effect U.S. price and income policy for the program commodities since 1933 and briefly traces the evolution of policy for each of these commodities since 1933. It also highlights the instruments currently in use for each program commodity. Although quite brief, this summary should give the reader a clear idea of the extent of price and income policy for American agriculture since 1930 and of the policy changes that have occurred over the past sixty years; the reader will thus be prepared for the analyses of various policy options presented in subsequent chapters. A full assessment of policy history for each commodity can be obtained by studying the more detailed accounts cited at the end of this chapter.

Program Commodities

The commodities for which price and income support is granted through legislative action are the feed grains (corn, grain sorghum, barley, oats, and rye), soybeans, wheat, rice, cotton, tobacco, sugar, peanuts, milk, wool and mohair, and honey. Some of these commodities get more of legislators' attention than do others. Furthermore, the policy instruments used differ considerably from commodity to commodity.

The relative importance of each of these commodities to U.S. agriculture is shown in Table 2.1. This table highlights several interesting facts. First, feed grains have increased in relative importance since 1940 at the expense of wheat, cotton, and tobacco. Second, the traditionally southern crops—cotton, tobacco, sugar, and peanuts—have stayed in the legislative picture since the 1930s in spite of an overall decline in their relative importance. Third, some commodities have received governmental support over the years in spite of the fact that they are of only minor significance in terms of total cash receipts or in terms of strategic importance—e.g., wool and honey. Finally, nearly 60 percent of the cash receipts from farm marketings are now derived from commodities for which no price or income support legislation exists.

A final point to keep in mind while reviewing past policy for the agricultural sector is that it has been directed to commodities rather than to people producing these commodities. This is an issue to which I will return in the final chapter of this book.

Policy Instruments

Price Support. Price support measures are intended to provide a floor below which market prices are not permitted to fall when reduced demand or increased supply or both would otherwise result in prices that are deemed to be too low. Nonrecourse loans and government purchases of farm commodities are the two instruments that have typically been used in tandem to support farm prices under these circumstances.

A nonrecourse loan is a loan from the Commodity Credit Corporation (CCC)[1] to which producers of specified commodities are entitled under

Table 2.1. Percentage of total cash receipts from farm marketings in the United States derived from program commodities, 1940–90

Commodity	1940	1950	1960	1970	1980	1985	1988	1989	1990
					(percentage)				
Feed grains	7.2	7.5	8.7	10.1	13.1	15.6	9.5	10.5	11.4
Soybeans	0.8	2.6	3.5	6.4	10.2	7.7	8.1	6.8	6.6
Wheat	5.3	7.2	6.9	3.6	6.3	5.5	4.2	4.5	4.0
Rice	0.4	0.7	0.7	0.9	1.1	0.7	0.7	0.5	0.6
Cotton	7.6	8.6	6.9	2.5	3.2	2.6	3.0	3.0	3.0
Tobacco	2.9	3.7	3.4	2.8	1.9	1.9	1.3	1.5	1.6
Sugar	0.9	0.7	0.7	1.1	1.5	1.0	1.3	1.2	1.2
Peanuts	na	0.8	0.5	0.8	0.4	0.7	0.7	0.7	0.7
Dairy products	18.1	13.1	13.9	12.9	11.7	12.5	11.7	12.2	12.1
Wool and Mohair	1.3	0.5	0.3	0.1	0.1	a	a	a	a
Honey	0.1	0.1	0.1	0.1	0.1	0.1	0.1	0.1	0.1
Total	44.5	45.5	45.6	41.3	49.6	48.3	40.5	41.0	41.3

Source: U.S. Department of Agriculture, *Agricultural Statistics,* various annual issues; and U.S. Department of Agriculture, 1989.

[a]Less than 0.1 percent.

na—not available.

whatever conditions of eligibility might be specified. The loan rate is specified by the extant legislation or by the secretary of agriculture according to rules given by the extant legislation. If a farmer can sell his crop for a price higher than the loan rate before the expiration date of the loan, he may, if he chooses, repay the loan plus the interest charges on the loan up to that moment and retake title to his commodity. If he has no such reason for redeeming the loan, he may simply let the loan mature (usually in nine months) and forfeit his crop to the CCC with no penalty and no other charges. The CCC can take no other recourse against the farmer if he defaults on his loan—the CCC simply takes title to the loan collateral.

The loan rate thus establishes a floor for the market price. If this price floor is above the equilibrium price, the market will not clear at the floor price and "surplus" production will result.[2] In order to protect this floor price then, the government must be willing to purchase the surplus production if there is no other means of disposing of it or of preventing it from happening. In the late 1950s and early 1960s, *direct government purchase of surplus commodities* was the principal means used to protect the farm price support system. Today such purchases are used to support farm prices only in the case of milk, honey, and sugar.

Direct Farm Income Support. Nonrecourse loans and government purchases protect farm incomes indirectly by maintaining farm prices at specified levels. Various direct income payment schemes are also used to support farm incomes. The latter have been more in vogue since the middle 1970s. One reason for this new trend relates to the fact that direct income payments are believed by many to be less market distorting than price support mechanisms and, therefore, are accompanied by lower social losses than are price supports. This issue will be discussed in more detail in Chapter 5.

Target prices and deficiency payments were first authorized under the Agriculture and Consumer Protection Act of 1973. When target prices are established and market prices are below target prices, farmers who meet the eligibility criteria (usually participation in whatever supply control programs also exist) are eligible to receive a direct income (deficiency) payment. The payment rate is equal to the difference between the target price and the higher of the average market price or the nonrecourse loan rate. The total deficiency payment is then determined as the product of (1) the payment rate, (2) the eligible acreage planted for harvest (the farm program acreage), and (3) the program yield (i.e., normal yield) established for that particular farm.

Marketing loans were first introduced by the Food Security Act of 1985.

They permit farmers who have a nonrecourse loan from the CCC to redeem that loan at less than the original loan rate if market prices subsequently fall below the loan rate. With marketing loans, the loan rate does not become a floor to the market price as would be the case with a fixed loan rate program. Hence the United States is not placed at a competitive disadvantage in relation to other countries producing the same commodity because of the loan rate policy.

The instrument of *loan deficiency payments* was also first authorized by the Food Security Act of 1985. A loan deficiency payment is in fact a subsidy in the form of a direct payment for *not* signing up for a CCC loan when eligible to do so. This type of payment can be made any time a marketing loan is in effect. The rate of payment is the same as the effective subsidy provided by a marketing loan. The advantage of such an instrument is that the CCC avoids the cost associated with accumulating and holding the commodity that it would sustain under the nonrecourse loan program.

An *incentive payment* is really the same as a deficiency payment. It is currently in force only in the case of wool. The United States is a deficit producer of wool, so the deficiency payment is actually made to encourage increased production. As implemented for wool, the payment rate increases with the quality of wool produced. Hence the wool incentive payment is not only made to encourage greater wool production but also to encourage higher quality wool production.

A *disaster payment* is a direct income payment made to producers whose agricultural output has been severely reduced due to a natural calamity such as hail, drought, or flood. It is available to farmers, though, only in areas where federal crop insurance is unavailable.

Generic certificates, first permitted under the Food Security Act of 1985, are used in lieu of cash to pay farmers for participating in various commodity programs. These certificates are called generic because they can be exchanged for any commodity the CCC has on hand. They have a fixed dollar face value and an eight-month life. They can be used to reacquire commodities pledged as collateral under the loan programs, they can be sold to others, or they can be returned to the CCC for cash. These certificates provide a mechanism for moving CCC stocks into commercial channels so that the loan rate no longer sets an effective price floor to the domestic market. Further, they reduce CCC stock buildups.

Supply Management. A variety of supply management tools have been implemented since the 1930s in an effort to prevent price-depressing surpluses and thus to help maintain farm incomes without incurring the huge federal costs associated with price support and government purchase programs or direct income payment mechanisms.[3]

Acreage allotments restrict the number of acres a farmer can plant. *Marketing quotas* restrict the amount of the commodity a farmer can sell. The two have generally been implemented at the same time, although this is not necessary. Indeed at the present time we have marketing quotas for burley tobacco and peanuts, but no acreage allotments for these two commodities.

Under a *cropland set-aside program* a farmer is required to set aside a specified percentage of the total *intended* planted acreage and devote this land to approved conservation practices in order to be eligible for nonrecourse loans and direct income payments.

An *acreage reduction program* is identical to a set-aside program except that the acreage reduction is based on a farmer's *actual previous* plantings rather than on *intended* plantings. Under this type of program, each farmer is assigned a base acreage which is calculated on the basis of past production. The reason for using actual previous plantings rather than intended plantings is to minimize the chance of a farmer enrolling land in the reduction program that would, if in fact planted, produce very small yields and thus not ordinarily be profitable to farm. Thus, this type of supply control program is intended to reduce "slippage" (see Glossary for a definition of *slippage*).

The Food Security Act of 1985 authorized an *underplanting provision* under which farmers planting between 50 and 92 percent (later changed to between 0 and 92 percent for wheat and feed grains) of their base acres to the program commodity and devoting the remaining base acres to a conserving use are eligible to receive deficiency payments on 92 percent of the base acreage. The 50/92 provision is available to cotton and rice producers, and the 0/92 provision is available to wheat and feed grain producers.

The Omnibus Budget Reconciliation Act of 1990 mandated a new program known as the *triple-base plan*, under which a producer's base acreage is to be divided into three portions: (1) program acres, (2) flexible acres, and (3) conserving-use acres. The conserving-use acres are subject to the restrictions that normally apply to acreage reduction program acres. The program acres must be planted to a program crop for which deficiency payments will be paid. The flexible acres can be planted to any program crop, oilseed crop, or nonprogram crop other than fruits and vegetables. Production on the flexible acres is not eligible for deficiency payments, nonrecourse loans, or marketing loans. This program applies to wheat, feed grains, cotton, and rice. Under the triple-base plan, 15 percent of the farmer's base acres must be "flexed." An additional 10 percent may be "flexed" without losing any base acres in subsequent years.

Under a *paid acreage diversion program* the government pays the farmer an annual rental fee in return for the farmer's agreement to place some of his land in soil-conserving uses or otherwise idle land from program crop production. When this type of program is authorized, farmers may or may not opt to participate. If they do not, they are not penalized for their choice. That is, they will still be able to take advantage of any income payment programs not tied to this choice or nonrecourse loan program currently in effect. Slippage can be minimized with this type of program through specific restrictions that can be imposed on the type of land enrolled in the program. The payment-in-kind (PIK) program implemented in 1983 was a type of paid acreage diversion program in the sense that participating farmers were paid in kind from CCC-owned commodities rather than in cash. A paid acreage diversion program is authorized by current legislation but at present is not implemented for any of the program crops.

Conservation advocates have long pushed for farm programs that would help protect the nation's soil resources from erosion caused by overplanting or mismanagement by farmers interested only in increasing their immediate incomes without regard to long-term erosion consequences. At times such programs have gained enough support to be enacted because they also promised to help combat burgeoning crop surpluses. The first such effort was made with the Soil Conservation and Domestic Allotment Act of 1936, which followed on the heels of the dust bowl. This act, though, took a fairly short-run view of the soil conservation issue in that it called for annual contracts. A second approach was the Soil Bank Program authorized by the Agricultural Act of 1956. This program involved three- to ten-year contracts with annual rental payments for land diverted. The third was the *conservation reserve program* authorized by the Food Security Act of 1985. Under this program, farmers are permitted to bid competitively to retire highly erodible cropland from production for ten to fifteen years in return for annual rental payments plus up to 50 percent of the cost of planting soil-conserving vegetation on the retired cropland. The program is administered by the Agricultural Stabilization and Conservation Service (ASCS) and was intended to take up to 45 million acres of highly erodible cropland out of production over the next several years.

The *Dairy Diversion Program* was authorized by the Dairy and Tobacco Adjustment Act of 1983 and operated for fifteen months during 1984–85. Under this program farmers who agreed to reduce their milk marketings by 5–30 percent below their base period production received $10 per hundredweight for each 100-pound reduction. This was the first attempt at supply control in the U.S. dairy industry.

The *Milk Production Termination Program* was authorized by the Food

Security Act of 1985. Farmers could bid competitively to liquidate their entire dairy operation and refrain from producing milk for the next five years. Selected for the program were those farmers who submitted the lowest bids until the targeted amount of reduction in milk production had been reached—twelve million pounds or about 10 percent of the previous year's production. Under this program, farmers whose bids were accepted were paid from $3.40 to $22.50 per hundredweight based on their most recent annual production record. To partially offset the cost of this program, remaining milk producers were assessed $.40 per hundredweight of milk sold.

Commodity Policy

This section provides a brief description of policy as it has evolved since 1933 to the present for each program commodity.[4] The 1933 legislation provides a convenient starting point for this review since it was the first major piece of legislation aimed at price and income support for agriculture. The Agricultural Adjustment Act of 1933 was passed in response to the economic hardships of farmers caused by low demand and prices during the Great Depression. Under this act, farmers were permitted to enter into contracts with the secretary of agriculture to adjust production of surplus commodities in return for direct payments financed by taxes levied on processors of those commodities. Further, the federal government was required to make mandatory price support loans on certain "basic" commodities. In 1936 this act was declared unconstitutional because of the contractual relation between farmers and the government and because the act attempted to finance the program through the collection of processing taxes.

In the discussion that follows, frequent reference is also made to the Agricultural Act of 1949. This act is part of "permanent" legislation for U.S. agriculture because it has not been superseded by subsequent legislation. Subsequent legislation (usually referred to as farm bills) has merely amended the 1949 act for a specified number of years. For the past several years, legislation with a five-year life has been passed, although it can be amended before five years has elapsed. For example, the farm bill passed in December of 1985 was amended in February of 1986!

It is crucial to remember that if a farm bill is not passed when the current amendment expires, policy for U.S. agriculture will revert back to the "permanent" legislation provided by the 1949 act. In general, this legislation calls for rather high support price levels and would likely involve much higher government costs than does current legislation.

The data in Tables 2.2–2.6 show the history of support, target, and market prices since the 1949 act took effect. The data in Tables 2.7–2.10 show the impact of supply control measures and the extent of government purchases of surplus production over this period. These data should be helpful in putting the text discussion in proper perspective and should be examined carefully while reading the text.

Feed Grains

CURRENT POLICY

- Nonrecourse loans
- Target prices/deficiency payments
- Acreage reserve program
- FOR program
- 0/92 underplanting provision
- Triple-base plan

Legislation in the 1930s brought feed grain producers assistance in the form of acreage controls and prices supported with government purchases of surplus production. The Agricultural Adjustment Act of 1933 aimed to restore the purchasing power of agricultural commodities to the 1909–14 level with voluntary acreage reduction programs and voluntary agreements between farmers and handlers. This act was declared unconstitutional in 1936, so alternative approaches were quickly sought. The Agricultural Adjustment Act of 1938 was passed to assist producers of corn and other program commodities to obtain nonrecourse loans at favorable loan rates. This legislation included mandatory nonrecourse loans, authority for marketing quotas if necessary, and loan rates set at a specified percentage of parity. The act also provided for payments to farmers for shifting land from "soil-depleting" to "soil-conserving" crops and for instituting approved conservation practices.

During the early and mid-1940s, policy shifted to encouraging production of farm commodities needed to meet wartime and postwar demands. Later, downward pressure on farm prices prompted the Agricultural Act of 1949, which set loan rates (support prices) at 90 percent of parity through 1954. Prices were supported with nonrecourse loans and direct government purchases.

Following the Korean War, feed grain surpluses resulting from high price supports again became a problem. Acreage allotments were implemented in an effort to reduce the surpluses. Planting within allotments was not mandatory, but the high price supports were available only to farmers who did plant within their allotments.

Table 2.2. Price supports, target prices, and market prices for program commodities, 1950–90

	Corn			Oats			Barley		
Year	Support or target price[a]	Loan rate	Avg. market price	Support or target price[b]	Loan rate	Avg. market price	Support or target price[a]	Loan rate	Avg. market price
				($/bu.)					
1950	**1.47**	**1.47**	**1.52**	...	**0.71**	**0.79**	**1.10**	**1.10**	**1.19**
1951	1.57	1.57	1.66	...	0.72	0.82	1.11	1.11	1.26
1952	1.60	1.60	1.52	...	0.78	0.79	1.22	1.22	1.37
1953	1.60	1.60	1.48	...	0.80	0.74	1.24	1.24	1.17
1954	1.62	1.62	1.43	...	0.75	0.71	1.15	1.15	1.09
1955	**1.58**	**1.58**	**1.35**	...	**0.61**	**0.60**	**0.94**	**0.94**	**0.92**
1956	1.50	1.50	1.29	...	0.65	0.69	1.02	1.02	0.99
1957	1.40	1.40	1.11	...	0.61	0.61	0.95	0.95	0.89
1958	1.36	1.36	1.12	...	0.61	0.58	0.93	0.93	0.90
1959	1.12	1.12	1.05	...	0.50	0.65	0.77	0.77	0.86
1960	**1.06**	**1.06**	**1.00**	...	**0.50**	**0.60**	**0.77**	**0.77**	**0.84**
1961	1.20	1.20	1.10	...	0.62	0.64	0.93	0.93	0.98
1962	1.20	1.20	1.12	...	0.62	0.62	0.93	0.93	0.92
1963	1.25	1.07	1.11	...	0.65	0.62	0.96	0.82	0.90
1964	1.25	1.10	1.17	...	0.65	0.63	0.96	0.84	0.95
1965	**1.25**	**1.05**	**1.16**	...	**0.60**	**0.62**	**0.96**	**0.80**	**1.02**
1966	1.30	1.00	1.24	...	0.60	0.67	1.00	0.80	1.05
1967	1.35	1.05	1.03	...	0.63	0.66	0.90	0.90	1.00
1968	1.35	1.05	1.08	...	0.63	0.60	0.90	0.90	0.91
1969	1.35	1.05	1.15	...	0.63	0.59	1.03	0.83	0.87
1970	**1.35**	**1.05**	**1.33**	...	**0.63**	**0.62**	**1.03**	**0.83**	**0.96**
1971	1.35	1.05	1.08	...	0.54	0.60	0.81	0.81	0.99
1972	1.41	1.05	1.57	...	0.54	0.72	1.10	0.86	1.21
1973	1.64	1.05	2.55	...	0.54	1.18	1.27	0.86	2.14
1974	1.38	1.10	3.02	...	0.54	1.53	1.13	0.90	2.81
1975	**1.38**	**1.10**	**2.54**	...	**0.54**	**1.46**	**1.13**	**0.90**	**2.42**
1976	1.57	1.50	2.49	...	0.72	1.56	1.28	1.22	2.25
1977	2.00	2.00	2.02	...	1.03	1.10	2.15	1.63	1.78
1978	2.10	2.00	2.25	...	1.03	1.20	2.25	1.63	1.92
1979	2.20	2.10	2.36	...	1.08	1.36	2.40	1.71	2.29
1980	**2.35**	**2.25**	**2.70**	...	**1.16**	**1.79**	**2.25**	**1.83**	**2.85**
1981	2.40	2.40	2.47	...	1.24	1.88	2.60	1.95	2.48
1982	2.70	2.55	2.55	1.50	1.31	1.49	2.60	2.08	2.18
1983	2.86	2.65	3.21	1.60	1.36	1.62	2.60	2.16	2.47
1984	3.03	2.55	2.63	1.60	1.31	1.67	2.60	2.08	2.29
1985	**3.03**	**2.55**	**2.23**	**1.60**	**1.31**	**1.23**	**2.60**	**2.08**	**1.98**
1986	3.03	1.92	1.50	1.60	0.99	1.21	2.60	1.56	1.61
1987	3.03	1.82	1.94	1.60	0.94	1.56	2.60	1.49	1.81
1988	2.93	1.77	2.55	1.55	0.90	2.61	2.51	1.44	2.80
1989	2.84	1.65	2.36	1.50	0.85	1.49	2.43	1.34	2.42
1990	**2.75**	**1.57**	**2.30**	**1.45**	**0.81**	**1.13**	**2.36**	**1.28**	**2.14**

Source: Ash and Hoffman 1989, Hoffman and Ash 1989, Mercier 1989; and U.S. Department of Agriculture, *Agricultural Statistics,* various annual issues.

[a]Target prices began in 1973. Prior to that time support prices were maintained by government purchases and/or supply control measures.

[b]Target prices for oats were first mandated by the 1981 legislation.

Table 2.3. Price supports, target prices, and market prices for program commodities, 1950–90

Year	Grain sorghum Support or target price[a]	Grain sorghum Loan rate	Grain sorghum Avg. market price	Rye Loan rate[b]	Rye Avg. market price	Soybeans Loan rate	Soybeans Avg. market price
				($/bu.)			
1950	**1.87**	**1.87**	**1.05**	...	**1.31**	**2.06**	**2.47**
1951	2.17	2.17	1.32	...	1.52	2.45	2.73
1952	2.38	2.38	1.58	...	1.72	2.56	2.72
1953	3.43	3.43	1.32	...	1.29	2.56	2.72
1954	2.28	2.28	1.26	...	1.21	2.22	2.46
1955	**1.78**	**1.78**	**0.98**	...	**1.06**	**2.04**	**2.22**
1956	1.97	1.97	1.15	1.27	1.16	2.15	2.18
1957	1.86	1.86	0.97	1.18	1.08	2.09	2.07
1958	1.83	1.83	1.00	1.10	1.02	2.09	2.00
1959	1.52	1.52	0.86	0.90	1.00	1.85	1.96
1960	**1.52**	**1.52**	**0.84**	**0.90**	**0.88**	**1.85**	**2.13**
1961	1.93	1.93	1.01	1.02	1.01	2.30	2.28
1962	1.93	1.93	1.02	1.02	0.95	2.25	2.34
1963	2.00	1.71	0.98	1.07	1.08	2.25	2.51
1964	2.00	1.77	1.05	1.07	1.04	2.25	2.62
1965	**2.00**	**1.65**	**0.98**	**1.02**	**0.98**	**2.25**	**2.54**
1966	2.05	1.52	1.02	1.02	1.06	2.50	2.75
1967	2.14	1.61	0.99	1.02	1.07	2.50	2.49
1968	2.15	1.62	0.95	1.02	1.02	2.50	2.43
1969	2.14	1.61	1.07	1.02	1.01	2.25	2.35
1970	**2.14**	**1.61**	**1.14**	**1.02**	**0.99**	**2.25**	**2.85**
1971	2.21	1.73	1.04	0.89	0.90	2.25	3.03
1972	2.39	1.79	1.37	0.89	0.96	2.25	4.37
1973	2.61	1.79	2.14	0.89	1.91	2.25	5.68
1974	2.34	1.88	2.77	0.89	2.51	2.25	6.64
1975	**2.34**	**1.88**	**2.37**	**0.89**	**2.36**	**2.35**	**4.92**
1976	2.66	2.55	2.03	1.20	2.47	2.50	5.58
1977	2.28	3.39	1.82	1.70	2.05	3.50	6.82
1978	2.28	1.90	3.43	1.70	1.99	4.50	6.28
1979	2.34	2.00	3.91	1.79	2.07	4.50	6.86
1980	**2.50**	**2.14**	**4.67**	**1.91**	**2.64**	**5.02**	**7.57**
1981	2.55	2.28	4.72	2.04	3.00	5.02	6.07
1982	2.60	2.42	4.00	2.17	2.39	5.02	5.71
1983	2.72	2.52	4.89	2.25	2.17	5.02	7.83
1984	2.88	2.42	4.15	2.17	2.08	5.02	5.84
1985	**2.88**	**2.42**	**3.45**	**2.17**	**2.05**	**5.02**	**5.04**
1986	2.88	1.82	2.45	1.63	1.49	4.77	4.78
1987	2.88	1.74	3.04	1.55	1.63	4.77	5.88
1988	2.78	1.68	4.05	1.50	2.52	4.77	7.42
1989	2.70	1.57	3.75	1.40	2.06	4.53	5.69
1990	**2.61**	**1.49**	**3.75**	**1.33**	**2.09**	**4.50**	**5.75**

Source: Crowder and Davison 1989; Lin and Hoffman 1989; and U.S. Department of Agriculture, *Agricultural Statistics*, various annual issues.

[a]Target prices began in 1973. Prior to that time support prices were maintained by government purchases and/or supply control measures.

[b]A loan rate for rye was first mandated by the 1955 legislation.

Table 2.4. Price supports, target prices, and market prices for program commodities, 1950–90

	Wheat			Rice			Cotton		
Year	Support or target price[a]	Loan rate	Avg. market price	Support or target price[b]	Loan rate	Avg. market price	Support or target price[c]	Loan rate	Avg. market price
	($/bu.)			($/cwt.)			(cents/lb.)		
1950	1.99	1.99	2.00	...	4.56	5.09	...	30.25	40.00
1951	2.18	2.18	2.11	...	5.00	4.82	...	32.36	38.00
1952	2.20	2.20	2.09	...	5.04	5.87	...	32.41	35.00
1953	2.21	2.21	2.04	...	4.84	5.19	...	33.50	32.00
1954	2.24	2.24	2.12	...	4.92	4.57	...	34.03	34.00
1955	2.08	2.08	1.98	...	4.66	4.81	...	34.55	32.00
1956	2.00	2.00	1.97	...	4.57	4.86	...	32.74	32.00
1957	2.00	2.00	1.93	...	4.72	5.11	...	32.31	30.00
1958	1.82	1.82	1.75	...	4.48	4.68	...	35.08	33.00
1959	1.81	1.81	1.76	...	4.38	4.59	...	34.10	32.00
1960	1.78	1.78	1.74	...	4.42	4.55	...	32.42	30.00
1961	1.79	1.79	1.83	...	4.71	5.14	...	33.04	33.00
1962	2.00	2.00	2.04	...	4.71	5.04	...	32.47	32.00
1963	2.00	1.82	1.85	...	4.71	5.01	...	32.47	30.00
1964	2.00	1.30	1.37	...	4.71	4.90	...	30.00	28.00
1965	2.00	1.25	1.35	...	4.50	4.93	...	29.00	21.00
1966	2.57	1.25	1.63	...	4.50	4.95	...	21.00	26.00
1967	2.61	1.25	1.39	...	4.55	4.97	...	20.25	22.00
1968	2.63	1.25	1.24	...	4.60	5.00	...	20.25	21.00
1969	2.77	1.25	1.24	...	4.72	4.95	...	20.25	22.00
1970	2.82	1.25	1.33	...	4.86	5.17	...	20.25	22.00
1971	2.93	1.25	1.34	...	5.07	5.31	...	19.50	28.00
1972	3.02	1.25	1.76	...	5.27	6.05	...	19.50	27.00
1973	3.39	1.25	3.95	...	6.07	11.00	...	19.50	45.00
1974	2.05	1.37	4.09	...	7.54	13.94	38.00	27.06	43.00
1975	2.05	1.37	3.56	...	8.52	10.12	38.00	36.12	51.00
1976	2.29	2.25	3.15	8.25	6.19	6.90	43.20	38.92	64.00
1977	2.90	2.25	2.29	8.25	6.19	7.94	47.80	44.63	52.00
1978	3.40	2.35	2.82	8.53	6.40	9.29	52.00	48.00	58.00
1979	3.40	2.50	3.51	9.05	6.79	9.05	57.70	50.23	62.00
1980	3.63	3.00	3.91	9.49	7.12	11.10	58.40	48.00	74.40
1981	3.81	3.20	3.69	10.68	8.01	11.90	70.87	52.46	67.10
1982	4.05	3.55	3.45	10.85	8.14	8.36	71.00	57.08	59.10
1983	4.30	3.65	3.51	11.40	8.14	8.31	76.00	55.00	65.30
1984	4.38	3.30	3.39	11.90	8.00	8.32	81.00	55.00	58.70
1985	4.38	3.30	3.08	11.90	8.00	6.53	81.00	57.30	56.80
1986	4.38	2.40	2.42	11.90	7.20	3.75	81.00	55.00	51.50
1987	4.38	2.28	2.57	11.66	6.84	7.27	79.40	52.25	63.70
1988	4.23	2.21	3.72	11.15	6.63	6.83	75.90	51.80	55.60
1989	4.10	2.06	3.72	10.80	6.50	7.35	73.40	50.00	63.60
1990	4.00	1.95	2.61	10.71	6.50	6.50	72.90	50.27	66.40

Source: Childs and Lin 1989; Harwood and Young 1989; Stults et al. 1989; and U.S. Department of Agriculture, *Agricultural Statistics,* various annual issues.

[a]Target prices began in 1973. Prior to that time support prices were maintained by government purchases and/or supply control measures.

[b]Target prices for rice were first mandated by the 1975 legislation.

[c]Target prices for cotton were first mandated by the 1973 legislation.

Table 2.5. Price supports, target prices, and market prices for program commodities, 1950–90

	Sugar			Wool		Honey	
Year	Market Stabilization price[a]	Loan rate[a]	Avg. market price[b]	Support rate	Avg. market price	Loan rate	Avg. market price
	(cents/lb.)		($/ton)	(cents/lb.)		(cents/lb.)	
1950	...	...	13.61	45.20	62.10	9.00	15.30
1951	...	...	14.10	50.70	97.10	10.00	16.00
1952	...	...	14.35	54.20	54.10	11.40	16.20
1953	...	...	13.94	53.10	54.90	10.50	16.50
1954	...	...	13.11	53.20	53.20	10.20	17.00
1955	...	...	13.51	62.00	42.80	9.90	17.80
1956	...	...	14.24	62.00	44.30	9.70	19.00
1957	...	...	13.49	62.00	53.70	9.70	18.70
1958	...	...	14.01	62.00	36.40	9.60	17.40
1959	...	...	13.50	62.00	43.30	8.30	17.00
1960	6.36	...	13.91	62.00	42.00	8.60	17.90
1961	6.40	...	13.46	62.00	42.90	11.20	18.00
1962	6.51	...	15.05	62.00	47.70	11.20	17.40
1963	6.61	...	14.42	62.00	48.50	11.20	18.00
1964	6.63	...	14.00	62.00	53.20	11.20	18.60
1965	6.80	...	14.08	62.00	47.10	11.20	17.80
1966	7.08	...	14.96	65.00	52.10	11.40	17.40
1967	7.25	...	15.67	66.00	39.80	12.50	15.60
1968	7.52	...	15.96	67.00	40.50	12.50	16.90
1969	7.91	...	14.72	69.00	41.80	13.00	17.50
1970	8.27	...	16.87	72.00	35.50	13.00	17.40
1971	8.69	...	17.45	72.00	19.40	14.00	21.80
1972	9.16	...	18.00	72.00	35.00	14.00	30.20
1973	10.52	...	31.68	72.00	82.70	16.10	44.40
1974	9.10	...	48.86	72.00	59.10	20.60	51.00
1975	10.38	...	27.60	72.00	44.70	25.50	50.50
1976	12.19	...	21.00	72.00	65.70	29.40	49.90
1977	13.50	13.50	24.20	99.00	72.00	32.70	53.00
1978	15.00	14.73	25.20	108.00	74.50	36.80	54.50
1979	15.00	13.00	33.90	115.00	86.30	43.90	59.00
1980	15.00	...	47.20	123.00	88.10	50.30	61.40
1981	17.44	16.75	29.20	135.00	94.40	57.40	63.20
1982	20.73	17.00	35.40	137.00	68.60	60.40	56.80
1983	21.17	17.50	37.00	153.00	61.30	62.20	54.40
1984	21.57	17.75	33.90	165.00	79.50	65.80	50.00
1985	21.50	18.00	33.80	165.00	63.30	65.30	47.50
1986	21.78	17.23	35.90	178.00	66.80	64.00	51.30
1987	21.76	18.00	38.20	181.00	91.70	63.00	50.30
1988	21.80	17.50	41.20	178.00	138.00	59.10	50.00
1989	21.95	21.80	43.10	177.00	124.00	56.40	49.80
1990	21.95	21.95	42.00	182.00	80.00	53.80	52.80

Source: Barry et al. 1990; Hoff and Phillips 1989; Lawler and Skinner 1989; and U.S. Department of Agriculture, *Agricultural Statistics,* various annual issues.

[a]Cents/lb. of raw sugar. Loan rates were established for the first time by the 1977 legislation.
[b]Price received for sugar beets.

Table 2.6. Price supports, target prices, and market prices for program commodities, 1950–90

	Peanuts			Tobacco				Milk	
Year	Support price on quota	Support price on nonquota[a]	Avg. market price	Support price on burley	Burley market price	Support price on flue-cured	Flue-cured market price	Support price on all milk	Avg. market price
	(cents/lb.)			*(cents/lb.)*				*($/cwt.)*	
1950	**10.80**	**...**	**10.90**	**45.70**	**49.00**	**45.00**	**54.70**	**3.07**	**3.16**
1951	11.50	...	10.40	49.80	51.20	50.70	52.40	3.60	3.85
1952	12.00	...	10.90	49.50	50.30	50.60	50.30	3.85	4.06
1953	11.90	...	11.10	46.60	52.50	47.90	52.80	3.74	3.48
1954	12.20	...	12.20	46.40	49.80	47.90	52.70	3.15	3.14
1955	**12.20**	**...**	**11.70**	**46.20**	**58.60**	**48.30**	**52.70**	**3.15**	**3.15**
1956	11.40	...	11.20	48.10	63.60	48.90	51.50	3.25	3.25
1957	11.10	...	10.40	51.70	60.30	50.80	55.40	3.25	3.27
1958	10.66	...	10.60	55.40	66.10	54.60	58.20	3.06	3.15
1959	9.68	...	9.60	57.20	60.60	55.50	58.30	3.06	3.17
1960	**10.06**	**...**	**10.00**	**57.20**	**64.30**	**55.50**	**60.40**	**3.06**	**3.25**
1961	11.05	...	10.90	57.20	66.50	55.50	64.30	3.40	3.36
1962	11.07	...	11.00	57.80	58.60	56.10	60.10	3.11	3.20
1963	11.20	...	11.20	58.30	59.20	56.60	58.00	3.14	3.21
1964	11.20	...	11.20	58.90	60.30	57.20	58.50	3.15	3.26
1965	**11.20**	**...**	**11.40**	**59.50**	**67.00**	**57.70**	**64.60**	**3.24**	**3.34**
1966	11.35	...	11.30	60.60	66.90	58.80	66.90	4.00	3.97
1967	11.35	...	11.40	61.80	71.80	59.90	64.90	4.00	4.06
1968	12.01	...	11.90	63.50	73.70	61.60	66.60	4.28	4.22
1969	12.38	...	12.30	65.80	69.60	63.80	72.40	4.28	4.45
1970	**12.75**	**...**	**12.80**	**68.60**	**72.20**	**66.60**	**72.00**	**4.66**	**4.70**
1971	13.42	...	13.60	71.50	80.90	69.40	77.20	4.93	4.86
1972	14.25	...	14.50	74.90	79.20	72.70	85.30	4.93	5.08
1973	16.43	...	16.20	78.90	92.90	76.60	88.10	5.29	6.20
1974	18.30	...	17.90	85.90	113.70	83.30	105.00	6.57	7.13
1975	**19.73**	**...**	**19.60**	**96.10**	**105.50**	**93.30**	**99.80**	**7.24**	**7.63**
1976	20.70	...	20.00	109.30	114.20	106.00	110.40	8.13	8.56
1977	21.53	...	21.00	117.30	120.00	113.80	117.60	9.00	8.71
1978	21.00	12.50	21.10	124.70	131.20	121.00	135.00	9.43	9.65
1979	21.00	15.00	20.70	133.30	145.20	129.30	140.00	11.13	11.10
1980	**22.75**	**12.50**	**25.10**	**145.90**	**165.90**	**141.50**	**144.50**	**13.10**	**12.00**
1981	22.75	12.50	26.90	163.60	180.70	158.70	166.40	13.30	12.73
1982	27.50	10.00	25.10	175.10	181.00	169.90	178.50	13.10	12.66
1983	27.50	9.25	24.70	175.10	177.30	169.90	177.90	13.10	12.61
1984	27.50	9.25	27.90	175.10	187.60	169.90	181.10	12.85	12.49
1985	**27.95**	**7.40**	**24.40**	**178.80**	**159.40**	**169.90**	**171.90**	**12.10**	**11.72**
1986	30.37	7.49	29.20	148.80	156.50	143.80	152.70	11.60	11.46
1987	30.37	7.49	27.70	148.80	156.30	143.50	158.60	11.25	11.37
1988	30.76	7.49	27.90	150.00	161.00	144.20	161.30	10.60	11.15
1989	30.79	7.49	27.80	153.20	167.10	146.80	167.30	10.70	12.38
1990	**31.57**	**7.49**	**34.10**	**155.80**	**167.00**	**148.80**	**167.00**	**11.10**	**12.34**

Source: Fallert, Blayney, and Miller 1990; Grise 1989; Schaub and Wendland 1989; and U.S. Department of Agriculture, *Agricultural Statistics,* various annual issues.

[a]Support price on nonquota peanuts initiated by the 1977 legislation.

Table 2.7. Acres harvested, total production, CCC ending stocks, and acres diverted by government programs for program commodities, 1950–90

	Corn				Grain sorghum			
Year	Acres harvested	Total production	CCC stocks	Acres diverted	Acres harvested	Total production	CCC stocks	Acres diverted
	(1,000)	*(mil. bu.)*		*(1,000)*	*(1,000)*	*(mil. bu.)*		*(1,000)*
1950	72,398	2,764	399	0.0	10,346	234	17	0.0
1951	71,191	2,629	389	0.0	8,544	163	1	0.0
1952	71,353	2,981	280	0.0	5,326	91	0	0.0
1953	70,738	2,882	362	0.0	6,295	116	22	0.0
1954	68,668	2,708	606	0.0	11,718	236	68	0.0
1955	68,462	2,873	758	0.0	12,891	243	76	0.0
1956	64,877	3,075	984	0.0	9,209	205	75	0.0
1957	63,065	3,045	1,046	0.0	19,682	568	294	0.0
1958	63,549	3,356	1,168	0.0	16,524	581	489	0.0
1959	72,091	3,825	1,285	0.0	15,402	555	560	0.0
1960	71,649	3,908	1,315	0.0	15,592	620	671	0.0
1961	58,449	3,626	810	19.1	10,957	480	646	6.1
1962	56,609	3,637	567	20.3	11,536	510	610	5.5
1963	59,227	4,019	815	17.2	13,326	585	613	4.6
1964	55,369	3,484	521	22.2	11,742	490	538	6.5
1965	55,392	4,103	249	24.0	13,029	673	383	7.0
1966	57,002	4,168	139	23.7	12,813	715	193	7.3
1967	60,694	4,860	182	16.2	14,988	755	192	4.1
1968	55,980	4,450	295	25.4	13,890	731	198	7.0
1969	54,574	4,687	255	27.2	13,437	730	156	7.5
1970	57,358	4,152	105	26.1	13,568	683	65	7.4
1971	64,123	5,646	160	14.1	16,142	868	48	4.1
1972	57,513	5,580	79	24.4	13,212	801	5	7.3
1973	62,143	5,671	7	6.0	15,700	923	0	2.0
1974	65,405	4,701	0	0.0	13,809	623	0	0.0
1975	67,625	5,841	0	0.0	15,403	754	0	0.0
1976	71,506	6,289	0	0.0	14,466	711	5	0.0
1977	71,614	6,505	4	0.0	13,797	781	5	0.0
1978	71,930	7,268	101	6.1	13,410	731	44	1.4
1979	72,400	7,939	260	2.9	12,901	809	46	1.2
1980	72,961	6,637	242	0.0	12,513	579	42	0.0
1981	74,524	8,119	280	0.0	13,677	876	42	0.0
1982	72,719	8,235	1,143	2.1	14,037	835	171	0.7
1983	51,483	4,175	201	32.2	10,001	488	103	5.7
1984	71,915	7,674	225	3.9	15,355	866	112	0.6
1985	75,224	8,877	546	5.4	16,782	1,120	207	0.9
1986	69,159	8.250	1,443	14.3	13,859	938	409	3.0
1987	59,505	7,131	750	23.0	10,531	731	464	4.1
1988	58,250	4,929	400	20.5	9,042	577	355	3.9
1989	64,703	7,525	363	10.8	11,103	615	341	3.3
1990	66,952	7,933	233	10.1	9,079	571	163	3.0

Source: Lin and Hoffman 1989; Mercier 1989; and U.S. Department of Agriculture, *Agricultural Statistics,* various annual issues.

Table 2.8. Acres harvested, total production, CCC ending stocks, and acres diverted by government programs for program commodities, 1950–90

	Wheat				Cotton			
Year	Acres harvested	Total production	CCC stocks	Acres diverted	Acres harvested	Total production	CCC stocks	Acres diverted
	(1,000)	(mil. bu.)		(1,000)	(1,000)	(mil. bales)		(1,000)
1950	61,607	1,019	160	0.0	17,843	10,014	76	0.0
1951	61,873	988	82	0.0	26,949	15,149	2	0.0
1952	71,130	1,306	292	0.0	25,921	15,139	236	0.0
1953	67,840	1,173	714	0.0	24,341	16,465	129	0.0
1954	54,356	984	971	0.0	19,251	13,697	1,661	0.0
1955	47,290	937	922	0.0	16,928	14,721	5,952	0.0
1956	49,768	1,005	808	0.0	15,615	13,310	4,829	1.1
1957	43,754	956	813	0.0	13,558	10,964	937	3.0
1958	53,047	1,457	1,084	0.0	11,849	11,512	984	4.9
1959	51,781	1,121	1,198	0.0	15,117	14,558	4,967	0.0
1960	51,896	1,357	1,277	0.0	15,309	14,272	1,678	0.0
1961	51,551	1,235	1,225	0.0	15,634	14,318	1,449	0.0
1962	43,541	1,094	1,075	10.7	15,569	14,867	3,750	0.0
1963	45,506	1,147	1,102	7.2	14,212	15,334	4,303	0.0
1964	49,762	1,283	800	5.1	14,055	15,145	6,557	0.5
1965	49,560	1,316	635	7.2	13,613	14,938	9,715	1.0
1966	49,613	1,305	299	8.3	9,553	9,557	6,677	4.6
1967	58,353	1,508	122	0.0	7,997	7,443	552	4.8
1968	54,765	1,557	100	0.0	10,159	10,926	24	3.3
1969	47,146	1,443	140	11.1	11,051	9,990	1,890	0.0
1970	43,564	1,356	277	15.7	11,155	10,192	262	0.0
1971	47,635	1,619	353	13.5	11,471	10,477	1	2.1
1972	47,303	1,546	355	20.1	12,984	13,704	0	2.0
1973	54,148	1,711	6	7.4	11,970	12,974	0	0.0
1974	65,368	1,782	1	0.0	12,547	11,540	0	0.0
1975	69,499	2,122	0	0.0	8,796	8,302	0	0.0
1976	70,927	2,142	0	0.0	10,914	10,581	0	0.0
1977	66,686	2,026	0	0.0	13,275	14,389	0	0.0
1978	56,495	1,776	48	9.6	12,400	10,856	0	0.3
1979	62,454	2,134	50	8.2	12,831	14,531	0	0.0
1980	71,125	2,381	188	0.0	13,215	11,018	0	0.0
1981	80,642	2,785	200	0.0	13,841	15,566	1	0.0
1982	77,937	2,765	190	5.8	9,734	11,864	396	1.6
1983	61,390	2,420	192	30.0	7,348	7,800	158	6.8
1984	66,928	2,595	188	18.6	10,379	13,000	123	2.5
1985	64,734	2,425	378	18.8	10,229	13,400	767	3.6
1986	60,723	2,092	602	21.0	8,468	9,700	73	4.2
1987	55,945	2,107	830	23.9	10,030	14,759	3	4.0
1988	53,189	1,812	283	22.5	11,948	15,412	50	2.2
1989	62,189	2,037	190	9.6	9,538	12,196	27	3.5
1990	69,353	2,739	117	7.1	11,708	15,500	0	1.9

Source: Harwood and Young 1989; Stults et al. 1989; and U.S. Department of Agriculture, *Agricultural Statistics,* various annual issues.

Table 2.9. Acres harvested, total production, CCC ending stocks, and acres diverted by government programs for rice, 1950–90

Year	Acres harvested	Total production	CCC stocks	Acres diverted
	(1,000)	*(1,000 cwt.)*		*(1,000)*
1950	**1,637**	**38,780**	**395**	**0**
1951	1,996	46,120	227	0
1952	1,997	48,190	8	0
1953	2,159	52,810	1,200	0
1954	2,550	64,110	18,445	0
1955	**1,826**	**55,840**	**27,374**	**0**
1956	1,569	49,440	12,555	0
1957	1,340	42,960	12,012	0
1958	1,415	44,740	9,455	0
1959	1,586	53,660	6,867	0
1960	**1,595**	**54,590**	**4,132**	**0**
1961	1,589	54,200	314	0
1962	1,773	65,930	1,860	0
1963	1,771	70,280	1,435	0
1964	1,786	73,160	1,044	0
1965	**1,793**	**76,300**	**621**	**0**
1966	1,967	85,050	232	0
1967	1,970	89,340	86	0
1968	2,353	104,240	6,325	0
1969	2,128	91,940	6,417	0
1970	**1,815**	**83,790**	**9,467**	**0**
1971	1,818	85,710	2,747	0
1972	1,818	85,420	148	0
1973	2,170	92,750	0	0
1974	2,531	112,630	4	0
1975	**2,818**	**128,370**	**19,214**	**0**
1976	2,480	115,580	18,721	0
1977	2,249	99,810	10,772	0
1978	2,970	133,200	8,300	0
1979	2,869	131,900	1,700	0
1980	**3,312**	**146,200**	**0**	**0**
1981	3,792	182,700	17,500	0
1982	3,262	153,600	22,320	422
1983	2,169	99,700	25,000	1,798
1984	2,802	138,800	44,300	790
1985	**2,492**	**134,900**	**43,600**	**1,240**
1986	2,360	133,400	8,700	1,480
1987	2,333	129,603	100	1,570
1988	2,900	159,900	0	1,090
1989	2,689	154,487	0	1,210
1990	**2,813**	**154,919**	**0**	**1,030**

Source: Childs and Lin 1989; and U.S. Department of Agriculture, *Agricultural Statistics,* various annual issues.

Table 2.10. Total production and ending stocks for program commodities, 1950–90

Year	Peanuts		Wool		Milk	
	Total production	CCC stocks	Total production	Total stocks	Total production	CCC stocks[a]
	(mil. lb.)		(1,000 lb. clean)		(mil. lb.)	
1950	**2,035**	**7**	**103,482**	**175,200**	**116,602**	**1,637**
1951	1,679	142	108,799	173,500	114,681	70
1952	1,356	92	111,288	205,000	114,671	200
1953	1,574	30	110,787	226,500	120,221	7,510
1954	1,008	0	112,480	242,700	122,094	10,519
1955	**1,548**	**37**	**115,092**	**249,500**	**122,945**	**5,517**
1956	1,607	151	115,518	186,100	124,860	1,951
1957	1,436	118	114,051	151,900	124,628	2,777
1958	1,814	196	116,251	125,500	123,220	984
1959	1,523	172	123,991	151,400	121,989	430
1960	**1,718**	**103**	**126,537**	**132,000**	**123,109**	**1,208**
1961	1,657	70	123,620	131,800	125,707	4,911
1962	1,719	105	117,645	118,300	126,301	7,818
1963	1,942	106	110,877	113,800	125,202	5,556
1964	2,099	65	101,283	103,300	126,967	973
1965	**2,390**	**89**	**96,098**	**118,900**	**124,173**	**538**
1966	2,416	114	93,040	117,800	119,892	46
1967	2,477	12	90,145	104,500	118,769	3,994
1968	2,547	0	84,618	117,300	117,234	3,020
1969	2,535	0	79,062	96,400	116,345	1,447
1970	**2,983**	**11**	**77,077**	**79,300**	**116,872**	**2,098**
1971	3,005	4	76,394	86,000	118,462	1,539
1972	3,275	24	83,691	71,200	119,904	2,005
1973	3,474	0	75,894	53,300	115,385	476
1974	3,668	552	69,370	51,500	115,553	310
1975	**3,847**	**958**	**63,114**	**47,500**	**115,326**	**124**
1976	3,739	0	58,661	41,600	120,269	410
1977	3,715	2	56,669	42,000	122,698	3,710
1978	3,952	0	54,353	48,500	121,461	4,254
1979	3,968	0	55,370	46,800	123,411	3,180
1980	**2,303**	**0**	**55,661**	**45,900**	**128,425**	**7,207**
1981	3,982	2	57,968	59,800	133,013	12,980
1982	3,440	0	56,036	58,400	135,505	15,451
1983	3,296	0	54,324	58,900	139,672	17,412
1984	4,406	0	50,409	51,600	135,450	11,767
1985	**4,123**	**0**	**46,433**	**50,600**	**143,147**	**9,105**
1986	3,701	0	44,790	46,800	143,384	8,702
1987	3,916	0	44,705	45,300	142,557	3,451
1988	3,981	0	47,116	55,900	145,152	4,313
1989	3,990	0	47,800	69,300	144,239	5,632
1990	**3,601**	**0**	**47,000**	**52,600**	**148,284**	**5,541**

Source: Fallert, Blayney, and Miller 1990; Lawler and Skinner 1989; and Schaub and Wendland 1989; and U.S. Department of Agriculture, *Agricultural Statistics,* various annual issues.

[a]Milk equivalent of dairy products.

In 1954 the Agricultural Trade Development and Assistance Act was passed in the hopes of channeling surpluses into foreign aid and thus increase effective demand for agricultural products. The Agricultural Act of 1956 established the Soil Bank Program which was designed to encourage farmers to divert cropland from production into conserving uses. This program was not as successful as had been hoped because farm yields on the land in production continued to increase and so surpluses continued.

The Emergency Feed Grain Program of 1961 was intended to put an end to the surpluses and to the buildup of government stocks. With this legislation a shift was made from total support through commodity loans to support with a loan rate combined with payments for acreage diversion. The program was voluntary and only participants in the acreage diversion program were provided support.

Beginning in 1963 and continuing into the early 1970s the major focus of policy for the feed grains was on price support loans at or near world prices, direct price support payments, and cropland diversion payments. This policy, although costly to the federal government, did bring surpluses under control, and farmers' incomes were sustained in part by generous government payments and in part by rapid productivity growth. In 1970 a set-aside acreage reduction program replaced the acreage diversion program in the hopes of giving farmers more flexibility of choice.

The Agriculture and Consumer Protection Act of 1973 represented another step in the direction of a more market-oriented policy. World demand for U.S. farm products was now at an all-time high, prices of farm products were quite favorable, and CCC stocks and thus government costs were minimal. Hence a different approach to farm programs was sought. Support prices based on parity calculations were replaced with target prices based on cost-of-production considerations. Income support was implemented via deficiency payments calculated as the difference between target prices and the maximum of loan rates or market prices when market prices fell below target prices. The 1973 act also introduced disaster payments. Deficiency payments, set-aside acreage, and disaster payments were applied to the base acreage allotments determined and apportioned by the secretary of agriculture. Additional plantings were not eligible for support, but no penalties were imposed.

The 1977 legislation basically continued that of 1973 and added a farmer-owned reserve (FOR) program, which was designed to reduce price instability and provide a more orderly grain reserve system. The FOR program was also intended to serve as a control on the holding of CCC inventories.

Under the Agriculture and Food Act of 1981 specific loan and target price minimums were mandated. Crop-specific acreage reduction programs were introduced, thus reviving the acreage allotment concept so that all planted acreage did not qualify for target price protection. Up until the 1981 act was passed, demand was strong and prices were favorable. Following 1981, however, surpluses again became a problem. In 1983 a PIK program was announced under which farmers received a payment in kind for additional acreage reduction instead of a cash payment. The PIK program together with the acreage reserve program and a paid land diversion program diverted nearly thirty-two million acres of cropland from feed grain production.

The Food Security Act of 1985 continued many provisions already in place but gave the secretary of agriculture additional discretion in carrying out policy. Major farm sector issues prior to passage of this act were high CCC stocks, low net farm incomes, and declining export market shares. The act thus put major emphasis on reducing loan rates to keep domestic prices at competitive world market levels and on increasing U.S. competitiveness in international markets.

Under this act nonrecourse loans were still available but the loan rate was to be set at between 75 and 85 percent of the average farm price for the preceding five years excluding the high- and low-price years. The secretary was permitted to reduce the loan rate an additional 20 percent to maintain competitiveness in world markets and to prevent CCC stock buildups.

Target prices and deficiency payments were also continued by the 1985 act, but target prices were mandated to decline over time. To receive deficiency payments, producers were required to comply with any acreage reduction program in effect. The act mandated an acreage reduction program when CCC carryover stocks reached a specified level. To further reduce production under such circumstances, the secretary was authorized to implement a voluntary paid land diversion program. The conservation reserve program was a third means available with which to reduce feed grain production.

The 1985 act provided for deficiency payments for feed grains not actually planted. A producer who planted as little as 8 percent of his base acreage and devoted the remainder to soil conserving uses could receive deficiency payments based on the assumption that he had planted up to 92 percent of his permitted acreage. This is the so-called 0/92 underplanting provision.

Several export programs designed to increase commodity exports and enhance the competitiveness of U.S. exports were continued. In addition, an Export Enhancement Program (EEP) and a Targeted Export Assistance Program (TEAP) were authorized by the 1985 act to help U.S. exporters compete in specific markets to which other countries subsidize exports or in general to help increase exports of agricultural commodities.[5]

Legislation passed in 1990 continued most of the provisions of the 1985 act, but some new features were added. The Food, Agriculture, Conservation, and Trade Act of 1990 provided for (1) target prices to be maintained at 1990 levels, (2) loan rates to be set at 85 percent of a five-year moving average of market prices, and (3) cross-compliance and offsetting-compliance (see Glossary for definitions) as a condition for program eligibility to be prohibited. The 1990 farm bill further provides for the establishment of acreage reduction program requirements on the basis of the stocks-to-use ratio rather than on the basis of carryover quantity alone.

Of additional significance are the key provisions of the Omnibus Budget Reconciliation Act of 1990 relating to agriculture. For feed grains this legislation mandated a 15 percent triple-base acreage reduction plan as outlined earlier in this chapter. Further, it requires deficiency payments on feed grains to be based on the prior twelve-month-average market price rather than the prior five-month average, and it reduces the 1991 acreage reduction percentage for feed grains to 7.5 percent. These provisions were implemented solely for deficit reduction purposes and are a clear instance of how federal budget pressures can and do directly influence farm policy choices.

Support prices are set for the feed grains other than corn at levels the secretary of agriculture deems "fair and reasonable" in relation to corn. A target price/deficiency payment program is not available for rye nor is a paid land diversion program. Nonrecourse loans, however, are available for rye as well as for the remaining feed grain crops.

Wheat

CURRENT POLICY

- Nonrecourse loans
- Target prices/deficiency payments
- Acreage reserve program
- FOR program
- 0/92 underplanting provision
- Triple-base plan

Wheat also suffered from surplus production and low prices in the 1930s, prompting Congress to provide wheat producers income support via nonrecourse loans, government purchases, acreage allotments and marketing quotas, export subsidies, and conservation incentives.

Between 1948 and 1961 nonrecourse loans and purchase agreements were the primary means of supporting wheat producers' prices and incomes. When surpluses accumulated, acreage allotments and marketing quotas were proclaimed, with penalties for exceeding allotments and quotas. A wheat land retirement program was instituted under the Agricultural Act of 1956. In 1962–63 a voluntary paid land diversion program was added.

Under 1964 legislation, acreage allotments were maintained but new features were added. All mandatory features of the old legislation were dropped, and fines and penalties for exceeding allotments were removed. The loan rate was lowered closer to world market levels. Only producers who planted within their allotment were eligible for price support loans. Incomes of producers who participated in the acreage reserve program received direct payments in the form of certificates so that the blend of market price and certificate value would be at about 80 percent of parity.

The 1971 legislation authorized a wheat allotment only for domestic food use. No limit was placed on wheat plantings. Allotments were only used for computing set-aside acreage requirements and marketing certificate payments.

The Agriculture and Consumer Protection Act of 1973 repealed the wheat certificate program and replaced it with the target price/deficiency payment program. This act also authorized disaster payments for wheat producers.

Under the Food and Agriculture Act of 1977, wheat producers received deficiency payments on their allotments regardless of how many acres of wheat they planted. At this time there was much less concern about surplus production although a set-aside program was also implemented. This act also made the FOR program available to wheat producers.

By 1981–82, wheat surpluses and low farm prices were again a problem, so supply management became an important issue. An acreage reserve program was introduced with diverted land required to be placed in approved conservation uses. In 1983 there were a 15 percent acreage reserve program, a 5 percent paid land diversion program, and a 10–30 percent PIK program under which participating producers could submit bids to idle their entire base in return for in-kind payments. Legislation enacted in 1984 mandated a 20 percent acreage reserve program and a 10 percent paid land diversion program for the 1984 and 1985 crops.

The Food Security Act of 1985 provided for price and income support and production control features identical to those for the feed grains. As was the case for feed grains, at the time of passage of the 1985 act the policy issues were high CCC stock levels and falling export market shares. Thus policies were adopted in an effort to increase the competitiveness of U.S. wheat, and export instruments were authorized with which to increase U.S. wheat exports.

Legislation passed in 1990 maintained most of existing policy for wheat and implemented the same new provisions for wheat outlined for feed grains with the exception that the 1991 wheat acreage reduction percentage was lowered to 15 percent by the budget bill. Again, the crucial issues Congress had to contend with in passing wheat legislation for 1990 were continuing wheat surpluses and federal budget pressures.

Rice

CURRENT POLICY

- Nonrecourse loans
- Target prices/deficiency payments
- Marketing loans
- Marketing certificate program
- Acreage reduction program
- 50/92 provision
- Triple-base plan

The Agricultural Adjustment Act of 1938 provided for nonrecourse loans, marketing quotas, acreage allotments, and direct payments to bring producer prices of rice up to parity. Nonrecourse loans were not offered, however, until 1941, at which time they became mandatory for farmers harvesting within their acreage allotment. Marketing quotas were to be proclaimed if total supply was estimated to exceed normal supply by 10 percent. The 1948 and 1949 acts continued this authority and set a mandatory price support at 90 percent of parity. The Agricultural Act of 1954 attempted to deal with the mounting surpluses of rice with flexible price supports. Marketing quotas were proclaimed and voted in by a farmer referendum for the 1955 crop. Marketing quotas and acreage allotments continued in effect through 1973.

The Agricultural Act of 1956 introduced the Soil Bank Program in an effort to further reduce rice acreage. This provision provided for an acreage reserve program that paid farmers to divert rice land and for a conservation reserve that paid farmers to retire rice land under long-term contracts of from three to ten years. These programs were not very effective, however, and the acreage reserve program ended in 1958 and the conservation reserve program ended in 1961.

In the early 1970s rice surpluses diminished and farm prices were quite favorable as exports increased. Accordingly rice marketing quotas were suspended for the 1974 and 1975 crops. The Rice Production Act of 1975 shifted rice production control from quotas and allotments to greater market orientation along the lines of the programs for other grains. A target price/deficiency payment program was established with allotments the payment base. Farmers could plant in excess of their allotment, but eligibility for loans and deficiency payments was restricted to the rice produced from allotted acres. This act also provided for an acreage set-aside program. The Food and Agriculture Act of 1977 continued these provisions and authorized a cash payment for diverting rice land to non–program crop uses.

The 1981 act allowed the secretary of agriculture to operate an acreage reserve and a paid land diversion program with no restrictions. The 1982 act mandated a 15 percent acreage reserve program and a 5 percent paid land diversion. The Food Security Act of 1985 raised the limit on the acreage reserve program to 35 percent and set a 30 million hundredweight target for carryover stocks.

The Food Security Act of 1985 provided price and income support for rice identical to that provided for wheat and feed grain producers. In addition it mandated marketing loans for the 1986–90 crops whereby producers may repay their nonrecourse loans at the prevailing world market price, at 50 percent of the 1986–87 loan rate, 60 percent of the 1988 loan rate, or 70 percent of the 1989–90 loan rate. The secretary of agriculture could require producers to purchase negotiable certificates redeemable for CCC-owned rice as a condition for repaying the loan at the lower level. The secretary of agriculture could also offer loan deficiency payments to producers who were eligible for loans but who did not take out loans. The various programs already in place designed to increase rice exports were continued. In addition rice exporters were assisted by the EEP and TEAP.

The CCC was mandated by the 1985 act to issue negotiable marketing certificates (generic certificates) to persons who entered into agreements with CCC to participate in the certificate program whenever the world price fell below the loan rate. These certificates could be redeemed for cash or for any CCC-owned commodities and were transferable.

The 1990 legislation maintains existing rice policy and implements the same new provisions as outlined for feed grains and wheat except that no change was specified in the acreage reduction requirement. In addition, marketing certificates issued to program participants when world price is below the loan rate can now be exchanged for any CCC-owned commodity, not just rice.

Soybeans

CURRENT POLICY

- Nonrecourse loans
- Marketing loans
- Loan deficiency payments
- Fee assessments

Soybean producers do not have a history of high price supports nor have they ever been subject to an acreage reduction program. Similarly there have been no acreage allotments or marketing quotas for soybeans. Such programs have never been pushed by soybean producers because up to now demand has been sufficient to clear the market at reasonable prices. Given the favorable demand situation, producers have frequently expanded soybean acreage onto land that could not be used for other program crops due to marketing quotas or allotments. On occasion Congress has encouraged this with extra incentives to wheat, feed grain, and cotton producers (e.g., in 1966, 1970, and 1973). Sometimes, however, soybean production has been discouraged by program provisions for other crops. Under the Food Security Act of 1985, for example, high support or target prices relative to loan rates for grains and cotton encouraged farmers to participate in the grain and cotton programs rather than expand soybean production when market demand was strong for the latter.

Following World War II, nonrecourse loans were in effect for soybeans at the discretion of the secretary of agriculture. Market prices were generally above the established loan rates, however, so there was a minimum of soybean loan activity. The late 1950s and early 1960s provided an exception. Even then, though, the oversupply situation did not become chronic.

A nonrecourse loan program for soybeans was mandated for the first time by the Food and Agriculture Act of 1977 and continued by the Agriculture and Food Act of 1981. Under the latter act, loan rates for soybeans were set at 75 percent of the simple average of prices received by farmers over the preceding five years excluding the high and low years. Neither target prices nor deficiency payments were available to soybean producers. These provisions were continued by the 1985 act.

The 1990 farm bill initiates a soybean marketing loan program and extends both the nonrecourse loan and marketing loan program to the remaining oilseed crops (canola or rapeseed, safflower seed, flaxseed, mustard seed, sunflower seed, and sesame seed). The 1990 farm bill thus creates a new category of program commodities; the six additional oilseeds.

As in the past there is no target price/deficiency payment or acreage reduction program available to oilseed producers. For federal deficit reduction purposes oilseed producers who participate in the loan program will be assessed a 2 percent loan origination fee.

Cotton

CURRENT POLICY

- Nonrecourse loans
- Marketing loan program
- Marketing certificate program
- Target price/deficiency payment
- Acreage reduction program
- Import quotas for upland cotton
- Loan deficiency payments
- 50/92 provision
- Triple-base provision

Mounting production surpluses and falling farm prices in the U.S. cotton industry led to legislation in the 1930s that provided for mandatory price support loans and marketing quotas keyed to acreage allotments. Acreage allotments and marketing quotas were used from 1938 to 1942. The Agricultural Act of 1948 provided for mandatory support of cotton at 90 percent of parity if producers approved marketing quotas. Subsequent legislation extended this level of support through the 1954 crop. Cotton was under marketing quota continuously from 1954 through 1970.

In the late 1950s and early 1960s, Congress had to face the reality that cotton surpluses were mounting and existing legislation provided no effective provision to deal with them. The Agricultural Act of 1964 authorized the secretary of agriculture to make payments to domestic handlers or textile mills in order to bring the price of cotton used in the United States down to the export price level. This essentially ended the two-price system that had been in effect since 1956. Also a domestic allotment program was authorized for 1964–65. Producers who planted within their allotment received a higher support. This act marked the beginning of voluntary programs for reducing cotton production.

The Food and Agricultural Act of 1965 was more market oriented than previous legislation, with price supports for all of the program commodities (except dairy) set near world market prices. Cotton was supported at 90 percent of estimated world price levels. Incomes of cotton farmers were maintained through payments based on the extent of participation in an acreage reduction program. Sale and lease of allotments within a state were now permitted. Beginning in 1966 cotton producers could receive a land rent for diverting cropland acreage to approved conserving uses.

By the end of 1970, the huge CCC inventory of cotton was gone, but the cost of accomplishing the feat (via land diversion) was very high, and the distribution of payments among different-sized cotton producers was decidedly uneven.

The Agricultural Act of 1970 continued the voluntary program for cotton, suspended marketing quotas, and provided for a cotton set-aside program. The set-aside payment for cotton producers was specified as the difference between the higher of 65 percent of parity or 35 cents a pound and the average market price for the first five months of the marketing year so long as this payment was at least 15 cents per pound. The cotton loan rate was set at 90 percent of the average world price for the previous two years.

For reasons indicated before, by 1973 a different approach to farm programs was sought. The result was a target price/deficiency payment program with loan rates set at or near world price levels. A cropland set-aside program was authorized for cotton but not needed.

The 1977 act gave the secretary of agriculture authority to adjust target prices based on a number of factors, including cost of production, but maintained loan rates for cotton at or near world market levels. Acreage allotments and bases were removed. The 1981 act provided for an acreage reduction program to control surpluses.

Increased stocks, depressed prices, and lower farm incomes led to implementation of a PIK program for the 1983 crop in addition to the existing acreage reduction and paid land diversion programs. Eligibility for program benefits and PIK program participation required growers to participate in the 20 percent acreage reduction program. A producer could idle up to an additional 5 percent of the base acreage in return for a cash diversion payment rate of 25 cents per pound. Farmers participating in the 20 percent acreage reduction program had the option of idling an additional 10–30 percent of their base acreage and receiving a PIK equal to 80 percent of the farm program yield. They also had the option of submitting sealed bids indicating the percentage of their farm program yield for which a PIK would be accepted for idling their entire base acreage.

The Food Security Act of 1985 mandated marketing loans for the 1987–90 cotton crops whereby if the world market price was less than 80 percent of the loan level, producers could repay their loan at a rate that the secretary of agriculture estimated would minimize loan forfeitures, accumulation of stocks, and storage costs and that would allow free marketing of cotton domestically as well as internationally. The secretary of agriculture was also authorized to offer loan deficiency payments. As with the other program crops, the emphasis of the 1985 act was on maintaining competitiveness in international trade.

The CCC must issue negotiable marketing certificates to first-handlers of cotton who have entered into agreements with the CCC to participate in the certificate program when the world market price is below the loan rate—that is, when the loan rate would otherwise prevent the United States from being competitive internationally. These certificates can be redeemed for cash, for any CCC-owned commodity, or sold to the highest bidder unless they are declared cotton-specific by the secretary.

The 1990 legislation maintains existing cotton policy and implements the same new provisions as outlined for feed grains, wheat, and rice except that no change was specified in the acreage reduction requirement. In addition, marketing certificates issued to first-handlers when world price is below the loan rate must be generic so that they are now automatically exchangeable for any CCC-owned commodity, not just cotton.

Wool and Mohair

CURRENT POLICY

- Direct or incentive payments
- Import duties
- Fee assessments

Price supports for wool first became mandatory in 1947 and for mohair in 1949. Price supports for these two commodities were set at between 60 and 90 percent of parity. Nonrecourse loans and government purchases were used to support these two commodities. The Agricultural Act of 1949 required that wool be supported at a price that would encourage an annual production of 360 million pounds. This goal was thought to require a support price set at the maximum of 90 percent of parity. Even so the production goal was not achieved.

The National Wool Act of 1954 authorized direct "incentive" payments as the method of supporting wool and mohair producer incomes and abandoned the loan program as CCC stocks soared to levels reaching one half of annual production. Under this program producers would receive as a direct payment the difference between the support price and the market price. This is the method still in use today. The 1954 act stated that "wool is an essential and strategic commodity which is not produced in quantities and grades in the United States to meet the domestic needs and that the desired domestic production of wool is impaired by the depressing effects of wide fluctuations in the price of wool in the world markets." This act also continued the tariff on wool imports and an equivalent ad valorem duty on wool thread and cloth of textile grade.

Wool and mohair payments, in contrast to those for other supported commodities, increase as the per unit value of the producer's wool or mohair increases. This payment is designed to encourage the production of a higher quality of wool and thus improve wool marketing. For example, the 1988 market price for shorn wool was $1.38 per pound and the 1988 support price was $1.78 per pound. Thus the payment rate in 1988 was 0.29 cents per pound ([1.78 − 1.38]/1.38) for every $1.00 of greasy wool marketed. Since the payment rate is applied to the dollar value of wool sales, the greater the price a producer receives for wool, the greater is the per pound incentive payment.

The 1990 act extends existing wool and mohair policy as contained in the 1954 act. For deficit reduction purposes, producers will be assessed a fee of 1 percent of their wool and mohair incentive payments.

Tobacco

CURRENT POLICY

- Transferable marketing quotas for burley and flue-cured
- Transferable allotments for flue-cured
- Nonrecourse loans
- Coresponsibility levy
- Fee assessments

The U.S. government has operated programs to support and stabilize tobacco prices since the early 1930s. Tobacco was designated a "basic" commodity in the Agricultural Adjustment Act of 1933 and cash payments were made to growers who restricted production. The Agricultural Adjustment Act of 1938 authorized marketing quotas for tobacco and a penalty for growers who exceeded their designated quota. Many changes have been made in the law since, but the marketing quota authority continues. The program is available for all kinds of tobacco except shade-grown wrapper and perique. Price supports have never applied to Pennsylvania filler. Cigar binder and Ohio filler first came under quotas in 1951. The Maryland crop was last supported in 1965.

Under the price support program, a loan rate (rather than a support price) is established for each grade of tobacco. If a buyer's bid price on any lot of tobacco is not above the loan rate for that grade, the eligible grower receives the loan rate instead of the bid price and the tobacco is acquired by a cooperative association. Under an agreement with the CCC, the cooperative arranges for receiving, redrying, packing, storing, and the eventual sale of the tobacco under loan.

Beginning in 1962, lease and transfer of flue-cured acreage allotments within counties were permitted. In 1965, poundage quotas were implemented for flue-cured tobacco. Poundage quotas were required for burley tobacco in lieu of acreage allotments in 1971. Poundage quotas for both types of tobacco removed incentives to increase yields at the expense of quality under acreage allotments.

Lease and transfer of all quotas became effective for burley tobacco in 1971. Producers were allowed to sell an amount of tobacco up to 110 percent of their quota, with marketings the following year to be reduced by the amount of marketings in excess of their quota or below their quota. Quotas plus or minus these over- or undermarketings are known as "effective" quotas.

A "four-leaf plan" was authorized between 1978 and 1983 in which growers were permitted to plant additional acreage if they agreed not to harvest the lower quality, four lower leaves of each stalk. This alleviated the stock buildup during this period when demand was weak. In 1984 this ruling was repealed.

In more recent years two major concerns of the tobacco industry and of policymakers were that the U.S. Treasury outlays for the tobacco program were becoming excessive, and the U.S. tobacco industry was losing its competitiveness in world markets because of escalating U.S. price supports. Thus in 1982 Congress passed the No-Net-Cost Tobacco Program Act, which required that to be eligible for price support, producers of all kinds of tobacco must contribute a "coresponsibility levy" to a fund to ensure that the loan program operates at zero net cost to the government except for administrative expenses. This legislation also gave the secretary of agriculture authority to reduce support rates for tobacco grades that are in excess supply so as to make U.S. prices more competitive.

The 1982 law also provided authority for owners of flue-cured allotments and quotas to sell these rights separately from the farms to which they are attached for use on other farms in the same county and to active tobacco producers. Corporations, utilities, educational and religious institutions, and other entities owning tobacco allotments but not significantly involved in farming were required to sell their allotments to active producers or people who planned to become active producers within the same county. If these rights were not sold by December 1, 1984, such allotment holders would forfeit their allotments. Other provisions of this act prohibited fall leasing of flue-cured quotas and stipulated that allotments on any flue-cured farm cannot exceed 50 percent of the farm's eligible cropland.

Legislation signed into law in 1983 provided among other things that the free lease and transfer of flue-cured tobacco quotas would be abolished in 1987 and that thereafter flue-cured quota owners can (1) grow the quota only on land to which the quota is assigned, (2) rent the quota to an active grower who will produce the crop on the land to which the quota is assigned, or (3) sell the quota to an active grower in the county. If sold, the seller must allow the buyer up to five years to pay for the quota. Also this legislation stipulates that beginning in 1984, no more than 15,000 pounds of burley quota can be transferred to a single farm. Previously the limit was 30,000 pounds.

Under 1990 legislation, participating tobacco producers will be assessed a loan origination fee of 1 percent of the loan value. All other provisions remain unchanged.

Honey

CURRENT POLICY

- •Nonrecourse loans
- •Price supports
- •Marketing loans
- •Fee assessments

When sugar rationing terminated at the end of World War II, honey prices fell to near pre-war levels, prompting Congress to act on behalf of the nation's beekeepers. The Agricultural Act of 1949 provided that honey along with the basic commodities should receive mandatory price support at between 60 and 90 percent of parity. Under the 1950 program, packers of honey signed contracts with the U.S. Department of Agriculture (USDA) under which they agreed to pay beekeepers 9 cents per pound (delivered to their packing plants) for all honey acquired from them that met requirements concerning cleanliness of the honey, its moisture content, and its flavor. USDA, in turn, agreed to accept all the honey contracting packers offered at the support price plus established charges for handling, storage, and any processing required by USDA. For the 1982 season a price support loan and purchase agreement program was instituted. This program has been operational in every season since.

Under the honey program, nonrecourse loans at the applicable price support rate on warehouse-stored and farm-stored honey are made available to beekeepers during the crop year on all honey produced during that year. If the market price fails to rise above the support price, beekeepers may cancel their loans by delivering honey of value equal to the loan value at the end of the loan period. Beekeepers who have not made use of the loan feature of the program may use the purchase option. Under this option, the CCC will buy at the support price any honey a beekeeper wishes to sell which is not already obligated to the CCC as loan collateral.

The Food Security Act of 1985 authorized the secretary of agriculture to offer producers marketing loans to maintain competitiveness with foreign suppliers. The new law also eliminated the parity formula for determining support prices and mandated a progressively lower support price as for the other program commodities. To further encourage a market-oriented approach, the CCC purchase agreement option was eliminated.

The 1990 farm bill provides that honey will be supported at not less than 53.8 cents per pound. Otherwise, program provisions specified by previous legislation remain unchanged. For deficit reduction purposes, honey producers will be assessed a fee of 1 percent of the level of price supports on all honey marketings.

Peanuts

CURRENT POLICY

- •Transferable marketing quotas
- •Two-tiered price support
- •Nonrecourse loans
- •Import Quotas
- •Fee assessments

Peanuts came under the production control and diversion provisions of the Agricultural Adjustment Act of 1933. The program included contracts with peanut growers obligating them to plant not over 90 percent of the 1933–34 planted acreage. Compliance ensured benefit payments for diverting peanuts into crushing for oil and meal. The Soil Conservation and Domestic Allotment Act of 1936 authorized payments to peanut farmers for voluntarily shifting acreage into soil-conserving legumes and hays. In 1937, four regional peanut growers' associations were organized to participate in the peanut diversion programs. These associations were authorized to buy up to a certain quantity of peanuts at prices established by the USDA. Storage costs and losses on surplus peanuts diverted to crushing were absorbed by USDA.

The Agricultural Adjustment Act of 1938 as amended in 1941 authorized marketing quotas for peanuts and made price supports mandatory at 50–75 percent of parity. Peanut marketing quotas were approved for the 1941–43 crops, with penalties for noncompliance.

The Agricultural Act of 1949 set support levels at 90 percent of parity for 1950 and between 75 and 90 percent of parity for subsequent years. Producers were to receive price supports only if acreage allotments and marketing quotas were in effect. Marketing quotas and acreage allotments have been in effect for peanuts ever since. Generally, the secretary of agriculture is directed to establish marketing quotas when supplies are excessive. Peanuts are an exception. Peanut marketing quotas must be proclaimed without regard to the supply situation. Farmers can vote out the quota in a referendum, but they never have.

To protect the domestic peanut price support program, the U.S. government has since 1953 set an annual import quota of 1.7 million pounds.

By 1977 the cost of the peanut program had become excessive. As a consequence the Food and Agriculture Act of 1977 implemented a two-price poundage quota program for peanuts while retaining the acreage allotment and price support programs. Under the two-price plan, each allotment holder was given a poundage quota. Producers could produce in excess of their quota within their acreage allotment, but the quantity for which they received the higher of the two price support levels was limited to the quota. Peanuts in excess of quota are referred to as "additionals."

Producers are required to have an allotment if they wish to grow and market peanuts. Under the 1977 act transfers of allotments within a county were allowed. Previously such transfer was permitted only if the secretary of agriculture approved.

Quota peanuts are grown mainly for the domestic market for edible uses and for seed. Quota peanuts could be contracted any time prior to harvest or placed under quota loan at harvest. Producers could contract for sale with a handler and the peanuts could be used only for crushing or export. Additional peanuts could also be delivered to buying points at harvest and placed under loan with the producers receiving the additional price support.

Once the peanuts were received and placed under loan, the producer no longer had control of them. Additionals received for loan could be used for crushing, export, or the domestic edible market. Use in the domestic edible market required the buyer to pay the handling costs plus 100 percent of the quota loan if purchased at the time of delivery or 105 percent of the quota loan if purchased after delivery or 107 percent of the quota loan if purchased on or after January 1 of the subsequent year. This provision, plus the import quota, ensured that the domestic market would not be undercut by the program.

The 1981 act maintained the two-tiered price support system and poundage quotas but suspended acreage allotments. Quota support prices were limited to quota holders and applied to the poundage quota. But since acreage allotments were removed, anyone was eligible to produce peanuts. Additionals, however, were eligible only for the lower of the two support prices, and they were subject to the marketing controls specified in the 1977 act.

The 1985 act provides for a penalty of 140 percent of the loan level for quota peanuts if a producer with quota markets in excess of his quota. Sale and lease of poundage quotas are still permitted but only within a county in the major peanut-producing states.

The peanut program is administered by three regional grower associations which act as agents for USDA. Pools are set up by area, by type of peanut, by quota peanuts and additionals, and by quality category. CCC profits from sales of additionals for domestic edible uses are reallocated to offset any losses CCC has incurred on quota peanuts within a given pool. At the end of the crop year, the CCC balances its books on the loan operations with the area grower associations. If a net surplus remains for a given pool, the surplus is returned to the associations for distribution as dividends to those growers who had peanuts under the loan in that pool. Net losses are absorbed by the government as a CCC budget expense.

The only change in peanut policy brought about by the 1990 legislation is that on all peanut marketings an assessment of 1 percent of the peanut support price (quota or additional support price, whichever applies) will be made.

Sugar

CURRENT POLICY

- Nonrecourse loans
- Government purchases
- Import duties
- County-by-county import quotas
- Fee assessments

The United States imposed tariffs on raw sugar in 1789 to help raise revenue for the federal government. The United States has maintained import duties on all imported sugar except during 1890–94, when there was a surplus in the federal treasury. This lapse in import duties was quickly followed by an influx of cheap imports from Europe, resulting in sugar prices in the United States falling below the cost of production for many producers. As a result Congress passed the McKinley Tariff Act of 1890, which became the first national legislation to recognize and encourage the new beet sugar industry in the United States. In 1894 a tariff was again levied on sugar imports, this time to protect the local industry rather than to generate revenue for the federal government.

The Jones-Costigan Sugar Act of 1934 provided for acreage restrictions and benefit payments to sugar beet and sugarcane growers in domestic areas and in the Philippines, paid for out of funds collected from a processing tax on sugar. This act also provided for a system of country-by-country quotas for controlling the supply of sugar from all sources, domestic and foreign.

The purpose of benefit payments was to give U.S. growers an incentive to limit their acreage in line with USDA-determined quotas. This program was voluntary so that benefit payments went only to those producers who adjusted planted acreages to their quota. The sugar production quota provisions were suspended in April 1942 but reinstated by the Sugar Act of 1948.

The Sugar Act of 1948 was permitted to expire in 1974 on the assumption that the sugar program was no longer needed. However, sugar surpluses quickly developed and prices fell to an average of 7.5 cents per pound by late 1976. Thus Congress included new sugar legislation in the Food and Agriculture Act of 1977. Under this act sugarcane and sugar beets were to be supported through loans or purchases at between 52.5 and 65 percent of parity. Processors were required to pay growers at least the support price specified by the program for average-quality sugar beets and sugarcane as long as growers met USDA minimum wages for field-workers.

Under the Agriculture and Food Act of 1981, a nonrecourse loan program was established with the loan rate for raw cane sugar of not less than 17 cents per pound in 1982, 17.5 cents per pound in 1983, 17.75 cents per pound in 1984, and 18 cents per pound in 1985. The secretary of agriculture was directed to support the price of domestically grown sugar beets at a level that is fair and reasonable in relation to the support level for sugarcane.

To minimize the risk of the CCC acquiring sugar during periods of low sugar prices (as occurred in 1977–78), a market stabilization price (MSP) was established for raw cane sugar above the purchase or loan rate. The MSP is considered by USDA to be the minimum market price required to discourage sale or forfeiture of any sugar to the CCC. The difference between the purchase or loan rate and the MSP covers the cost of freight and related marketing expenses for raw sugar, the interest required to redeem a loan, and an incentive factor to encourage processors to sell sugar in the marketplace rather than to sell or forfeit it to the CCC. To maintain domestic market prices at or near the MSP, the secretary of agriculture estimates the domestic demand for and supply of sugar and then imposes country-by-country import quotas accordingly.

Continuing low world prices for sugar plus the legal restriction that prohibited more than a 50 percent ad valorem duty on sugar imports prevented the secretary from adequately protecting the CCC against sugar forfeitures in late 1981. Thus in early 1982 the president modified the import fee system then in effect and established a new system of country-by-country import quotas because he determined that excessive imports of cheap foreign sugar were not in the interest of domestic producers. These provisions are continued by the Food Security Act of 1985. The 1985 act also gives the secretary additional authority to operate the sugar program "at no cost to the Federal Government." This clause pressures him to maintain tight import quotas and high domestic prices so as to minimize CCC stock accumulations.

The 1990 farm bill establishes provisions for marketing allotments for sugar whenever imports are estimated to be below 1.25 million short tons. Otherwise, sugar policy remains essentially unchanged. For deficit reduction purposes first processors of cane and beets will be assessed a marketing fee of 1 percent of the loan rate.

Dairy

CURRENT POLICY

- Price support on raw milk
- Government purchases
- Federal Marketing orders establish minimum class prices on grade A Milk
- Import quotas
- Assessments for excessive government purchases
- Fee assessments

U.S. dairy legislation in force today, as is the case for most all farm legislation, evolved out of policy formulated in the 1930s. This legislation is aimed at stabilizing the domestic dairy industry and ensuring U.S. dairy farmers an "adequate" income. The policy is implemented via price supports administered by the CCC, classified pricing and pooling through federal milk marketing orders, and very tight import controls.

A support price for producer milk is established by the secretary of agriculture based on guidelines provided by the extant agricultural legislation. This is a price goal to be sought, not a guarantee. Purchase prices for butter, nonfat dry milk, and cheddar cheese are also established by the secretary of agriculture with a view toward maintaining the targeted support price for raw milk. If processors cannot find commercial buyers for all the butter, nonfat dry milk, or cheddar cheese they produce for at least the announced purchase price, the CCC is obligated to purchase these products at the specified purchase prices. When the market price is above the announced purchase price and thus, at least theoretically, the handler buying price for producer milk is above the announced support price for raw milk, the CCC is authorized to sell on the open market any butter, nonfat dry milk, and/or cheese stored in government warehouses.

Nearly 80 percent of the fluid-grade (grade A) milk[6] produced in the United States is regulated by one of forty-four milk marketing orders. These orders establish minimum prices to be paid by handlers for (1) fluid grade milk used for fluid purposes and (2) fluid grade milk diverted to manufacturing uses. These minimum prices were until 1985 established on the basis of a formula specified by the market administrator based on testimony taken at public hearings. In all cases the minimum prices were based on a competitive pay price for manufacturing-grade milk (the average price paid farmers by processing plants in Minnesota and Wisconsin for grade B milk), which in turn is heavily influenced by the support price for producer milk. Milk marketing orders also establish a pooling procedure so that the actual price farmers receive is a weighted average of the minimum price for each class of milk. The Food Security Act of 1985 specified by legislation the differential to be added to the average price received by Minnesota-Wisconsin producers for grade B milk to determine the minimum class I price to be paid by handlers in each of the forty-four milk marketing orders.

Imports of dairy products are restricted by quotas authorized under section 22 of the Agricultural Act of 1949. Currently, positive quotas exist for twelve categories of cheese, chocolate, buttermilk, skimmed and whole milk, dried cream, evaporated milk, and dry milk. In recent years imports of dairy products have amounted to the equivalent of about 2 percent of total U.S. milk production. Hence U.S. import quotas serve to effectively insulate the U.S. dairy industry (both farming and processing) from foreign competition.

Until recent times the United States had no need for milk supply control measures on a national scale. A combination of factors has, however, led to quite burdensome surpluses at present: sustained high support prices since the early 1970s, low feed prices, and failure of the 1983–84 dairy assessment program to reduce supplies. The dairy assessment program was authorized by the Omnibus Budget Reconciliation Act of 1982. Under this program the secretary of agriculture had the authority to deduct 50 cents per hundredweight from the proceeds of commercial sales of all milk by producers if support purchases were estimated to be 5 billion pounds milk equivalent or more during the current fiscal year, and an additional 50 cents per hundredweight if over 7.5 billion pounds milk equivalent. The dairy assessment program resulted in a reduction in the effective support price of $1.50 over the period for which it was in force.

To further deal with the surplus problem, in 1983 Congress enacted a dairy diversion program, under which contracts were offered to producers who agreed to reduce their marketings by 5–30 percent in return for a payment of $10 per hundredweight reduction in milk production. Any further assessments under the previous program were to be credited toward the diversion payment. Finally, in 1986 the Milk Production Termination Program (Whole-Herd Buyout Program) was put into operation. Dairy producers were offered the opportunity to submit a bid to sell to the government their right to produce milk for the next five years.

The Food Security Act of 1985 further provided for an automatic downward adjustment of the milk support price at the beginning of the year by 50 cents per hundredweight if the secretary of agriculture determined that CCC purchases of dairy products required to maintain the announced support price would exceed 5 billion pounds of milk equivalent. An upward adjustment of 50 cents per hundredweight was required if CCC purchases were projected to be less than 2.5 billion pounds of milk equivalent.

Under the Food, Agriculture, Conservation, and Trade Act of 1990, the price support for milk containing 3.67 percent milk fat can be adjusted up or down depending on anticipated CCC purchases of dairy products for the ensuing year. The support price may not be reduced below $10.10 per hundredweight, although if CCC purchases are anticipated to reach or exceed a critical level when the support price is already at $10.10, farmers could be assessed a special fee to be determined by the secretary of agriculture. The new farm bill also expressly prohibits a dairy termination program similar to that authorized by the 1985 farm bill. As in the case of other program commodities for which no other means of reducing federal budgetary costs were available, dairy farmers will be assessed a fee of 5 cents per hundredweight in 1991 and 11.25 cents per hundredweight thereafter, with the proviso that these assessments can be refunded to any producer who does not increase production over the previous year's level.

What Has Been Learned

Some sixty years of involvement with price and income policy for agriculture has taught several lessons. The first lesson is that in the absence of effective supply control, prices supported above equilibrium levels lead to surpluses! The one notable exception is in wool and mohair. In this case, however, the United States is a large deficit producer. It will take much higher wool and mohair price supports to encourage additional production of these products. A second lesson is that controlling the acreage of one crop is not sufficient. Cross-compliance provisions are necessary so that when substitutions in production are relatively easy, surpluses or shortages of another crop are avoided.

A third lesson is that it is difficult to change the direction of policy. Changing emphasis from a policy of farm income support to some alternative such as encouraging resource adjustments has received limited attention. Other policy thrusts, such as soil conservation, are only made when convenient. That is, when surpluses accumulate, conservation programs appear so that plantings of program crops can be reduced. But when the market becomes tight and surpluses disappear, as was the case from the mid-1970s to the early 1980s, farmers are encouraged to plant fencerow to fencerow with little regard for soil conservation. The more market-oriented policies of the late 1960s and early 1970s were very difficult to get through Congress. The difficulty here was compounded by the state of the agricultural economy. In 1973 the final push in this direction was achieved largely because farm incomes were high and government costs of the farm program were low. By 1983, however, the situation was reversed and huge farm subsidies once again appeared.

This brief historical review has also embedded in it lessons related to pricing or legislating ourselves out of international markets. High prices for any commodity encourage the search for new sources of supply and/or substitute products. High U.S. support prices for cotton, sugar, peanuts, and tobacco have encouraged other countries to increase their production of these commodities since the 1930s. Certainly, supporting the price of cotton at levels higher than equilibrium spurred the development of synthetic fibers. Supporting the price of sugar in the United States at prices three to four times world market prices has encouraged the development of glucose, dextrose, and most recently high-fructose corn syrup.

Other policy decisions can also have important consequences. In 1980 the United States imposed an embargo on shipments of wheat, feed grains, soybeans, and selected other products to the USSR in retaliation for the USSR invasion of Afghanistan. The lesson here is that other countries producing a surplus of these commodities were quite willing and able to supply the USSR. The embargo not only failed to deter the USSR from invading Afghanistan, it cost the United States a short-run market for grains as well as Treasury dollars to compensate farmers for this loss. Whether or not it resulted in market losses over the longer run is still debatable (see U.S. Department of Agriculture 1986). Sluggish recovery of exports in the face of a significant depreciation of the U.S. dollar suggests that there were significant longer run market losses as well.

Notes

1. See the Glossary for a description of the CCC and its functions. A short description of each of the policy instruments discussed here will also be found in the Glossary.

2. *Surplus production* as used throughout this book will mean the excess of commercial production over commercial use plus commercial storage at prevailing market prices. It is that portion of current production that is purchased by government. It would not exist in the absence of government intervention which encourages producers to produce more than will clear the market at current market prices.

3. Not included in this listing is the farmer-owned reserve (FOR) program (see the Glossary for a complete description of the FOR program). The FOR program has elements of a supply control program. More fundamentally, however, the FOR program is a safety-net program designed to maintain a supply of grain for emergency purposes and a stabilization instrument designed to reduce the instability of prices of commodities for which FOR is available. What impact the program has is obviously dependent upon how it is implemented. Salathe, Price, and Banker (1984) argue that the program has had a positive impact on commodity prices and farm incomes but has not significantly reduced price variability (for a more detailed discussion of price stabilization programs see Chapter 11).

4. This description draws heavily on the background reports for the 1990 farm legislation prepared by Economic Research Service and referenced at the end of this chapter.

5. See Chapter 9 for a full description of these programs and an assessment of their consequences.

6. Milk that meets the sanitary requirements established by the U.S. Public Health Agency for drinking milk. All other milk is termed grade B, or manufacturing-grade, milk.

Suggested Readings and References

Ash, Mark, and Linwood Hoffman. 1989. "Barley: Background for 1990 Farm Legislation." U.S. Department of Agriculture. Economic Research Service. Staff Report no. 89-65. December.

Barry, Robert D., Luigi Angelo, Peter J. Buzzanell, and Fred Gray. 1990 "Sugar: Background for 1990 Farm Legislation." U.S. Department of Agriculture. Economic Research Service. Staff Report no. 90-06. February.

Becker, Geoffrey S. 1986. "Fundamentals of Domestic Commodity Price Support Programs." Congressional Research Service, Library of Congress. Report no.

86-128 ENR. June.

Childs, Nathan W., and William Lin. 1989. "Rice: Background for 1990 Farm Legislation." U.S. Department of Agriculture. Economic Research Service. Staff Report no. 89-49. November.

Cochran, W. W. and Mary Ryan. 1976. *American Farm Policy, 1948–73*. Minneapolis: University of Minnesota Press.

Crowder, Brad, and Cecil Davison. 1989. "Soybeans: Background for 1990 Farm Legislation." U.S. Department of Agriculture. Economic Research Service. Staff Report no. 89-41. September.

Fallert, Richard F., Don P. Blayney, and James J. Miller. 1990. "Dairy: Background for 1990 Farm Legislation." U.S. Department of Agriculture. Economic Research Service. Staff Report no. 90-20. March.

Glaser, Lewrene K. 1986. "Provisions of the Food Security Act of 1985." U.S. Department of Agriculture. Economic Research Service. Agricultural Information Bulletin no. 498. April.

Grise, Verner N. 1989. "Tobacco: Background for 1990 Farm Legislation." U.S. Department of Agriculture. Economic Research Service. Staff Report no. 89-48. October.

Harwood, Joy L., and C. Edwin Young. 1989. "Wheat: Background for 1990 Farm Legislation." U.S. Department of Agriculture. Economic Research Service. Staff Report no. 89-56. October.

Hoff, Frederic L., and Jane K. Phillips. 1989. "Honey: Background for 1990 Farm Legislation." U.S. Department of Agriculture. Economic Research Service. Staff Report no. 89-43. September.

Hoffman, Linwood A., and Mark Ash. 1989. "Oats: Background for 1990 Farm Legislation." U.S. Department of Agriculture. Economic Research Service. Staff Report no. 89-46. September.

Lawler, John V. and Robert A. Skinner. 1989. "Wool and Mohair: Background for 1990 Farm Legislation." U.S. Department of Agriculture. Economic Research Service. Staff Report no. 89-62. November.

Lin, William, and Linwood Hoffman. 1989. "Sorghum: Background for 1990 Farm Legislation." U.S. Department of Agriculture. Economic Research Service. Staff Report no. 89-67. December.

Mercier, Stephanie. 1989. "Corn: Background for 1990 Farm Legislation." U.S. Department of Agriculture. Economic Research Service. Staff Report no. 89-47. September.

Salathe, Larry, J. Michael Price, and David E. Banker. 1984. "An Analysis of the Farmer-Owned Reserve Program, 1977–82." *American Journal of Agricultural Economics* 66 (1): 1–11.

Schaub, James D., and Bruce Wendland. 1989. "Peanuts: Background for 1990 Farm Legislation." U.S. Department of Agriculture. Economic Research Service. Staff Report no. 89-61. November.

Stults, Harold, Edward H. Glade, Jr., Scott Sanford, and Leslie A. Meyer. 1989. "Cotton: Background for 1990 Farm Legislation." U.S. Department of Agriculture. Economic Research Service. Staff Report no. 89-42. September.

U.S. Department of Agriculture. 1986. "Embargoes, Surplus Disposal, and U.S.

Agriculture: A Summary." Economic Research Service. Agricultural Information Bulletin no. 503. November.
_____. 1989. "Economic Indicators of the Farm Sector: National Financial
Summary." Economic Research Service. ECIFS 8-1. September.

3 The Basis for Governmental Legislation for Agriculture

U.S. government legislation has almost always been favorable to agriculture. This legislation has for the most part been designed to accomplish a variety of developmental and/or income distribution objectives. Further, this legislation has been promulgated and sustained on the basis of considerable support, or at least lack of resistance, from the American populace. The inquisitive student of agriculture, politics, or economics will ask, Why has this been the case and why does it continue? Is agriculture such a key element of the national economy that it must be protected by this kind of government support? Is the political power of the various agricultural interests strong enough to explain such favorable legislation? Are prices and incomes in American agriculture chronically and unacceptably low so that generous taxpayer support of the sector is justified? Are the characteristics of agricultural production and/or product marketing sufficiently troublesome or unique as to call for special support? Do these characteristics constitute special cases of market failures discussed earlier which would justify corrective measures in the form of price and income legislation for agriculture? Are conditions in the agricultural sector today such that would call for less government intervention or for legislation of a different kind?

Clear answers to the above set of questions will not be found in this book. In fact, most of these questions have answers that are partly determined by values and beliefs held by the public, as we learned in Chapter 1, and cannot be generated by scientific methods or pure reasoning alone. Nevertheless, the questions raised here are of critical importance to a study of policy for agriculture. How they are answered by the body politic will shape future policy for this sector. It is important, then, to have at hand background information relevant to such issues before proceeding further.

Importance of Agriculture

Producer of Food. There is no disagreement over the fact that food is essential to the health and well-being of a nation's citizens. Consequently, many argue, the agricultural sector must be generously supported and protected so as to preserve for consumers a ready access to an adequate supply of food not only in times of natural calamities and war but also in times of peace. Indeed most U.S. agricultural legislation explicitly states that one of its aims is to maintain an adequate supply of pure and wholesome food for consumers at reasonable prices.

In some countries this is translated into a national goal of "food self-sufficiency." A land of such abundance as is the United States has rarely if ever had to worry greatly about food self-sufficiency, although it does worry about large variations in food supplies from day to day or month to month or year to year. In the more densely populated countries like Japan and those in Western Europe where citizens still have vivid memories of the ravages of war and of war-caused food shortages, food self-sufficiency is a much more prominent national objective.

In addition to an adequate and stable supply of food, a nation's citizens want assurances that their food-producing sector is as efficient as possible so that food costs are kept to a minimum. To assess the efficiency of the food sector is a difficult task. Comparative statistics such as presented in Table 3.1 are often used for this purpose. Based on these data it would appear that the United States has few if any equals. U.S. consumers allocate about 10.4 percent of their total personal consumption expenditures to food purchases, whereas in other countries consumers allocate much larger amounts of their consumption expenditures to food purchases.

The data in Table 3.1 give some insight into the efficiency of food production, but they by no means tell the whole story and can be grossly misused. For example, in 1986, 53.3 percent of the personal consumption expenditures of Indians went for food purchases. We might expect food to be relatively less expensive in India than in the United States because Indians do not process their food as highly as do Americans, nor do they add as many packaging services to food as do Americans. At the same time, however, India's per capita gross national product is only about one sixtieth that of the United States.[1] Given the low per capita income of Indians, then, one should not be surprised that the percentage shown in Table 3.1 for India is so high.

Regardless of the relevance of the comparisons shown in Table 3.1, one must question whether or not the type of farm policy we have had in the United States has resulted in the most efficient food-producing industry we could have had. The policies discussed so far and to be discussed later in

Table 3.1. *Food expenditures as a percentage of total personal consumption expenditures in selected countries, 1970 and 1986*

Country	1970	1986
Australia	18.3	15.5
Austria	26.1	17.5
Belgium	24.1	17.7
Canada	15.1	11.5
Colombia	33.4	31.0[a]
Denmark	20.6	16.4
Finland	24.0	18.8
France	22.0	16.8
Ghana	53.6	53.6[b]
Greece	35.5	34.6
Hungary	33.3	27.9[c]
India	63.0	53.3
Italy	32.2	21.5
Japan	26.4	18.8
Kenya	41.3	41.4[d]
Mexico	34.9	31.9[d]
Netherlands	22.0	14.4
Norway	24.4	18.9
Philippines	51.9	51.5
South Korea	47.4	36.3
Spain	33.2	26.2
Sweden	20.8	17.9
Switzerland	21.8	20.2
Thailand	46.4	35.7
USSR	35.0[e]	28.0
United Kingdom	19.5	13.7
United States	14.3	10.4
West Germany	22.0	16.8
Zimbabwe	24.3	12.7[a]

Source: Korb 1987; and Korb and Cochrane 1989.
[a]Data for 1983.
[b]Data for 1978.
[c]Data for 1982.
[d]Data for 1980.
[e]Data for 1975.

this book have had very little to do with the food-*processing* sector so it is difficult to argue that farm policy has caused inefficiencies here. But, as subsequent chapters of this book will show, government programs for agriculture probably have led to some inefficiencies in production of farm products. Nevertheless, food still appears to be very reasonably priced in the United States.

Agriculture's Role in the National Economy. Another line of argument in support of favorable legislation for agriculture relates to this sector's alleged importance to the general economy. Such maxims as "As agriculture goes, so goes the economy" have long been used to express this idea. The agricultural interests in several states (ranging in diversity from Iowa to Pennsylvania) proudly point to the fact that on a value-added basis, agriculture is their most important industry. This is even more poignant if *agriculture* is defined to include, in addition to farming, the food-processing and food-retailing sectors as well as the farm input manufacturing and retailing sectors. It is on the basis of such statements and pride that ensuring the viability of farming as an economic activity has been deemed to be in the interest of society as a whole, even if the legislation required to do so favors the farm sector at the expense of the urban sector.

As those who have studied the development process in detail agree, most economic activity in a newly developing country centers around agriculture. In the initial stages of development, farmers not only till the soil to produce the various agricultural products, they also provide their own inputs. That is, they raise their own animals that serve as a source of power, produce the grain and hay required to grow and sustain these animals, utilize the waste from these same animals for fertilizing their land, produce their own lumber by felling and hewing homegrown trees, build their own homes and farm buildings, make their own tools for working the land (e.g., plows and harrows), produce their own clothes from wool or flax grown on the farm, make their own shoes and animal harnesses from the hides of animals they raise, harvest their own grain with the aid of cradles they have constructed themselves, etc.

As the country matures, however, industries emerge to supply more and more of the farmer's input needs. A power and machinery industry develops to supply tractors and plows and mechanical harvesters; a chemical industry develops to supply farmers with commercial fertilizer, pesticides, and herbicides; a feed industry develops to produce specially prepared animal and poultry feeds; a transportation sector develops so farmers can obtain inputs from the industrial centers and transport products to central markets more efficiently and effectively; a shipping industry develops so that agricultural products can be marketed in international commerce.

Similarly in the early stages of a country's development, farmers process the products they produce on the farm. Their primary concern is to provide for their own family's food needs. Any surplus food processed will be sold to the few nonfarmers that exist. As development proceeds apace, however, nonfarmers become so numerous that farmers cannot supply all of their own food needs *and* the food needs of nonfarm families. Since nonfarm families have their own obligations, they do not all wish to process their food from raw products they buy from farmers—rather, they want to buy food already processed. Hence a specialized food-processing sector develops. This sector assumes the responsibility of processing agricultural produce into a form that the consumer demands, at a time that the consumer demands it, and at a place that the consumer demands it.

In the early stages of the development of America, agriculture was most assuredly the dominant industry. In the beginning of the nineteenth century as development of lands west of the Atlantic seaboard was in full swing, about 72 percent of the gainfully employed workers in this country were employed in farming (Table 3.2), and about 40 percent of the national income originated in primary agriculture. One may easily and justifiably conclude that during this period "as agriculture went, so went the economy."

But as Table 3.2 also clearly shows, these percentages declined quite rapidly, so that today only 2.8 percent of the gainfully employed workers in the United States are employed in farming, and only 1.7 percent of the gross private product originates in farming! While farming is unquestionably an important activity, one can no longer say that it is *the* most important sector economically, and it is no longer *the* sector around which all other economic activity in the country revolves.

Agriculture's Political Base

The proportion of the nation's population living on farms in the United States was most certainly near 90 percent in 1800, had been reduced to 35 percent by 1910 and to 23 percent by 1940, and has since dwindled to about 2 percent (Table 3.3). One might infer from this trend that there will now be greater difficulty in securing and maintaining legislation favorable to farmers than was the case during the first half of the twentieth century. This conclusion would presumably be based on the notion that there will be greater success in the political arena of rent seeking the greater is the number of rent seekers with a common purpose (see Chapter 1).

There are two problems with this argument. First, the farm sector is still successful in securing favorable legislation as judged by the pervasiveness of programs providing protection and direct income aid to farm people. Second, while the number of people living on farms as a proportion of the total population has declined precipitously, the proportion of the nation's residents living in rural areas has not. In 1800, over 90 percent of the population lived in rural areas. By 1900 this proportion had declined to 60 percent and by 1960 to 30 percent. Slightly more than one quarter of the nation's population still lives in rural areas and this proportion has been

Table 3.2. *Farmworkers as a percentage of all gainfully employed workers and gross product originating in primary agriculture as a percentage of total gross private product, United States, 1800–90*

Year	Farm employment	Product originating in farming
1800	na	39.5
1820	71.9	34.4
1840	68.6	34.6
1860	59.0	30.8
1880	49.3	20.7
1900	37.7	20.9
1920	25.7	12.3
1940	16.9	6.4
1960	10.7	3.9
1980	3.7	2.4
1985	2.7	1.9
1990	2.8	1.7

Source: U.S. Bureau of the Census 1975; and Council of Economic Advisers 1990.
na—not available.

Table 3.3. *Percentage of U.S. residents living in rural areas and living on farms in rural areas, 1800–90*

Year	Rural residents	Farm residents
1800	93.9	na
1820	92.8	na
1840	89.2	na
1860	80.2	na
1880	71.8	43.8
1900	60.3	39.3
1920	48.8	30.2
1940	43.5	23.2
1960	30.1	8.7
1980	26.2	2.7
1985	26.1	2.2
1990	25.8	1.9

Source: U.S. Bureau of the Census 1975; and U.S. Department of Agriculture, *Agricultural Statistics*, various annual issues.
na—not available.

quite stable over the last decade. This is a substantial proportion of the total population. One might well suppose that all of these residents—not just those living on farms—would be interested in securing legislation favorable to agriculture. The livelihood of most of the rural residents in the United States undoubtedly depends on a strong agricultural sector. They will most likely argue for and support favorable legislation for primary agriculture.

Nevertheless, some argue there are signs that the relative political strength of the agricultural sector is waning. Knutson, Penn, and Boehm (1983, 84–85) see three such signs:

1. New public interest groups, recognizing the impact of food and agricultural policy on the people and interests they represent, are continuously injecting new ideas into legislative and executive decision processes. Frequently, they are even willing to challenge adverse decisions in the courts.

2. The members of the agriculture committees of the Congress are no longer responsive only to the agriculture establishment. The new interest groups have members on these committees who will not only hear them out but will either plead their case directly or will give them an opportunity to plead their case.

3. USDA is no longer in full control of the food and agriculture policy decisions in the executive branch. These decisions are viewed as being too important to be left to USDA. They may have important foreign policy implications, budget implications, balance-of-trade implications, or food price implications. On specific issues, the secretary of state, director of the Office of Management and Budget, or the secretary of the treasury may be more important and influential in a policy decision than the secretary of agriculture.

Another factor of perhaps even more importance here is that the so-called southern bloc in the Senate and House has lost much of its power of past eras as a result of the new system of electing committee chairpersons (see Hardin 1986). The consequences of this development are that the proagriculture legislators from the South wield much less power in forging agricultural legislation today than in previous times.

Farm Prices and Incomes

Prices of Farm Products and of Farm Inputs. The data in Table 3.4 show that between 1950 and 1990 real prices received for corn, wheat, milk, and eggs have declined by 72, 76, 35, and 64 percent, respectively. The index of prices of all farm commodities relative to the general price level declined by 51 percent over this period. The index of prices paid for farm inputs relative to the general price level, on the other hand, has not declined greatly over this period. In fact, real prices paid were

Table 3.4. Prices of selected farm commodities and of farm inputs deflated by the consumer price index, 1950–90

Year	Corn	Wheat	Milk	Eggs	All commodities[a]	Prices paid[b]
	($/bu.)	($/bu.)	($/cwt.)	(¢/doz.)		
1950	**6.31**	**8.30**	**16.14**	**150.6**	**232.4**	**153.5**
1951	6.38	8.12	17.62	183.5	253.8	157.7
1952	5.74	7.89	18.30	157.0	237.7	158.5
1953	5.54	7.64	16.14	178.7	209.7	149.8
1954	5.32	7.88	14.76	136.1	200.7	148.7
1955	**5.04**	**7.39**	**14.96**	**147.4**	**190.3**	**149.3**
1956	4.74	7.24	15.22	144.5	183.8	147.1
1957	3.95	6.87	14.98	127.8	181.5	149.5
1958	3.88	6.06	14.29	133.2	190.3	148.8
1959	3.61	6.05	14.30	107.9	178.7	147.8
1960	**3.38**	**5.88**	**14.22**	**121.6**	**175.7**	**148.6**
1961	3.68	6.12	14.11	119.1	173.9	147.2
1962	3.71	6.75	13.54	111.9	178.8	149.0
1963	3.63	6.05	13.40	112.7	173.2	147.1
1964	3.77	4.42	13.39	109.0	167.7	148.4
1965	**3.68**	**4.29**	**13.43**	**107.0**	**171.4**	**149.2**
1966	3.83	5.03	14.85	120.7	179.0	151.2
1967	3.08	4.16	15.03	93.7	164.7	149.7
1968	3.10	3.56	15.06	97.7	160.9	146.6
1969	3.13	3.38	14.96	109.0	158.0	144.4
1970	**3.43**	**3.43**	**14.72**	**100.8**	**154.6**	**141.8**
1971	2.67	3.31	14.49	77.5	153.1	143.2
1972	3.76	4.21	14.52	73.9	162.7	148.3
1973	5.74	8.90	16.08	118.2	220.7	159.9
1974	6.13	8.30	16.90	107.9	213.0	164.3
1975	**4.72**	**6.62**	**16.26**	**97.4**	**187.7**	**165.4**
1976	4.38	5.54	16.98	102.5	179.3	167.0
1977	3.33	3.78	16.04	91.7	165.0	165.0
1978	3.45	4.33	16.26	80.1	176.4	165.6
1979	3.25	4.83	16.53	80.3	159.8	170.8
1980	**3.28**	**4.75**	**15.78**	**68.3**	**145.6**	**167.5**
1981	2.72	4.06	15.14	69.4	152.9	165.0
1982	2.64	3.58	14.08	61.7	137.8	162.7
1983	3.22	3.52	13.63	61.3	135.5	161.6
1984	2.53	3.26	12.95	69.6	136.7	157.8
1985	**2.07**	**2.86**	**11.85**	**53.1**	**119.0**	**151.5**
1986	1.37	2.21	11.41	56.2	112.2	145.1
1987	1.71	2.26	11.04	48.3	111.8	142.6
1988	2.15	3.14	10.36	44.6	116.7	143.7
1989	1.90	3.00	10.94	55.6	118.5	143.5
1990	**1.76**	**2.00**	**10.50**	**54.8**	**114.8**	**140.8**

Source: U.S. Department of Agriculture. *Agricultural Statistics,* various annual issues.
Note: The deflator used was the consumer price index for all commodities based on 1982–84 = 100.
[a]Index of prices received by farmers (1977 = 100).
[b]Index of prices paid by farmers (1977 = 100).

generally higher during the 1973–83 period than at any time during the 1950s.

The downward trend in real prices received has not been free of fluctuations since 1950. Real prices received for almost all farm commodities rose sharply in the 1970s as world demand increased and as a variety of weather phenomena around the globe reduced world grain production. In the second half of the 1970s, a steep and continuous downtrend in real farm prices again set in. Similarly, real prices paid increased sharply in the early 1970s in response to increased demand for inputs, higher land prices, and higher energy prices. The subsequent downtrend in real prices paid was delayed somewhat and was not nearly as sharp as was the downtrend in real prices received.

It is fairly clear that farmers have been caught in a rather severe price-cost squeeze over the course of the past four decades. How have they survived this phenomenon? They have survived, in large part, by producing a greater level of output per unit of input, again for practically every farm commodity. Some of the most dramatic yield increases have occurred in corn, wheat, milk, and egg production (see Table 3.5), but large yield increases have also occurred in the production of other feed grains, rice, cotton, potatoes, and fruits and vegetables, as well as in animal production. This, together with the fact that profit margins per unit of output have shrunk, has encouraged farmers to increase the average size of their farm operation so as to increase the number of units on which to make a profit and in this way maintain a standard of living to which the farm family has been accustomed.

It is interesting to note here that great yield increases have occurred in milk production and we have not yet begun commercial use of one of the more highly publicized products of the "biotechnology" era—bovine somatotropin, which is a hormone naturally produced in the pituitary of the cow that stimulates milk production and that can now be produced commercially in the laboratory. Many individuals have looked at the new research area yielding biotechnology as something that will revolutionize agriculture and cause huge agricultural adjustment problems. It might well be said, however, that a huge technological revolution in agriculture is nothing new, nor are the adjustment problems associated with such revolutions.

Demand, Supply, and Technology. All empirical studies of the demand for farm products conclude that demand for these products is both income and price inelastic (Table 3.6). An income elasticity of 0.2 means that a 10 percent change in per capita income of consumers

Table 3.5. Selected crop and animal yields and ratio of index of all output to index of all inputs, 1950–90

Year	Wheat	Corn	Sorghum	Soybeans	Rice	Cotton	Milk	Eggs per layer	People fed per worker	Ratio of all output to all inputs
	(bu.)	*(bu.)*	*(bu.)*	*(bu.)*	*(cwt.)*	*(lb.)*	*(lb.)*			
1950	**17**	**38**	**23**	**22**	**237**	**269**	**5,242**	**172**	**15**	**60**
1951	16	37	19	21	231	270	5,259	175	16	61
1952	18	42	17	21	241	280	5,300	178	17	63
1953	17	41	18	18	245	325	5,467	183	18	64
1954	18	39	20	20	251	342	5,580	184	19	65
1955	**20**	**42**	**19**	**20**	**306**	**417**	**5,772**	**192**	**20**	**68**
1956	20	47	22	22	315	409	6,020	197	22	68
1957	22	48	29	23	321	388	6,175	199	23	68
1958	27	53	35	24	316	466	6,413	202	23	74
1959	22	53	36	24	338	462	6,707	207	24	75
1960	**26**	**55**	**40**	**23**	**342**	**447**	**6,977**	**209**	**26**	**78**
1961	24	62	44	25	341	440	7,232	210	27	78
1962	25	64	44	24	372	458	7,396	212	28	79
1963	25	68	44	24	397	518	7,564	213	29	82
1964	26	63	42	23	410	517	7,964	217	31	81
1965	**27**	**74**	**52**	**25**	**426**	**527**	**8,089**	**218**	**35**	**85**
1966	26	73	56	25	432	480	8,296	218	38	82
1967	26	80	50	25	454	447	8,664	221	41	85
1968	28	79	53	27	443	516	8,955	220	42	88
1969	31	86	54	27	432	434	9,281	220	44	88
1970	**31**	**72**	**50**	**27**	**462**	**439**	**9,669**	**218**	**45**	**87**
1971	34	88	54	28	471	438	9,948	223	47	95
1972	33	97	61	28	470	507	10,185	227	48	94
1973	32	91	59	28	427	520	9,946	227	49	95
1974	27	72	45	24	445	441	10,229	230	49	90
1975	**31**	**86**	**49**	**29**	**456**	**453**	**10,279**	**232**	**50**	**98**
1976	30	88	49	26	466	465	10,863	235	50	99
1977	30	91	57	31	444	520	11,156	235	53	100
1978	31	101	55	29	448	420	11,147	239	56	102
1979	34	110	63	32	460	544	11,438	240	60	106
1980	**33**	**91**	**46**	**27**	**441**	**400**	**11,938**	**242**	**61**	**101**
1981	35	109	64	30	482	540	12,260	243	65	116
1982	35	113	59	32	471	585	12,334	243	68	117
1983	39	81	49	26	460	510	12,643	247	72	99
1984	39	107	56	28	495	601	12,193	245	76	118
1985	**37**	**118**	**67**	**34**	**541**	**629**	**13,248**	**247**	**81**	**128**
1986	34	119	68	33	565	550	12,828	247	88	125
1987	38	120	69	34	556	706	13,574	248	89	124
1988	34	85	64	27	551	619	14,145	251	88	117
1989	33	116	55	32	575	614	14,244	250	87	130
1990	**39**	**118**	**63**	**34**	**551**	**635**	**14,642**	**252**	**88**	**131**

Source: U.S. Department of Agriculture, *Agricultural Statistics,* various annual issues.
[a]The index of all farm output and the index of all farm inputs both equal 100 in 1977.

Table 3.6. Domestic price and income elasticity at the retail level for selected farm products

Product	Price elasticity	Income elasticity
All food	−0.23	0.18
Corn meal	−0.22	0.06
Bread	−0.15	0.00
Rice	−0.32	0.06
Beef	−0.64	0.29
Pork	−0.41	0.13
Lamb and mutton	−2.63	0.57
Veal	−1.72	0.59
Eggs	−0.32	0.05
Chicken	−0.78	0.18
Fresh milk	−0.35	0.20
Cheese	−0.46	0.25
Potatoes	−0.31	0.12
Apples	−0.72	0.14

Source: George and King 1971.

will increase the demand for farm products by 2 percent. Such small responses to increases in income means that growth in demand for farm products is largely dependent upon the rate of growth of population. With population growth rates slowing down everywhere and most certainly in the United States, this is not a happy prospect for U.S. farmers. Indeed it is one of the reasons why there is so much attention being paid at present to promote the sale of additional agricultural products, particularly at home.

Price elasticity of demand for farm products in the aggregate appears also to be in the vicinity of 0.2 (in absolute value). The price elasticity of demand varies considerably among farm commodities, but there are few food products with an elasticity of 1.0 or more (in absolute value). This is most significant to the sector since it means that for any agricultural product, an increase in output will result in a decrease in gross farm revenue.[2]

The supply of most farm commodities is also inelastic in the short run (Table 3.7), although not as inelastic as demand. Supply elasticity is typically highest for the livestock products with short production cycles (e.g., eggs and broilers and hogs) since the production of these commodities can be altered within a few months. Supply elasticity tends to be higher for crops which are produced as a sideline enterprise and planted on only a small portion of total acreage. Potatoes, some vegetables, and some fruits fall into this category. Short-run supply elasticities tend to be lower for wheat and

*Table 3.7. Short-run elasticities of supply
for selected agricultural products*

Product	Supply elasticity
Aggregate farm output	0.2
Feed grains	0.4
Wheat	0.3
Soybeans	0.5
Potatoes	0.8
Tobacco	0.4
Fruits	0.2
Beef	0.5
Hogs	0.6
Poultry	0.9
Eggs	1.2
Milk	0.3

Source: Tweeten 1979.

feed grains, which occupy a large proportion of the cropland and are grown in areas where alternatives are limited. Supply elasticities also tend to be low for beef and fruit because these commodities have relatively long production cycles.

These demand and supply characteristics have significant implications for farmers. First, we note that when demand is inelastic, gross farm income falls as output increases. This is by no means unique to agriculture. But the agricultural sector is also characterized by a large number of producers no one of whom is large enough to influence price by output restrictions or by inventory control as can their counterparts in an oligopolistic manufacturing industry. These two factors taken together do make agriculture somewhat distinctive and constitute a major argument for public policy designed to provide for supply control in this sector.

The demand and supply characteristics of agriculture when coupled with farmers' tendency to rapidly adopt output-increasing technology also have strong implications for agriculture. Cochrane (1958) has developed what he calls "the treadmill" thesis to explain the fact that there is continuous pressure in agriculture to expand output in spite of the fact that this means reduced gross revenue for the industry. Cochrane's argument proceeds along the following lines. Given the competitive structure within which farming takes place, the individual farmer recognizes that he cannot influence the price received for his product. His method of achieving increased income is by adopting cost-reducing and output-enhancing technology. For the early adopters this practice is successful, as their per unit costs are lowered by use of the new technology. Unfortunately, as more and more farmers adopt this output-increasing technology, the industry supply curve shifts to the right, market prices fall, aggregate output increases, and given an inelastic demand, gross farm income falls. The reduced farm price puts a price-cost squeeze on the farmers who are using the old technology, thus forcing them to adopt the new technology in order to reduce their costs and stay competitive. Hence, aggregate output increases further, and thus farm prices and gross revenues fall even further. When the process is completed, "Mr. Average Farmer is right back where he started, as far as his income position is concerned. [It is in this sense that Mr. Average Farmer] is on a treadmill with respect to technological advance" (Cochrane 1958, 96).

There is no stopping or slowing down this process so long as there is new technology available and so long as farmers' asset positions are such that they can continue to finance adoption of the new technology. That is, only a financial crisis in agriculture that renders farmers incapable of purchasing the new technology will bring to a halt the advance of farm output.

Farm Income. To examine the income position of farm people relative to that of nonfarm people, I have constructed a measure of the average *money* income of farm households which can be compared to the average money income of all U.S. households (Table 3.8). This income measure for farm households includes money income from all sources—farm and off-farm sources as well as government payments.[3] It excludes all *nonmoney* income—the monetary value of food produced and consumed on the farm, changes in the value of inventories, and estimated rental value of farm dwellings.

The data presented in Table 3.8 indicate that average money income of farm households was comparatively low during the early 1950s, lower still during the mid- and late 1950s, but on a par with or even above that of the general population during the latter half of the 1960s and throughout the 1970s. During the early 1980s, average money income of farm households again fell below that of the general population but had recovered by 1985. Farm households' average money income was well above that of the general population in 1972–76 and again in 1985–90.

Over the last nearly three decades, the number of farms with annual gross sales of $40,000 or more has increased nearly sixfold while the number of farms with annual gross sales of less than $10,000 has decreased threefold (Table 3.9). Operators of most of the relatively small farm units though are not dependent on farming alone for their livelihood. By 1970 and still today, the smallest farms earned on average over 100 percent of

Table 3.8. Money income of farm households and of all households in the United States, 1950–90

Year	Money income Farm households	Money income All U.S. households	Farm as a % of all U.S. households
1950	**$2,487**	**$4,205**	**59.2**
1951	2,872	4,506	63.7
1952	2,852	4,703	60.6
1953	2,815	4,888	57.6
1954	2,650	4,828	54.9
1955	**2,551**	**5,038**	**50.6**
1956	2,683	5,340	50.2
1957	2,713	5,392	50.3
1958	2,889	5,462	52.9
1959	2,789	5,793	48.1
1960	**3,927**	**5,979**	**65.7**
1961	4,660	6,159	75.7
1962	4,810	6,298	76.4
1963	5,450	6,554	83.2
1964	5,879	6,818	86.2
1965	**6,356**	**7,110**	**89.4**
1966	7,624	7,693	99.1
1967	7,734	7,989	96.8
1968	8,189	8,760	93.5
1969	9,418	9,544	98.7
1970	**10,254**	**10,001**	**102.5**
1971	9,823	10,383	94.6
1972	12,328	11,286	109.2
1973	17,854	12,157	146.9
1974	18,204	13,094	139.0
1975	**15,694**	**13,779**	**113.9**
1976	16,463	14,922	110.3
1977	14,866	16,100	92.3
1978	17,954	17,730	101.3
1979	18,798	19,554	96.1
1980	**18,434**	**21,063**	**87.5**
1981	17,412	22,787	76.4
1982	19,608	24,309	80.7
1983	20,959	25,609	81.8
1984	21,861	27,464	79.6
1985	**35,338**	**29,066**	**121.6**
1986	36,601	30,759	119.0
1987	43,264	32,144	134.6
1988	44,562	37,017	120.4
1989	43,172	39,500	109.3
1990	**44,018**	**41,400**	**106.3**

Source: Computed from data in U.S. Department of Agriculture. "Economic Indicators of the Farm Sector," Economic Research Service, various annual issues.

their family income from off-farm sources—that is, they were losing money on their farming operations! Through the 1980s, midsized farms—those with annual gross sales of $40,000–$99,999—earned on average 40 percent or more of their family income from off-farm sources! Even some of the commercial-sized farms (e.g., those with annual sales of $100,000–$199,999) now earn on average about 20 percent of their family income from off-farm sources. In total, 59 percent of the income of farm families came from off-farm sources in 1980, 50 percent in 1985, and almost 45 percent in 1988.

Table 3.9 also shows the percentage of farm operator family income derived from direct government payments. Whether the narrowing of the farm-nonfarm income gap can be attributed exclusively to farm programs is debatable. Certainly the importance of off-farm sources of income to farm families in the 1980s relative to earlier decades appears to have been a more important contributing factor for the smaller farms. In the 1980s, farm families with annual farm sales of $40,000 or less from farm produce would have been destitute had they been dependent solely on income from farm production and from direct government payments. Farm families with annual sales of $40,000–$199,999 also appear to rely more heavily on off-farm income than on direct government payments. Only the largest farms (those with annual sales of $500,000 or more) appear in some years to be more dependent on government payments than on off-farm income. In the latter cases, however, the farm operation generates a very substantial family income so that these units can survive quite nicely in the absence of *both* off-farm income *and* direct government payments.

All in all then, it is clear that much progress has been made toward farm-nonfarm income equality. On the majority of U.S. farms, off-farm income has been a much more important factor accounting for this progress toward equality than has direct government payments. This, together with the fact that the larger farms probably need no income support to maintain a standard of living comparable to nonfarm families, suggests that the need for farm income enhancement is much less evident today than it was during the 1950s and early 1960s.

It is well known that farm income is much more variable than is nonfarm income because of the riskiness of the farming business. It is also well known that the distribution of income is more skewed in farming than in other occupations and that there is a greater incidence of poverty in agriculture (as measured by annual money income) than elsewhere. Indeed, all this may suggest the need for some income protection for farmers. On the other hand, cash or money income is only one measure of the relative well-being of farm families. Wealth is another. On the basis of wealth, I conclude that farm families are in a superior position! A recent report (U.S. Department of Agriculture 1988, 52) indicates that the "farm wealth of

Table 3.9. Number of farms, farm family income, and percentage of income from off-farm sources and from direct government payments by sales classes, 1960–88

Annual gross sales[a]	Number of farms					Average farm operator family income[b]				
	1960	1970	1980	1985	1988	1960	1970	1980	1985	1988
	(1,000)					*($1,000)*				
Under $5,000	2,466	1,695	931	811	751	3.6	7.4	15.9	24.6	30.7
$5,000–$9,999	660	372	312	289	279	5.0	8.8	15.7	15.1	27.4
$10,000–$19,999	497	362	289	284	274	6.7	11.3	14.3	15.3	25.7
$20,000–$39,999	227	302	282	243	251	11.2	16.5	13.1	21.2	31.7
$40,000–$99,999	90	165	355	341	320	16.5	28.1	18.5	29.4	43.4
$100,000–$199,999	23	36	166	221	216	31.4	55.9	39.2	60.5	88.7
$200,000–$499,999	na	13	81	78	76	na	102.5	81.1	136.4	178.1
$500,000 or more	na	4	24	26	30	na	662.8	476.5	656.7	780.5
All farms	3,963	2,949	2,440	2,293	2,197	5.1	12.1	23.9	37.1	52.8

	Percentage of family income from nonfarm sources[c]					Percentage of family income from government payments				
	1960	1970	1980	1985	1988	1960	1970	1980	1985	1988
Under $5,000	69.0	95.5	109.0	106.0	102.0	2.0	4.6	0.3	0.3	1.2
$5,000–$9,999	31.5	62.0	105.0	101.0	92.2	4.4	12.2	1.0	2.4	3.7
$10,000–19,999	18.9	37.1	103.0	96.4	80.7	4.8	16.4	1.4	5.2	9.1
$20,000–$39,999	14.9	20.4	85.4	76.1	61.2	4.4	16.6	4.0	13.5	18.1
$40,000–$99,999	11.0	14.1	47.5	47.0	33.8	5.2	15.3	6.3	19.0	26.0
$100,000–$199,999	0.0	15.1	23.4	21.1	19.9	4.2	13.2	4.3	18.0	23.8
$200,000–$499,999	na	0.0	13.5	8.6	9.1	na	9.1	3.0	14.0	18.0
$500,000 or more	na	0.0	4.2	4.7	3.3	na	4.7	0.8	4.6	5.2
All farms	42.4	49.4	59.4	50.1	44.5	3.5	10.4	2.2	9.1	12.5

Source: U.S. Department of Agriculture 1989.

[a]Includes cash receipts from the sale of farm produce, government payments to farmers, and other farm-related income.

[b]Includes government payments to farmers, the value of farm products consumed on farm, and the rental value of farm dwellings in addition to commercial farm sales and income from off-farm sources.

[c]Percentages in excess of 100 here indicate that net income from the farm operation was negative.

na—not available.

farm operator households averaged about $225,000 in 1986. The average wealth of U.S. households in 1984 was $78,739. The wealth of farm operator households is greater than the wealth of U.S. households at all levels of income."

Characteristics of Agricultural Production

Dependence on Nature. No industry is as vulnerable to the vagaries of nature as is agriculture. The most dramatic way in which weather has affected agriculture in the past is undoubtedly through drought. And the most memorable drought in recent history was that of 1936, which led to the dust bowl in the Great Plains and Midwest. Not quite as severe were the droughts of 1953 and 1974, both of which had a major impact on wheat, cotton, and corn yields; and the droughts of 1970, 1980, 1983, and 1988, which affected feed grain and soybean yields.

Other weather patterns can also severely impact agricultural production. For example, tropical storm Agnes in 1972, which caused severe flooding in Pennsylvania and elsewhere in the East, resulted in rather significant reductions in milk production per cow as pastures and hay fields were flooded. Farmers are also subjected to hail damage, weather-related crop diseases, early frosts, as well as weather-related insect infestations. The following passage from the 1975 *Economic Report of the President* is instructive in terms of putting the weather factor in perspective:

> In the spring of 1974 there seemed to be good reason to expect excellent U.S. grain production even if weather conditions were to be somewhat below average. Much field preparation had been completed the previous fall. Surveys showed that farmers were planning increases in their plantings because of favorable prices and the removal of Government acreage diversion programs. Fertilizer supplies were tight, but they exceeded the previous year; and efforts were under way to minimize bottlenecks in production and distribution. Then wet weather delayed spring plantings—which itself slightly reduced yields and made crops more vulnerable to early frosts—and prevented some fields from being planted at all. But the summer's dry and hot weather was the major setback. Preliminary official estimates of the feed grain crop fell from 234 million (short) tons in March to 215 million tons in July, and then to 175 million tons in August, when the first survey based on actual yield estimates became available. Significant though smaller reductions occurred for wheat (from 2.1 billion bushels in March to 1.8 billion bushels in August) and soybeans (from 1.5 billion bushels in March to 1.3 billion bushels in August). Severe frosts in September and early October further damaged the feed grain and soybean crops. (Pp. 163–64)

As if weather conditions in the United States were not enough, U.S. farmers are impacted as well by weather conditions in other parts of the world. In 1972 the USSR purchased huge quantities of U.S. wheat and corn because of a particularly bad harvest due to drought in that year. The USSR has had droughts in previous years, but never before had the USSR purchased such huge quantities on international markets in an attempt to make up for the shortfall. In the same year much of South and Southeast Asia had a poor grain crop, forcing countries in this region of the world to acquire additional grain supplies on the world market. A relatively good harvest of grains around the world was recorded in 1973, but in 1974 there was again a poor grain harvest in the USSR. In addition, in 1973 demand for soybeans increased markedly due to the continuing poor catch of anchovies (and other fish used to produce fish meal needed for commercially prepared animal feedstuffs) off the coast of Peru. Here the problem was El Niño.[4]

These events of the 1970s which originated outside the United States were fairly quickly and strongly felt by U.S. farmers. Further, they were welcomed by U.S. farmers because they led to increased export demand and high prices. Indeed these events encouraged U.S. farmers (with government sanction) to reclaim most of the land idled by previous farm programs. In other years, of course, export demand and prices will be down because of particularly good harvests in other countries of the world. Good harvests in other countries will thus work to the detriment of U.S. farmers. Further, considerable time will be required for U.S. farmers to scale back production (by idling land again), as was the case following the boom years of the mid-to-late 1970s.

Long Production Lags. Another characteristic of agricultural production is that long periods of time elapse between production decisions and production realization. The midwestern grain farmer must make a commitment to plant corn in May, but the benefits of this commitment are not reaped until late fall or early winter. Further, corn is harvested only once a year. Hence this output must be stored to meet a continuous demand throughout the year. The dairy farmer who this year decides to expand the herd by keeping more than the usual number of replacement heifers will not see milk output increase until two years into the future. In the meantime there is no certainty as to the productive potential of these replacement heifers even though they must be fed and cared for until they begin producing milk regardless of the price of milk or of the price of feed. The apple, peach, orange, or grape producer must wait even longer between production decisions involving the planting of trees or vines and production realization.

This means that farmers cannot adjust output upward or downward very rapidly. Rather, once committed by spring planting, once replacement animals are in place, or once the trees are planted, the farmer must wait for the next production phase to begin in order to make significant adjustments in output. Obviously there are limits to this argument. If, for example, output prices fall so low that the marginal cost of completing the production process exceeds the expected price of the product, the rational farmer will abandon production at that point. Nevertheless, there is in most cases a range of prices over which output is relatively fixed during one production phase.

Price Instability. In a normally functioning market free of governmental interference and imperfectly competitive elements, commodity prices may vary from year to year more or less in a random fashion due to rightward or leftward shifts in either the supply or the demand schedule. As we have already seen, the supply schedule in agriculture can shift due to either weather or technology or both. Similarly the demand schedule can shift as tastes change or as per capita incomes change or as prices of close substitutes change (see Tomek and Robinson 1972).

The variability of prices due to natural phenomena is the more frequent concern however. This variability cannot be easily predicted or anticipated. It is precisely for this reason that price instability in agriculture is of concern to many, for this situation can lead to one or more of the following results: (1) the business manager may be unwilling to make the needed investments because of the uncertainty of future earnings, (2) lenders may be unwilling to provide loan funds at a reasonable cost because of the high risk, or (3) the business manager's costs may rise because under uncertainty he must spend more time and/or money to discover or anticipate what future prices will be.

As I noted before, since both supply and demand for most agricultural commodities is price inelastic at least in the short run, small shifts in either of these two schedules will lead to quite sizable changes in price. This coupled with the fact that agricultural output is quite sensitive to the vagaries of nature leads many people to conclude that prices of agricultural commodities are highly variable. This in turn points to justification for governmental support for agriculture in an effort to protect farmers from price instability.

In Table 3.10 are shown indexes of variation for prices of a variety of agricultural and nonagricultural commodities for each of four time periods since 1950. These coefficients were computed after first extracting a straight-line trend from the price data so that the resulting coefficients would not be biased by trend.[5] Relatively short time periods were used so that any changes in variability over time could be observed. This procedure has the obvious disadvantage of giving undue weight to a year when a quite

Table 3.10. *Index of variability of prices received by farmers for selected agricultural products and producer prices of selected nonagricultural commodities, 1950–89*

Commodity	1950–59	1960–69	1970–79	1980–89
Supported agricultural commodities				
Wheat	3.9	10.3	34.7	14.1
Rice	6.7	3.0	31.7	22.1
Corn	4.2	5.9	25.4	18.4
Oats	6.5	4.2	24.4	23.9
Barley	7.2	7.0	28.4	18.3
Grain sorghum	13.2	5.8	21.7	17.2
Rye	10.3	5.2	26.2	19.2
Soybeans	5.4	6.5	16.3	17.5
Cotton	5.5	8.6	13.0	8.3
Tobacco, burley	5.3	5.9	5.4	5.7
Peanuts	6.3	2.3	4.6	5.4
Sugar beets	3.3	4.6	36.1	13.8
Wool	20.1	10.4	30.6	25.8
Honey	4.6	4.0	16.3	5.5
All milk	6.5	5.8	5.0	2.6
Nonsupported agricultural commodities				
Potatoes	28.4	26.6	25.6	14.3
Steers, choice	15.3	6.6	12.4	5.9
Barrows and gilts	14.3	11.3	17.3	9.9
Lambs	14.9	6.4	9.7	9.4
Choice veal calves	16.9	7.7	21.4	13.9
Eggs	10.3	7.3	14.4	9.4
Broilers	6.6	6.2	14.1	7.1
Turkeys	4.6	8.3	14.0	12.0
Tomatoes, processing	6.8	9.4	12.7	3.4
Beans, snap	5.3	2.4	12.0	3.0
All apples	18.5	13.7	13.0	12.8
All oranges	19.5	24.1	21.3	15.7
All lemons	13.3	16.8	20.5	34.7
All grapes	20.2	11.6	15.4	16.7
Sour cherries	19.7	37.8	73.5	45.1
All pecans	22.3	26.8	20.6	12.5
Nonagricultural commodities				
Copper, electrolytic	17.4	7.4	12.1	21.3
Tin	10.5	11.5	35.8	11.5
Bituminous coal	4.4	4.2	10.0	4.3
Crude oil	2.7	1.5	17.1	18.2
Douglas fir	23.6	25.6	31.3	43.7

Source: U.S. Department of Agriculture, *Agricultural Statistics,* various annual issues; U.S. Department of Commerce, *Statistical Abstract of the United States,* various annual issues.
Note: Coefficients of variation estimated from detrended data (see text).

large price change occurred. This does not appear to be a major problem for the commodities analyzed here. Further, this procedure confines the particularly turbulent periods of the early to mid-1970s and the 1980s each to one time period rather spreading them over two or more time periods.

The first lesson to be learned from these data is that agricultural commodities *are* characterized by considerable price variability. Most farm commodities experienced greater levels of price variability during the period of food and energy shortages (the 1970s) and its aftermath. This was true of both supported and nonsupported commodities, although less so for the latter than for the former. Interestingly enough, even though farm policy has been continuous over this entire period for wheat, rice, the feed grains, cotton, wool, and sugar, prices of these commodities have *not* been stabilized. On the other hand, prices of tobacco, peanuts, and milk have remained relatively stable throughout the entire 1950-89 period. This can undoubtedly be attributed to the fact that the latter commodities have been effectively isolated from foreign competition and/or have been protected with very tight price and production controls.

A second lesson to be learned from these data is that there are a number of agricultural producers successfully operating in the face of considerable price variability but *without* price and income support or protection from foreign competition. Consider, for example, potatoes, hogs, and the fruits included in Table 3.10. Also there are industries that have managed to maintain reasonably stable prices on their own—for example, the broiler, egg, tomato, and snap bean industries.

A final lesson to be learned from Table 3.10 is that agriculture is not the only industry subjected to price variability. Note in particular the indexes of variability for Douglas fir and to a lesser extent for tin and copper.

Capital Considerations

HIGH CAPITAL REQUIREMENTS. The amount of capital required to get into farming relative to the value of output produced in any given year is quite large. In dairy, for example, assets per cow are today on the order of $3,000–$4,000, yet annual cash sales per cow are in the vicinity of $1,200–$1,500. Table 3.11 shows the capital turnover ratio in agriculture compared with that in manufacturing. The capital turnover ratio is simply the ratio of gross sales in the case of agriculture or the value of shipments in the case of the manufacturing industries and the retail sector to beginning-year land, building, and machinery assets. It suggests nothing about profitability or efficiency. It will vary considerably from year to year and particularly in

Table 3.11. Capital turnover ratio in agriculture and selected manufacturing industries, United States, 1977 and 1982

Industry	Capital turnover ratio	
	1977	1982
Farming Sector		
All farms in the U.S.	0.22	0.19
Illinois grain farms	0.11	0.09
Illinois hog farms	0.19	0.20
Illinois beef farms	0.23	0.25
Pennsylvania dairy farms (20–29 cows)	0.43	0.46
Pennsylvania dairy farms (90–100 cows)	0.51	0.67
Manufacturing Sector		
Meat packing (SIC 2011)[a]	14.52	16.15
Sausages (SIC 2013)	8.43	7.66
Poultry dressing (SIC 2016)	8.50	6.83
Creamery butter (SIC 2021)	15.01	21.65
Cheese (SIC 2022)	10.58	10.20
Ice cream (SIC 2024)	4.55	4.38
Fluid milk (SIC 2026)	7.35	7.81
Bread (SIC 2051)	4.19	4.20
Weaving mills, cotton (SIC 2211)	2.42	1.86
Sawmills (SIC 2421)	2.81	1.73
Pulp mills (SIC 2611)	0.86	0.76
Pharmaceutical preparations (SIC 2834)	3.37	3.44
Nitrogen fertilizers (SIC 2873)	1.03	1.07
Petroleum refining (SIC 2911)	4.14	4.89
Men's footwear (SIC 3143)	9.65	8.99
Blast furnaces and steel mills (SIC 3312)	1.30	0.93
Farm machinery and equipment (SIC 3523)	5.10	3.07
Motor vehicles and car bodies (SIC 3711)	11.36	5.11
Retail Trade		
Retail food stores (SIC 51)	na	8.64
Eating and drinking places (SIC 58)	na	3.10

Source: U.S. Department of Agriculture 1989; U.S. Bureau of the Census 1977, 1982; Dum 1977; Grisley 1982; Wilken and Kesler 1978; and Wilken et al. 1983.

Note: Capital turnover ratio = value of gross sales divided by beginning year assets.

[a]SIC—Standard Industrial Classification code number.

na–not available.

response to the business cycle or, in general, as capacity utilization varies. Further, it will vary from one industry to another because it is dependent upon the length of the production phase. One would, for instance, expect this ratio to be high for the creamery butter industry since plants in this industry receive a new supply of milk for processing several times a week if not daily.

Nevertheless, the ratios shown in Table 3.11 do indicate the importance of capital relative to cash sales generated in the different industries. It is, thus, instructive to compare the capital turnover ratio for all of farming and for a few farm types on the one hand and for selected manufacturing industries and the retail food business on the other. Of the manufacturing industries shown in Table 3.11, only two (pulp mills and blast furnaces and steel mills) show a ratio of less than 1 and in both cases it is near 1. In farming, on the other hand, this ratio is less than 0.3 for all but dairy farms! Clearly farming is much more capital intensive than are most of the manufacturing industries and the retail food sector. Dairy farming appears to be somewhat less capital intensive than grain, hog, or beef cattle farming. But even in dairying much more capital is needed relative to sales than in most industries of the manufacturing and retail sectors.

Among other things, the comparisons afforded by the data in Table 3.11 point out that entry into agriculture is likely much more difficult than is entry into other industries because of the high capital requirements relative to the gross revenue this capital will generate. Prospective farmers are expected to generate their own capital. They do not have the luxury that nonfarm businesses do of relying on stockholders for their equity capital.

CAPITAL MOBILITY. Conventional economic theory assumes that markets are such that there is only one price for any given asset and that an entrepreneur can purchase additional quantities of that asset or sell some portion of that asset he now owns at the same price. That is, conventional theory assumes perfect mobility of factors. G. L. Johnson (1958; Johnson and Quance 1972) was one of the first applied economists to recognize that in the farm sector at least, the real-world situation is much more complicated. Johnson observed that many specialized farm assets have an acquisition price that is generally much higher than their selling prices (or salvage values) and that this leads to a considerable range over which output price can vary before changes are made in asset use and therefore in quantity of output.

A variety of assets can be so characterized. Durable assets such as fruit trees, fencing, farm buildings, and drainage tile with salvage values near zero or even negative (e.g., there will be a cost associated with digging up fruit trees and disposing of them unless this cost can be completely recovered by selling the wood for firewood) are fixed over wide ranges of prices for the industry and the individual farm although they may not be fixed for individual farm enterprises. That is, tiling and fences are useful regardless of whether the farmer is growing corn or soybeans, silos are useful regardless of whether the farmer is producing corn silage or haylage, etc. Specialized durable assets such as corn pickers, grain combines, hay balers, and forage choppers have a very low salvage value since such machines have little use outside agriculture. These assets are probably fixed over wide ranges of prices for the industry and relatively fixed for individual farm enterprises. A hay baler, for example, is not very useful in corn or soybean production nor is a corn picker useful in sunflower production. Assets such as these will be used to produce specialized products until completely depreciated, almost without regard to product prices.

Thus, over a wide range of output prices and in the short to intermediate run at least, aggregate supply for farm products is quite inelastic. Some resources fixed to the industry tend to be less fixed on individual farms and much less so for individual commodities. Land has a low salvage value in nonfarm uses in most areas and it responds very little to output prices. In Chapter 10 we will see that cropland supply is very inelastic in both the short run and in the long run. Nevertheless, there is little to prevent shifts in crops or livestock grown on that relatively fixed input. This explains why supply tends to be more elastic for individual farm products than for the aggregate of farm output.

While certain assets in agriculture are clearly fixed in the above sense, "asset fixity" is not unique to agriculture. Furthermore, as we will see in the final chapter of this book upon returning to this issue, asset fixity does not appear to be a serious problem to farmers in the long run judging by the extent of actual adjustments farmers have made in the recent past. Hence, it is not clear that asset fixity can be used as a basis for justifying any but short-run policy aids.

Labor Use and Mobility. Farming is certainly a much less labor intensive industry than it was fifty or a hundred years ago, but relative to most nonagricultural industries, it is still quite labor intensive. Nevertheless, much of the new technology farmers have adopted to increase productivity has resulted in the substitution of nonlabor inputs for labor. This, in turn, has meant an excess of labor in agriculture. A perfectly competitive labor market would have permitted this excess labor to move out of agriculture with little difficulty.

Historically there have been severe impediments to the adjustment of farmworkers out of agriculture and into nonfarm jobs in order to prevent the persistence of low returns. These impediments have included lack of education and skills, lack of knowledge about nonfarm job opportunities, employment barriers created by organized labor, and lack of availability of nonfarm jobs in the area where these surplus labor resources could feasibly relocate.

For the most part these types of impediments have by now been removed or at least relaxed so that much of the surplus labor in agriculture has migrated to the nonagricultural sector. Nevertheless, there are still instances in which nonfarm jobs requiring the skills possessed by surplus agricultural workers are not available. In these instances, the excess labor resources cannot move out of agriculture.

Furthermore, reallocating labor resources to a different type of agricultural production is never easy and quite often simply not possible. It is relatively easy for midwestern grain farmers to shift from corn production to soybean production. It is quite another matter for dairy farmers to shift from milk production to corn production or even to shift from milk production to beef production. The intensity of resource use in dairy and beef are vastly different, so that with the same resources, Wisconsin dairy farmers simply cannot produce the same level of gross (or net) receipts with beef as they can with dairy. As Malone (1983) suggests, the on-farm alternatives for these dairy farmers are quite limited. In the long run, excess resources will exit if output prices fall or fail to increase. In the short run, however, some stickiness should be expected.

These kinds of problems are of paramount concern in the principal dairy regions of this country at the moment. We currently have a surplus of milk, and the obvious solution is to move resources out of dairy production. Unfortunately, as may well be reflected by the relatively low participation rate among Upper Midwest and Northeast dairy farmers in the dairy herd buyout program authorized by the Food Security Act of 1985, the alternatives available to dairy farmers in these regions are quite limited. At the present time in many rural communities there is a lack of jobs available for dairy farmers with which to supplement or replace their farm income.

Market Structure

Industrialization of the Agricultural Sector. Recent reports have documented some striking structural changes bearing directly on the agricultural sector (see, e.g., Schertz et al. 1979; Bergland 1981). The trend toward larger and fewer farms documented in Table 3.9 is in large part attributed to industrialization of agriculture and/or specialization. Not only have farmers specialized, but some tasks previously done by farm operators are relegated to other sectors. Farmers were once largely self-sufficient, providing their own horsepower, fuel, seed, and other raw material and financing their operation from internal equity capital. Not so today. In 1950, for example, the ratio of purchased inputs to farm-supplied inputs was 0.36 (based on a ratio of quantity indexes) (see Table 3.12). In 1990 that same ratio was 1.22. Along with specialization came a substitution of purchased inputs for labor as farmers sought to increase productivity and

Table 3.12. Trends in use of inputs in the agricultural sector, United States, 1950–90

Year	Ratio of purchased to home-grown inputs	Ratio of chemical to labor inputs	Ratio of machinery to labor inputs	Tractor horsepower[a]
1950	**0.36**	**0.07**	**0.27**	**26**
1951	0.37	0.08	0.31	29
1952	0.39	0.10	0.34	30
1953	0.39	0.11	0.37	32
1954	0.40	0.11	0.38	34
1955	**0.41**	**0.12**	**0.38**	**36**
1956	0.43	0.13	0.40	39
1957	0.46	0.14	0.42	42
1958	0.49	0.15	0.46	44
1959	0.52	0.17	0.46	46
1960	**0.56**	**0.18**	**0.47**	**47**
1961	0.58	0.21	0.48	51
1962	0.61	0.23	0.49	54
1963	0.64	0.28	0.51	55
1964	0.67	0.31	0.54	58
1965	**0.71**	**0.34**	**0.56**	**59**
1966	0.75	0.42	0.62	62
1967	0.78	0.52	0.66	62
1968	0.79	0.56	0.69	65
1969	0.81	0.62	0.73	68
1970	**0.82**	**0.67**	**0.76**	**69**
1971	0.87	0.75	0.81	67
1972	0.88	0.78	0.78	71
1973	0.90	0.83	0.83	67
1974	0.90	0.84	0.84	67
1975	**0.87**	**0.78**	**0.91**	**67**
1976	0.93	0.96	0.98	68
1977	1.00	1.00	1.00	67
1978	1.05	1.07	1.04	77
1979	1.11	1.24	1.05	87
1980	**1.06**	**1.28**	**1.05**	**85**
1981	1.08	1.34	1.02	84
1982	1.06	1.27	0.99	85
1983	1.09	1.05	0.92	100
1984	1.11	1.30	0.93	90
1985	**1.11**	**1.35**	**0.94**	**91**
1986	1.12	1.36	0.96	95
1987	1.16	1.42	0.94	102
1988	1.18	1.48	0.96	101
1989	1.24	1.61	0.96	98
1990	**1.22**	**1.60**	**0.96**	**97**

Source: U.S. Department of Agriculture, "Economic Indicators of the Farm Sector: National Financial Summary," Economic Research Service, various annual issues.

[a]Horsepower per 100 acres of principal crops planted.

capitalize on economies of scale. Between 1950 and 1990, for example, the ratio of chemical use to labor use increased from 0.07 to 1.60, and the ratio of machinery use to labor use increased from 0.27 to 0.96 (based again on a ratio of quantity indexes). Tractor horsepower used per 100 acres of principal crops planted increased from 26 to 97 over this same period. Today farmers are heavily dependent on the nonfarm sectors for their inputs. Economic conditions affecting these nonfarm sectors quickly trickle down to affect the farm sector as well. Thus, farm policy cannot be developed independent of nonfarm considerations and policies.

Use of Coordination Devices. Production, processing, distribution, retailing, and consumption activities must in some way be coordinated so that proper signals are sent through the economic system to guide output and consumption decisions. In a perfectly competitive system this coordination is done through the open market—by the "invisible hand." When operative, this system is highly effective, does not stifle innovativeness and progress, and prevents abuse of one market participant by another. It is operative, however, only when there are numerous buyers and sellers who interact for the purpose of exchanging information and establishing a truly "competitive" price.

Other coordinating devices, though, are used extensively in agriculture (see Table 3.13). These include grower-processor contracts and vertical integration. Contractual agreements take on a variety of forms depending on the commodity. Basically they specify the nature of the buyer-seller relationship. Some specify the date of product delivery, some the price at which the transaction is to be made, and some the type of production practices to be employed. In vertically integrated operations, decision making at more than one stage in the marketing channel (e.g., apple production and apple processing) is in the hands of a single firm. Highly perishable commodities that use land and capital intensively are prone to be produced under contract or vertical integration.

Formal coordination as opposed to open-market coordination of production and processing ranges from 100 percent for sugar beets and cane to less than 1 percent for hay and forage (Table 3.13). Formal coordination is quite extensive in milk, broilers, turkeys, fruits, and vegetables. About 16 percent of cattle feeding involves some type of formal coordination.

In highly perishable commodities, contracts that are entered into before the crops are planted can control quantity and timing of production as well as eliminate expensive storage and handling operations. Contracts reduce price risks and ensure a market. However, formal coordinating devices require farmers to relinquish some (or all) of their independent decision-

Table 3.13. Percentage of output produced under contract or vertical integration, selected commodities, 1960–80

Commodity	Production and marketing contracts[a]			Vertical integration[b]		
	1960	1970	1980	1960	1970	1980
Crops						
Feed grains	0.1	0.1	7.0	0.4	0.5	0.5
Hay and forage	0.3	0.3	0.5	0.0	0.0	0.0
Food grains	1.0	2.0	8.0	0.3	0.5	0.5
Oil-bearing crops	1.0	1.0	10.0	0.4	0.5	0.5
Seed crops	80.0	80.0	80.0	0.3	0.5	10.0
Fresh vegetables	20.0	21.0	18.0	25.0	30.0	35.0
Processing vegetables	67.0	85.0	83.1	8.0	10.0	15.0
Potatoes	40.0	45.0	60.0	30.0	25.0	35.0
Citrus fruits	60.0	55.0	65.0	20.0	30.0	35.0
Sugar beets	98.0	98.0	98.0	2.0	2.0	2.0
Sugarcane	24.4	31.5	40.0	75.6	68.5	60.0
Livestock						
Fed cattle	10.0	18.0	10.0	3.0	4.0	6.0
Hogs	0.7	1.0	1.5	0.7	1.0	1.5
Fluid milk	95.0	95.0	95.0	2.9	2.1	1.4
Manufacturing milk	25.0	25.0	25.0	2.0	1.0	1.0
Eggs	5.0	20.0	45.0	10.0	20.0	44.0
Broilers	93.0	90.0	89.0	5.4	7.0	10.0
Turkeys	30.0	42.0	62.0	4.0	12.0	28.0
Total farm output	15.1	17.2	22.9	3.9	4.8	7.4

Source: Manchester 1985.
[a]Agreements between farmers and processors, dealers, or others, including cooperatives.
[b]Production and marketing activities performed by the same firm.

making responsibility to larger and more economically powerful nonfarm firms. Furthermore, when these devices are used extensively, there are too few buyers and sellers using open markets to establish a competitive price. It is thus difficult for the market to reflect all of the forces of supply and demand. The limited market information available under these circumstances is often insufficient to serve as a base for rational resource allocation decisions.

The fact that formal coordination devices such as these are used extensively in agriculture is not necessarily to be condemned. They arose out of a perceived need by the private sector to resolve a structural or behavioral inadequacy in the marketplace. The proliferation of these devices, however, might be suggestive of the need for new concerns by policymakers—concerns about the adequacy or fairness of contract terms, concerns about the adjudication of contract disputes, consequences of the resulting "thinness" of open markets, etc.

Erosion of Infrastructure. Food-processing and retailing firms are fairly numerous. These firms are becoming larger at the expense of small local firms that are no longer able to compete. Larger firms are national or regional multiproduct firms that do not depend solely on any one production area for raw materials. On the contrary, they obtain their supplies anywhere they can get the volume and quality necessary to support a nationwide or regionwide marketing program. With ready access to markets thus reduced, small-scale producers for local markets are at a serious competitive disadvantage. Also, due to the decline in transportation services (particularly rail) some rural areas do not have ready access to production inputs or product markets.

Of perhaps greater significance is the fact that for a production activity to be viable at all in a particular area, it must be undertaken on a large enough scale such that support services and processing capacity can be provided at an economically justifiable scale. Given the scale economies in fluid milk processing, for example, a plant processing less than 50 million pounds of milk per year is not cost competitive. Thus, a minimum of 4,500 good producing cows must be in the region supplying such a plant. A study by Schneider et al. (1973) suggests that the minimum number of cows needed to support a feed mill is in the range of 26,000 or more—more than the dairy cow population of some states! Similar constraints exist for the production of such commodities as broilers, eggs, vegetables, and fruits.

Here again one should not necessarily despair over these results but should ask what if any new concerns should policymakers address. For the most part, researchers and the industry probably have too little information about the causes and consequences of these trends to be very helpful. Major issues here are whether this phenomenon restricts the adjustment possibilities of farmers, and whether legislators can assist farmers caught up in this situation.

Is the Support Justified?

Agriculture is (for the most part at least) the only sector capable of producing the raw material needed for food. On this count, the sector deserves special attention to ensure a continued supply of one of humanity's basic necessities and to ensure that this need is provided at least cost. Given science's increasing capability to produce synthetic food products, it is interesting to speculate what the future holds in store in terms of resources needed to produce food. If, for example, single-cell proteins are developed that are processable into food at low cost or technologies that break down cellulose fiber into low-cost food ingredients are perfected, the resource requirements needed in the future to produce food will be vastly different from those of today. For the foreseeable future, however, the agricultural sector as currently structured will be the principal source of raw material for food.

There would appear to be no basis for the argument that agriculture needs protection because it is the key sector around which the entire economy revolves. This may be true in certain portions of the developing world, but it is no longer true in the United States.

The political power of agricultural interests has in past years been fairly substantial and this undoubtedly explains some of the favorable support for agriculture. However, some see increased difficulty in securing favorable legislation for agriculture in the future. This is probably less a function of the declining numbers of farm people and/or changes in basic values of the population toward agriculture than it is to the changing nature of the political process itself.

Few industries are as vulnerable to the vagaries of nature at home as well as abroad as is agriculture. This characteristic leads to high levels of output and/or price variability and is, of course, beyond the control of farmers. Prices of agricultural products are thus subject to a great deal of variability and this leads to uncertainties concerning resource allocation decisions in both production and consumption. Nevertheless, farm policy has not been particularly effective in stabilizing prices of the major crops. Furthermore, several of the nonsupported agricultural commodities and a few nonagricultural commodities exhibit price variability as great or greater than that of supported crops and these industries still remain viable. Justification for past and current price policy for agriculture on the basis that it stabilizes farm prices is therefore weak. This is not to say that protection against price variability in agriculture is unwarranted.

Most production processes in agriculture involve long lags between production planning and production realization. Impediments to the short-run mobility of labor and fixed assets are characteristic of this industry. In the longer run, however, resources do adjust, as I will discuss at some length in Chapter 12. Agriculture is a very capital intensive industry—not only is it difficult to move capital from one enterprise to another or to move capital investments out of agriculture, entry into agriculture is difficult because of the high capital requirements. While not in itself justification for special treatment, these attributes deserve special attention since they tend to put limits on rapid adjustments of resources that are often called for in today's world.

In general, domestic demand for agricultural products is highly inelastic.

This means that rightward shifts in the supply curve, resulting in increased market supplies and lower farm prices, will bring forth reduced gross revenues from the sale of agricultural products. Nevertheless, there are strong incentives for farmers to increase output as new cost-reducing and output-enhancing technologies become available. Since agriculture is also probably the closest we come to a purely competitive industry (in the sense that it has a large number of producers no one of which is large enough to influence price), these attributes have led to considerable support for public policy designed to accomplish what farmers cannot do collectively for themselves—control supply.

Some trends over the last four decades are important in thinking through future policy for the agricultural sector. First, the need for income enhancement for farm families is less evident today than in days past. In part this can be attributed to the fact that many of the excess labor resources have migrated out of agriculture into sectors where these resources can earn a more competitive return. In part also this can be attributed to the fact that many farm families with insufficient resources to earn a competitive income from farming alone can, for the most part, find off-farm jobs with which to supplement their farm income. Second, the sector has become more industrialized and specialized so that it is now more closely interlinked and dependent upon other sectors of the economy. The extensive use of coordination devices such as vertical integration and contracts has alleviated the pressure for supply control measures in some parts of the agricultural sector and has reduced some of the risks inherent in agricultural production. At the same time, however, these devices have generated added burdens in the area of pricing. Finally, the changing structure of agriculture and of the agricultural input industries in many areas of the country has led to an erosion of the infrastructure serving farmers. This, in turn, threatens the survivability of the traditional family farm.

Notes

1. Per capita gross national product in 1986 was $17,480 in the United States but only $290 in India (World Bank 1988).

2. Let $e = (dq/dp)(p/q)$ be the price elasticity of demand. I assume this elasticity to be negative. Then since total revenue equals price times quantity, the elasticity of total revenue with respect to a quantity change is given by $e_y = (dqp/dq)(q/pq)$ $= (dp/dq)(q/p) + (p/p) = 1 + e^{-1}$. Hence if e is less than 1 in absolute value (i.e., if demand is price inelastic), e_y will be negative, so that when price falls and quantity increases (as will be the case if the demand curve is downward sloping), gross revenue will fall.

3. The series shown in Table 3.8 revises a similar series first estimated by Economic Research Service and reported in U.S. Department of Agriculture 1988. A consistent estimate of off-farm income of farm families is only available starting in 1960. For the 1950–59 period, off-farm income of farm families is estimated based on the personal income series estimated from 1934 to 1979 by Economic Research Service and reported in U.S. Department of Agriculture 1980.

4. El Niño is a warm ocean current that causes severe changes in animal life on both land and sea in this region. El Niño is an irregularly recurring phenomenon of varying severity. In fact, the El Niño of 1982 had much more severe impacts than that of 1972 not only off the coast of Peru but in other parts of the world as well, including along the California coast. It is interesting to note that subsequent to the 1972 El Niño, cheaper synthetic sources of the amino acids provided by fish meal became available and were approved by the Food and Drug Administration so that expensive fish meal is no longer a necessity in commercial animal feeds.

5. The coefficients were calculated as 100 times the square root of

$$\frac{\Sigma_1^n \left[(p_i - \hat{p}_i)^2 / \hat{p}_i \right]}{(n - 1)}$$

where n is the number of observations in the time series on p, and $\hat{p}_i$ is the estimated value of p_i based on a regression of p on a variable representing time. If all of the p_i lie on the regression line, the coefficient so computed is 0, indicating no variability about the trend line. If there is no trend, so that $\hat{p}_i = \bar{p}$ (the mean of p over all n observations) for all i, then the coefficient so computed is equal to the coefficient of variation calculated as shown in any standard statistics text.

Suggested Readings and References

Bergland, Bob S. 1981. *A Time to Choose: Summary Report on the Structure of Agriculture.* Washington, D.C.: U.S. Department of Agriculture.

Cochrane, Willard W. 1958. *Farm Prices: Myth and Reality.* Minneapolis: University of Minnesota Press.

———. 1979. *The Development of American Agriculture: A Historical Analysis.* Minneapolis: University of Minnesota Press.

Council of Economic Advisers. 1990. *Economic Report of the President.* Washington, D.C.: U.S. Government Printing Office.

Dum, Samuel A. 1977. "Pennsylvania Dairy Farm Business Analysis." Pennsylvania State University Cooperative Extension Service. Farm Management 58.

George, P. S., and G. A. King. 1971. "Consumer Demand for Food Commodities in the United States with Projections for 1980." University of California Giannini Foundation Monograph no. 26. March.

Grisley, William. 1982. "Pennsylvania Dairy Farm Business Analysis." Pennsylvania State University Cooperative Extension Service. Farm Management 65.

Hallberg, Milton C. 1988. "The U.S. Agricultural and Food System: A Postwar Historical Perspective." Northeast Regional Center for Rural Development,

Pennsylvania State University. Publication no. 55. October.

Hardin, Clifford M. 1986. "Congress Is the Problem." *Choices,* Premiere edition, 6–10.

Hathaway, Dale E. 1983. *Government and Agriculture.* New York: Macmillan.

Johnson, G. L. 1958. "Supply Function—Some Facts and Notions." In Earl O. Heady et al., eds. *Agricultural Adjustment Problems in a Growing Economy.* Ames: Iowa State College Press.

Johnson, G. L., and C. L. Quance, eds. 1972. *The Overproduction Trap in U.S. Agriculture.* Baltimore: Johns Hopkins University Press.

Knutson, Ronald D., J. B. Penn, and William T. Boehm. 1983. *Agricultural and Food Policy.* Englewood Cliffs, N.J.: Prentice-Hall.

Korb, Penni. 1987. "Comparing International Food Expenditures." *National Food Review* (U.S. Department of Agriculture, Commodity Economics Division. Economic Research Service) NFR-38, pp. 18–21.

Korb, Penni, and Nancy Cochrane. 1989. "World Food Expenditures." *National Food Review* U.S. Department of Agriculture, Economic Research Service NFR-41, pp. 26–29.

Malone, John W., Jr. 1983. "Opportunities for Resource Adjustment in Northeast Dairy Production." In *Implications of Reduced Milk Prices on the Northeast Dairy Industry,* edited by M. C. Hallberg and R. L. Christensen. Department of Agricultural Economics and Rural Sociology, Pennsylvania State University. A.E.& R.S. no. 167.

Manchester, Alden C. 1985. "The Farm and Food System: Major Characteristics and Trends." In *The Farm and Food System in Transition: Emerging Policy Issues.* no. 1. East Lansing: Michigan Cooperative Extension Service, Michigan State University.

Schertz, Lyle P., et al. 1979. *Another Revolution in U.S. Farming?* Washington, D.C.: U.S. Department of Agriculture.

Schneider, Lee D., et al. 1973. "Issues in Agricultural Land-Use Management in New Jersey." Department of Agricultural Economics and Marketing. Rutgers University. Special Report 17. February.

Tomek, William G., and Kenneth L. Robinson. 1972. *Agricultural Product Prices.* Ithaca: Cornell University Press.

Tweeten, Luther. 1979. *Foundations of Farm Policy.* 2d ed., rev. Lincoln: University of Nebraska Press.

U.S. Bureau of Census. 1975. *Historical Statistics of the United States: Colonial Times to 1970.* Washington, D.C.: Bureau of the Census.

_____. 1977. *Census of Manufacturers.* Washington, D.C.: Bureau of the Census.

_____. 1982. *Census of Manufacturers.* Washington, D.C.: Bureau of the Census.

U.S. Department of Agriculture. 1980. "Economic Indicators of the Farm Sector: Income and Balance Sheet Statistics, 1979." Economic Research Service. Statistical Bulletin no. 650. December.

_____. 1988. "Economic Indicators of the Farm Sector: Farm Sector Review, 1986." Economic Research Service. ECIFS 6-3. January.

_____. 1989. "Economic Indicators of the Farm Sector: National Financial Summary, 1988." Economic Research Service. ECIFS 8-1. September.

Wilken, D. F., and R. P. Kesler. 1978. "Summary of Illinois Farm Business Records." University of Illinois Cooperative Extension Service Circular 1162. August.

Wilken, D. F., R. P. Kesler, C. E. Cagley, and Irene Chow. 1983. "Summary of Illinois Farm Business Records." University of Illinois Cooperative Extension Service Circular 1214. August.

World Bank. 1988. *World Development Report 1988*. International Bank for Reconstruction and Development. New York: Oxford University Press.

II

The Benefits
and Costs
of Farm Policy

4

Benefits to Farmers of Government Agricultural Programs

Farm policy as conceived and implemented in the 1930s and as continued to the present was and is intended to be of benefit first and foremost to farmers. This is a well-accepted fact and requires no apologies even though many supporters of farm programs appear often to be on the defensive and in search of other farm program beneficiaries. Thus, a first task of this chapter is to examine the extent to which farmers and farm families have been aided by the farm programs in place.

Farmers though are not a homogeneous group. Some grow wheat, corn, or cotton, while others produce potatoes, fruits, or vegetables. Some emphasize animal production and grow crops for a source of feed for the animals, while others raise no animals at all. Some are so small in terms of value of farm products produced that they must rely on nonfarm jobs to supplement the income generated from the farm in order to sustain the farm family. Some operate million-dollar farm units. Some own the land they farm, while others rent most, if not all, of the land they farm. Diversity of this nature prompts a second question to be explored in this chapter: To what extent are America's farmers treated equally or unequally by the farm programs in place?

Contribution to Farm Family Income

Farm price and income support programs were initiated at a time when farm incomes were severely depressed—in the late 1920s and early 1930s. Even through most of the 1950s, money income of the farm household was estimated to be only about 50–60 percent of that of the average U.S. household (see Chapter 3). By the 1970s, however, the gap between farm and nonfarm money income had narrowed significantly so that in most recent years, money income per farm household has been *higher* than has the money income of the average U.S. household.

In Chapter 3, I suggested that off-farm sources of income had been more important than direct government payments in narrowing this gap. However, we must also consider whether farm policy has impacted prices of farm products and therefore farm revenues. Indeed, the acreage control features of farm policy have reduced production of the affected commodities and in turn have resulted in higher market prices than would have prevailed had these policy tools not been used. This is illustrated graphically in Figure 4.1, where curve DD represents the market demand curve, curve SS represents the aggregate supply curve under no government interference, and curve S'S' represents the aggregate supply curve under an acreage reduction program.

Since the amount of land that can be used under an acreage reduction program is restricted, any given quantity of output must be produced with more nonland inputs. The farmer is thus forced to produce a given level of output with a nonoptimal combination of resources (i.e., to produce less efficiently). Consequently, the marginal cost of production is higher. Since this is true for all producers who voluntarily participate in the acreage reduction program, the aggregate supply curve shifts to the left, as indicated by curve S'S' in the diagram.

As Figure 4.1 shows, equilibrium output will decline to Q' as intended by the program implemented. As we saw in Chapter 3, the demand for agricultural commodities is quite inelastic. Hence a rise in price and a corresponding fall in quantity produced can be expected to lead to an increase in aggregate gross farm income. This illustrates that just because

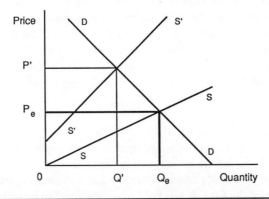

Figure 4.1. Impact of an acreage reduction program on market price and quantity of output

farm programs following the mid-1960s have operated so as to keep loan rates near market-clearing price levels (as I suggested in Chapter 2) does not mean that farm programs have had no impact on market prices and farm revenues. It is hardly satisfying to argue that farm policy has maintained loan rates near market-clearing price levels when simultaneously farm policy has resulted in elevated market-clearing price levels as a result of the leftward shift of the supply curve.

Obtaining quantitative estimates of the extent to which farm programs enhance prices is far from an exact science. To do so one must make assumptions about the slopes and curvatures of the relevant supply and demand functions, the independent variables to be included in these functions, farmers' propensity to participate in the farm programs, how the supply curves will shift when acreage reduction programs are removed, the length of lags involved in the various response relations, etc. Once the appropriate assumptions have been made and the relations estimated, one can determine the price enhancement impacts using a fully integrated, econometrically estimated model of the agricultural sector that is to be solved for equilibrium prices and quantities with the aid of a computer. Actual prices and quantities are then compared to estimated equilibrium prices and quantities to infer the consequences of the farm policies in place. This is the approach used by a recent Economic Research Service (ERS) team (see O'Brien and Fulton 1985) in which projections were made for 1986 to 1990 of (1) the impact of reverting back to permanent legislation (see Glossary) for agriculture and (2) the impact of removing *all* price and income support mechanisms in agriculture. The ratios of prices projected for 1986 under the permanent legislation scenario to those projected under the free market scenario are

Wheat	1.89
Feed grains	1.21
Rice	1.69
Cotton	1.55
Tobacco	1.19
Peanuts	1.85
Sugar	1.72

Using a less quantitatively sophisticated approach, Gardner (1981) compared actual market prices for 1978–79 with his best estimate of what prices would have been had the government programs then in place been removed. The reported price ratios from Gardner's analysis are

Wheat	1.18
Feed grains	1.06

Rice	1.02
Cotton	1.02
Tobacco	1.33
Peanuts	1.40
Sugar	1.88

We should expect the Gardner estimates to be lower than the ERS ratios not only because of the different projection years employed but more importantly because of the different policies analyzed. For example, "permanent" legislation for agriculture analyzed by the ERS team calls for price supports at 90 percent of parity and no supply controls. Gardner, on the other hand, studied the impact of removal of all government programs existing at the time of his analysis. Farm policy existing in 1978–79 was certainly favorable to farmers, but it was not nearly as favorable as permanent legislation would have been.

One could debate the merits of these estimates for some time. It seems clear, though, from these and similar studies that have been conducted more recently (see Baker, Hallberg, and Blandford 1989) that the farm programs currently in place do cause commodity prices to be substantially higher than they would otherwise be. Thus, in addition to direct income payments (deficiency payments, disaster payments, land diversion rental payments, etc.) farmers benefit via higher prices in the marketplace, which, given inelastic demand curves, translates into increased farm incomes.

Distribution of Benefits from Farm Programs

Distribution by Size of Farm. Several researchers have examined the distribution of farm program benefits in the form of direct payments across farms of different sizes. All of these studies come to the same general conclusion—the distribution of benefits is highly skewed in favor of the larger farms (for a comprehensive review of these studies, see Johnson and Short 1983).

This is not a very surprising conclusion. The farmer with the greater volume of produce to market will clearly receive greater total benefits from payments made on the basis of number of bushels (pounds or bales) produced than will the farmer with a smaller marketable output.

Similarly, the larger farmer stands to gain substantially more than does the smaller farmer from payments received for participating in a paid land diversion program. The disparity in benefits here will only be tempered by payment limits currently in force. Furthermore, as Johnson and Short point out, a higher proportion of the larger and more specialized farmers than of the smaller farmers participate in the farm programs.

It is instructive to examine the distribution of direct payments in some detail using the most recent data available. Direct government payments received by farmers include deficiency payments made in cash, cash or in-kind payments for reducing or diverting cropland acreages, disaster payments, and miscellaneous payments such as wool price support payments and dairy diversion or buyout payments. Not included are any monetary benefits farmers receive from the nonrecourse loan programs.

The set of data shown in the top one third of Table 4.1 suggests that the distribution of direct government payments to farmers is far from even across different-sized farms when size of farm is measured by annual gross sales. Further, the distribution appears to have become progressively more uneven over the last two and one-half decades. In 1960, the proportion of direct government payments was relatively evenly distributed among the first five sales classes shown in Table 4.1. In 1988, on the other hand, nearly one third of the direct payments went to those farms with annual gross sales of between $100,000 and $249,999, whereas only 8.3 percent of the payments went to those farms with annual gross sales of $19,999 or less.

Much has been made of this seemingly unequal distribution of payments in recent years, and farm policy has been on the receiving end of much criticism as a result. Three additional considerations, however, might suggest tempering the criticism. First, it is clear that comparing the distribution of payments among farm sizes as measured by annual gross sales is hazardous at best. A farm with $40,000 gross sales in 1960 may have been a relatively large farm during that year, whereas a farm with $40,000 gross sales in 1988 would be a quite small unit. This point can be made more dramatic by comparing *net farm income* generated by farms in a given sales group for a given year with the *average money income* of all families in the United States for the same year. In Table 4.2, I provide such a comparison for three of the sales classes of farms. From this table it is clear that a farm big enough to generate an income from farm operations sufficient to provide the farm family a standard of living comparable to or better than this family's nonfarm counterpart in 1960 (i.e., a farm with gross sales of $20,000–$39,999) was a mere weekend hobby in 1980–88!

Second, as suggested before, it should hardly be surprising that the distribution of direct government payments is skewed toward the larger producers (as measured by annual gross sales), since payment rates are based on volume produced. This point can be observed fairly clearly from the first set of data shown in Table 4.1. Further, this distribution has become more highly skewed in the direction of larger farmers over the years.

Third, a more appropriate measure of the equality or inequality of payment distribution is afforded by examining the distribution of government payments per farm as a percentage of gross income per farm. The distribution shown in the lower one third of Table 4.1 gives a realistic impression of the association between direct payments and the earning capacity of the farm. Clearly some inequality still exists, but this comparison suggests that the inequality may not be as severe as many believe it to be.

Table 4.1. Distribution of direct government payments and gross farm income by farm size class, 1960–88

Annual gross sales	1960	1965	1970	1975	1980	1985	1988
Percentage distribution of government payments							
Under $5,000	25.7	20.1	15.6	9.4	3.7	0.9	1.9
$5,000–$9,999	20.5	16.1	10.8	6.7	4.0	1.4	2.0
$10,000–$19,999	22.7	25.5	18.1	10.1	4.6	3.0	4.4
$20,000–$39,999	15.8	19.8	22.1	17.3	11.4	9.0	10.0
$40,000–$99,999	11.0	12.5	19.1	28.5	32.1	24.0	25.0
$100,000–$249,999	4.3	6.0	7.2	13.0	21.9	31.3	31.5
$250,000–$499,999	na	na	3.8	7.9	15.1	19.4	16.8
$500,000 and over	na	na	3.3	7.1	7.2	10.3	8.4
Percentage distribution of gross farm income[a]							
Under $5,000	15.9	11.1	9.8	5.3	4.2	3.8	3.3
$5,000–$9,999	14.9	10.1	6.5	3.4	2.7	2.5	2.3
$10,000–$19,999	20.6	17.1	11.7	6.1	4.0	3.7	3.4
$20,000–$39,999	17.7	19.5	18.4	11.4	6.9	5.3	5.2
$40,000–$99,999	14.7	18.2	20.9	24.0	18.4	15.7	13.7
$100,000–$249,999	16.2	24.0	10.3	15.8	18.4	23.1	21.2
$250,000–$499,999	na	na	8.0	13.2	18.8	18.2	17.2
$500,000 and over	na	na	14.4	20.8	26.6	27.7	33.7
Government payments per farm as a percentage of gross income per farm							
Under $5,000	3.0	10.4	10.7	1.5	0.7	1.1	5.1
$5,000–$9,999	2.6	9.1	11.1	1.7	1.3	2.7	7.3
$10,000–$19,999	2.1	8.5	10.4	1.4	1.0	3.9	11.2
$20,000–$39,999	1.7	5.8	8.1	1.3	1.4	8.1	16.6
$40,000–$99,999	1.4	4.0	6.2	1.0	1.5	7.5	15.8
$100,000–$249,999	0.5	1.4	4.7	0.7	1.0	6.5	12.9
$250,000–$499,999	na	na	3.2	0.5	0.7	5.1	8.5
$500,000 and over	na	na	1.6	0.3	0.2	1.8	2.2

Source: U.S. Department of Agriculture 1989.
[a]Excludes government payments but includes all other farm-related income.
na—not available.

Table 4.2. Net income from farm sources as a percentage of average money income of all households in the United States, 1960–88

Annual gross sales	1960	1965	1970	1975	1980	1985	1988
$20,000–$39,999	152	121	104	47	7	8	18
$40,000–$99,999	231	213	198	124	41	35	47
$100,000–$249,999	503	465	401	303	134	127	135

Source: U.S. Department of Agriculture 1989; and U.S. Bureau of the Census, *Current Population Reports,* various issues.

Those who see too much inequity associated with medium to large farms getting large government payments have argued for unconditional limits on the size of government payments on the basis that such farms have little need for income support over some minimum level. Accordingly payment limits of $55,000 were introduced in 1970; $20,000 for wheat, feed grains, and cotton combined in 1973; $55,000 for rice in 1975; $52,250 for rice in 1978; $50,000 for rice in 1979; $40,000 for wheat, feed grains, and cotton in 1978; $45,000 for wheat, feed grains, and cotton in 1979; and $50,000 for all commodities since 1980. The $50,000 limit does not include the value of nonrecourse loans received, marketing loan payments, or Findley payments (see Glossary for definitions of these terms). The 1981 and 1985 legislation called for disaster payment limits of $100,000. The Omnibus Budget Reconciliation Act of 1987 placed a $250,000 limit on *all* program payments from 1987 through 1990. Payment limits were further reduced by the Food, Agriculture, Conservation, and Trade Act of 1990 (see *payment limitation* in Glossary).

Those who argue for no payment limits or very high payment limits counter that if we impose payment limits, then we will not get those farmers who in fact produce most of the agricultural output to participate in the government programs in place. Note that 72 percent of the output by value in 1988 was produced by the largest 15 percent of the farms—those farms with annual sales of $100,000 or more (see Tables 4.1 and 3.9). If the objective of farm policy is to reduce farm surpluses, then we should attempt to ensure that those farmers who produce most of the output will participate in the farm programs in place. Restricting the amount of direct payments they can receive may discourage them from participating.

Distribution by Region. The distribution of payments across states is also far from even. In Table 4.3 we see that in 1988, government payments as a percentage of gross farm income were greatest in the wheat and feed grain areas of the country, followed closely by the

Table 4.3. *Direct payments to farmers per farm and as a percentage of gross farm income by state, 1988*

State	Direct payments per farm	Net income per farm	Direct payments as a % of gross farm income
Southeastern States			
Alabama	$2,385	$20,142	4.0
Florida	780	70,110	0.5
Georgia	3,551	27,978	4.2
South Carolina	2,958	12,442	5.9
Delta States			
Arkansas	7,018	31,920	7.5
Louisiana	5,761	18,324	8.7
Mississippi	5,774	22,162	8.5
Appalachian States			
Kentucky	1,659	9,634	5.1
North Carolina	2,147	25,004	2.9
Tennessee	1,543	8,454	5.0
Virginia	1,340	12,396	2.7
West Virginia	581	3,033	3.2
Northeastern States			
Connecticut	725	44,675	0.6
Delaware	3,500	68,800	1.7
Maine	1,000	14,616	1.4
Maryland	2,669	25,969	2.9
Massachusetts	406	30,116	0.5
New Hampshire	594	21,406	1.0
New Jersey	1,301	43,892	1.2
New York	2,117	13,932	3.0
Pennsylvania	1,180	16,220	1.7
Rhode Island	130	69,481	0.1
Vermont	831	16761	1.2
Corn Belt States			
Illinois	15,614	13,250	18.2
Indiana	8,328	8,400	13.3
Iowa	15,561	18,785	15.4
Missouri	4,154	8,185	9.9
Ohio	4,493	9,019	9.3
Lake States			
Michigan	5,411	10,745	9.3
Minnesota	11,260	16,418	14.7
Wisconsin	4,993	16,382	7.5
Northern Plains States			
Kansas	12,290	23,025	10.7
Nebraska	18,819	36,169	11.2
North Dakota	21,346	11,119	27.1
South Dakota	14,171	22,314	15.0
Southern Plains States			
Oklahoma	4,120	15,903	6.9
Texas	6,178	19,632	8.9

Table 4.3. Continued

State	Direct payments per farm	Net income per farm	Direct payments as a % of gross farm income
Mountain States			
Arizona	$9,605	$70,259	3.6
Colorado	10,275	27,509	6.3
Idaho	7,413	27,058	6.2
Montana	15,720	9,850	22.5
Nevada	2,423	15,615	2.5
New Mexico	5,100	20,864	4.9
Utah	2,887	14,038	4.6
Wyoming	4,236	11,079	4.5
Pacific States			
Alaska	2,903	19,032	4.8
California	3,989	72,119	1.9
Hawaii	86	36,796	0.1
Oregon	2,679	25,534	3.7
Washington	5,471	33,516	5.3

Source: U.S. Department of Agriculture 1989.

Delta States. Farmers in the vegetable and fruit and northeastern dairy areas of the country get precious little direct government payments.

This is not to say, of course, that dairy farmers receive no government assistance. Quite clearly the nation's dairy producers have been handsomely supported via price supports (particularly through the 1980s, as was pointed out in Chapter 2) and import controls. Further, most dairy farmers do produce feed grains and thus are eligible for direct payments if they elect to participate in the programs available to them. Participation rates in the feed grain programs in the Northeast are generally quite low since farmers in this region are more interested in producing feed for their dairy cows. They would participate in an acreage reduction program only if they anticipate that the deficiency payments they would receive for doing so would exceed the amount they would have to pay other farmers to obtain the grain needed to feed their livestock—grain they could no longer produce on their own farm.

Hence the distribution shown in Table 4.3 must be interpreted with care and with a knowledge of the type of agriculture in each state or area. It is apparent, though, that the distribution of payments among farmers in states with similar production patterns is not uniform, nor is the distribution of payments uniform across different commodity groups.

Distribution by Commodity. Table 4.4 demonstrates that for most commodities, most government payments go to the medium-to-large sized producers as measured by cropland acreages.[1] In the case of corn and oats, the farms with smaller acreages receive a substantial portion of the payments, however. This is undoubtedly due to the fact that on farms producing feed grains, not all of the available acreage is planted to one crop. This is in sharp contrast to the situation on farms growing wheat, cotton, and rice.

Table 4.5 indicates how direct payments were distributed by size of payment and by commodity in 1985. Here we see that the majority of direct payments went to producers who received modest to small payment amounts. Note, however, the glaring exception afforded by rice and cotton. For these two crops, well over 50 percent of the total payments distributed went to producers who received government payments of $40,000 or more each!

Table 4.4. Percentage of total direct payments received by producers on farms with different cropland acreages by commodity, 1982

Total cropland acreage	Percentage of farms receiving direct payments							
	Wheat	Corn	Barley	Sorghum	Oats	Cotton	Rice	Total
0–69	3.1	4.4	1.7	5.8	7.5	4.9	4.5	4.2
70–139	5.0	10.2	3.4	7.7	16.6	5.2	5.5	6.0
140–219	5.7	12.6	4.5	8.4	12.7	5.2	6.8	6.9
220–259	2.6	5.9	2.2	4.0	7.5	2.8	3.9	3.4
260–499	15.6	26.7	15.4	20.6	19.0	17.9	21.9	19.3
500–999	25.9	22.8	27.8	24.2	18.7	31.3	29.5	27.8
1,000–1,499	15.4	8.2	16.8	12.4	5.6	17.0	13.4	14.5
1,500–1,999	8.9	3.8	9.3	6.4	4.4	8.2	6.6	7.4
2,000–2,499	5.5	1.9	5.6	3.6	2.8	3.6	3.3	3.8
2,500 and over	12.3	3.5	13.3	6.9	5.2	3.9	4.6	6.7
Total	100.0	100.0	100.0	100.0	100.0	100.0	100.0	100.0

Source: Cate and Becker 1984.

Benefits of Government Programs
to Resource Owners

Assume existing policy increases the price received by farmers for a crop such as corn beyond the competitive equilibrium level without restricting output. In Figure 4.2 this is illustrated by the price increase from P_e to P_s. Farmers, responding along their supply curve, will then be encouraged to increase their production of corn beyond

Table 4.5.   Percentage of total direct payments received by producers by size of payment and commodity, 1985

Crops for which payments were received	Size of payment						
	$50,000 or more	$40,000–$49,000	$30,000–$39,000	$20,000–$29,000	$10,000–$19,000	Under $10,000	Total
Wheat only	3.6	9.1	8.1	12.5	22.7	44.0	100.0
Corn only	3.8	3.5	5.3	10.3	24.9	52.2	100.0
Rice only	39.5	15.3	11.3	11.8	11.3	10.8	100.0
Cotton only	41.1	12.8	10.9	10.8	10.4	14.0	100.0
Feed grain only	3.9	3.8	5.5	10.0	23.0	53.7	100.0
Wheat and feed grains	5.5	9.2	10.2	15.8	26.1	33.2	100.0
Rice and other	26.3	22.5	15.5	15.3	12.6	7.8	100.0
Cotton and other	25.1	17.7	13.7	14.8	14.6	14.1	100.0
All other combinations	44.5	20.6	12.4	9.5	8.4	4.6	100.0
All crops	12.8	9.5	9.2	13.0	21.4	34.1	100.0

Source: U.S. General Accounting Office 1987.

the equilibrium output level (Q_e) to level Q_s, as shown in Figure 4.2. From the theory of the firm, we know that the *derived* demand for an input is a function of the price of this input, the price of all other inputs, *and* the price of the output. In fact, the derived demand for an input is normally an *increasing* function of the price of the output. Hence, we can normally expect a policy that supports the price of the program commodity above the equilibrium level with no supply control to cause a rightward shift of the demand curve for an input. As Figure 4.3 indicates, this will result in an increase in the equilibrium price of the input from P_e to P', *and* in the equilibrium quantity of the input from Q_e to Q'.

This is a very important result. It shows first of all how price supports can impact the price farmers receive and thus also farm receipts. More important, for our present purposes, it shows that price policy for agriculture can have an impact on the owners of the inputs used to produce farm output. In the remainder of this section I demonstrate that the owners of different farm inputs will fare differently depending on the supply characteristics of their respective industries.

Hired Workers. Assume the initial result of price supports of the type under consideration here is that hired farmworkers receive higher wages out of the increased receipts farm operators obtain from the government price supports.[2] The pool of available workers is so large that we would expect a major clamoring for the higher wage rate farm

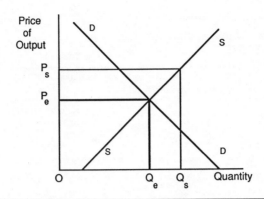

Figure 4.2. Impact of price support policy with no supply control

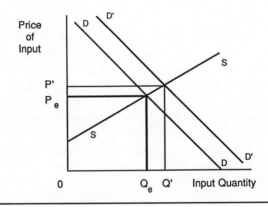

Figure 4.3. Impact of income enhancement policy on an input market with elastic supply

operators are now willing to pay. Thus it would not take much of an increase in wages to attract additional workers into agriculture. In order to retain workers in the remaining sectors of the economy, wage rates everywhere would have to rise by the same amount that they do in agriculture. The overall long-term effect on equilibrium wage rates then would likely be very small if perceptible at all. This conclusion hinges on the assumption that hired workers are fairly mobile and can easily shift from one sector to another in search of jobs. As we saw in Chapter 3, this is a fairly realistic assumption for today's world, although it has not always been the case.

Operator Labor. Similarly we would expect no appreciable impact on farm operator labor returns in the long run. Certainly in the short run, increased farm commodity receipts would be reflected in increased operator labor returns. In the longer term, however, these gains could be expected to be dissipated throughout the agricultural sector as adjustments are made. In the year following the favorable agricultural program, for example, additional effort would be diverted toward producing the supported crop by those who previously did not produce this crop. This action could be expected to diminish the higher labor returns to the supported crop and perhaps raise the labor returns on nonsupported crops slightly. In the end, labor returns would tend to be equated in all crops but probably at only a slightly higher level. Thus, in the end there would be only a slight, if any, effect on farm operator labor returns. Again the key here is that adjustments in the employment of farm operator labor are possible and highly probable.

Input Suppliers. Producers of purchased inputs—fertilizers, herbicides, machinery, and other manufactured goods used in farm production—also stand to gain in the short run but not in the long run. The price of these inputs would definitely be bid up in the short run because in a period of a year or less input supplies will tend to be fixed. Over the longer term, however, suppliers of these inputs would increase their output in response to the increased short-run returns. As the increased output is put on the market, prices would tend to fall again until nonnormal profits in these industries are driven out. Indeed, prices of these inputs can be expected to be driven back down nearly to levels that existed before the farm programs were put into effect. Again the possibility, and indeed high probability, of adjustments being made by the agricultural input industries is a key factor here.

Landowners. Consider what happens, though, to a resource that cannot easily be diverted to other uses. Land is a prime example. When the price of the supported commodity increases, some land that had previously been devoted to nonagricultural uses may be brought into production of the supported commodity. The possibilities here are fairly limited, however, because of the special soil and climatic requirements of agricultural production. Some land previously used to produce other agricultural crops may also be shifted into the supported commodity as the output price ratios change. Here again, though, the possibilities are fairly limited. The situation in this case is more like that depicted in Figure 4.4,

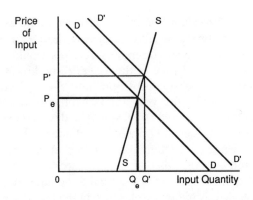

Figure 4.4. Impact of income enhancement policy on an input market with inelastic supply

where the supply of land is very inelastic. There will be very little change in the quantity of the input supplied but a great deal of change in the price of this input.

Land that is already being used to produce the supported commodity will remain in production of this commodity. This land merely earns a surplus—that is, an economic rent. This rent will not be dissipated, because there are no claims on it by significant amounts of other land. If the supported commodity uses only a small proportion of the total land available for all agricultural production, the economic rent will be small. Obviously the gains to land will be greater, the greater the number of agricultural crops supported in this way.

If the price support program results in the expected value of one acre of corn production being $10 higher, then the annual rental value of land will be increased by $10 per acre. When one buys an acre of land, one obtains the rights to a future flow of rental payments. If the program that results in an increased rental value of $10 per acre can be expected to continue indefinitely, the additional value of land is the discounted present value of the additional $10 per acre in annual rental payments. If this indefinite stream of rents rises with inflation and if the real rate of interest is 6 percent, then the increase in value would be $167 per acre ($10/0.06). In this way, then, farm program benefits get "capitalized" into the value of the land resource.

In the case of the land resource, the key factor is that adjustments in land use are not as likely to occur as in the case of other inputs, so the resource under question earns an economic surplus or rent. The owners of this resource are the primary beneficiaries rather than farm laborers. Farm operators who rent land similarly do not benefit from the program. This is a crucial concern when considering the distributional consequences of farm programs. Approximately one third of the farm land is owned by nonfarm operators (although many of these are retired farm operators). About 37 percent of farm land is rented. Interestingly enough, a very high percentage of the farm operators in the lower sales classes are full owners. Boxley (1985) found that 75 percent of the land operated by farmers with annual cash sales of less than $20,000 was owned by these same operators.

New Asset Created. One of the criticisms levied against input restriction programs such as acreage reduction or paid land diversion programs is that they induce less efficient production (see Figure 4.1) as farmers are forced to alter the mix of inputs required to produce the commodity in question. An alternative supply control measure used for tobacco and peanuts in the United States and for milk, tobacco, eggs, and turkeys in Canada is a producer quota that limits marketings of the supported commodity. In this case producers are permitted to produce the lower output with no restrictions on input use. Hence the supply curve will not be shifted from its original position—that is, production efficiency need not necessarily be altered.

The economic consequences of a marketing quota program are illustrated in Figure 4.5. With no supply control or price support program, equilibrium price is at P_e and equilibrium quantity is at Q_e. Under a marketing quota program, marketings are restricted to Q', and consumers

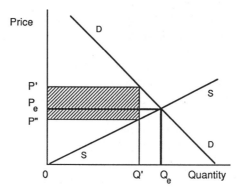

Figure 4.5. Economic impact of a marketing quota program

bid up the price of the more limited supply to P'. Producers are, of course, very pleased with price P'. However, they would have been willing to produce Q' even had price been at P''. In the aggregate, then, producers receive a "surplus" of an amount equal to (P' − P'') × Q', or the shaded area in the diagram, over the course of a given production period.

As in the case of an acreage reduction program, the owners of these "rights" to produce and market a crop (i.e., the holders of marketing quotas) also have the "rights" to the future (annual) flow of their portion of the shaded area in Figure 4.5. If these marketing quotas can be leased out or sold on the open market, they accumulate value. Hence a new asset is created. The value of this asset provides a market measure of the expected future benefits of the quota program.

Recent estimates of the market value of quotas in the United States and Canada are as follows (see Gardner 1981 and World Bank 1986):

Tobacco quota in North Carolina	$1.20 per pound
Peanut quota in Georgia	$0.06 per pound
Milk quota in Canada	$3,500 per cow
Tobacco quota in Canada	$1.50 per pound
Egg quota in Canada	$23 per hen
Turkey quota in Canada	$0.54 per pound

The original recipients of these quotas are, of course, the recipients of a "windfall." If the quota program is subsequently terminated, the final owners of the quotas will be net losers by the amount of the final value of the quota. Strong resistance to termination of quota programs can thus be expected.

It is, of course, possible to adopt a quota program under which quotas are neither leasable nor salable—that is, they remain with the original recipient until he or she sells the farm. In this case, the quota value accumulates (is capitalized) into the value of the land and accrues to the landowner. This is generally considered the least desirable way of implementing a quota program for two reasons. First, it prevents a more efficient producer from bidding the quota away from a less efficient producer. Thus with a nontransferable quota program, production inefficiencies are not only probable, they are highly likely. Second, this option locks in a regional production pattern that may not be in society's long-term interests from either an efficiency or a developmental standpoint. For these reasons the preferred method of implementing a quota program from an efficiency point of view is to permit unrestricted exchange of the quotas between farmers regardless of their location.

Summary

The discussion of this chapter has highlighted several issues related to the beneficiaries of farm programs. In the case of the smaller farms, there is precious little gained from direct government payments. I previously concluded that for the majority of the relatively small farms, the need for income support is not great because these units derive most of their family income from off-farm sources. This is not to say we should ignore or, worse yet, abandon the small, multiple-job-holding farm family. Rather, it is to say that traditional farm programs designed to give income support to farm families are of little consequence to farm families on the smaller farms, so we must look to other policy options or approaches to ensure their economic viability or to enhance their income position.

By the same token, direct government payments are of little importance to the very largest farms, because on these farms most of the family income comes from the farming operation itself. But large farmers do benefit from the farm programs to the extent that the prices they receive are higher than would otherwise be the case. And given the volumes they produce, the price enhancement features of the farm programs in place translate into sizable farm income gains. The small farmer also benefits from this price enhancement, but the small farmer has such a small volume of produce to sell that price enhancement, like direct payments, is of little consequence.

The midsized farm families probably have the greatest need for income support. Happily, these are the farms for which the distribution of direct payments appears to be concentrated. That is, direct payments are decidedly *not* equally distributed across different-sized farms, but this inequality may well be justifiable in terms of the relative income needs of farm families.

The distribution of direct payments by farm size, though, is not consistent among producers of different commodities. The evidence suggests that the very large cotton and rice producers are the recipients of a greater share of the direct payments than can be justified on the grounds of income needs. This is particularly true in view of the fact that the cotton and rice programs also serve to enhance the prices cotton and rice producers receive for their output.

Finally, in assessing who benefits from farm programs, it is important to note that landowners and those who own marketing quotas, if quotas exist, benefit more than do hired workers, renter-operators, or input suppliers. This, then, is another source of inequity in the farm programs of today. Interestingly enough, this attribute tends to work to the advantage of small farmers in that the majority of the small farmers are owner-operators.

Notes

1. Unfortunately, a similar distribution for a more recent year is not available. We can, however, be reasonably certain that the same general pattern shown in Table 4.4 would emerge using more recent data.

2. This and the next three subsections borrow heavily from Gardner 1981.

Suggested Readings and References

Baker, Derek, Milton Hallberg, and David Blandford. 1989. "U.S. Agriculture under Multilateral and Unilateral Trade Liberalization—What the Models Say." Department of Agricultural Economics and Rural Sociology, Pennsylvania State University. A.E. & R.S. no. 200. January.

Boxley, Robert F. 1985. "Farmland Ownership and the Distribution of Land Earnings." *Agricultural Economic Research* 37 (4): 40–44.

Cate, Penelope C., and Geoffrey S. Becker. 1984. "Federal Farm Programs: A Primer." Library of Congress Congressional Research Service 84-232ENR.

Gardner, Bruce L. 1981. *The Governing of Agriculture*. Lawrence: Kansas Regents Press of Kansas.

Johnson, James D. and Sara D. Short. 1983. "Commodity Programs: Who Has Received the Benefits?" *American Journal of Agricultural Economics* 65 (2): 912–21.

O'Brien, Patrick, and Thomas Fulton. 1985. "Possible Economic Consequences of Reverting to Permanent Legislation or Eliminating Price and Income Supports." U.S. Department of Agriculture. Economic Research Service. Agricultural Economic Report no. 526. January.

U.S. Department of Agriculture. 1989. "Economic Indicators of the Farm Sector: State Income Summary, 1988." Economic Research Service. ECIFS 8-2. October.

U.S. General Accounting Office. 1987. "Farm Programs: 1985 Payments and Crop Loans by State." Fact Sheet for the Chairman, Committee on the Budget, U.S. Senate, GAO/RCED-87-155FS. July.

World Bank. 1986. *World Development Report 1986*. The International Bank for Reconstruction and Development. New York: Oxford University Press.

5 Costs of Government Agricultural Policy

Infanger, Bailey, and Dyer (1983) argue that the budget process has become an instrument for fundamental farm policy change and, in particular, was the determining factor in the austerity of the 1981 farm legislation. In effect they argue that as the cost of food and agricultural programs accelerates or reaches a critical level, Congress is prompted to alter policy in such a way as to reduce this drain on the federal budget. Indeed one can point to a variety of discussions on Capitol Hill and elsewhere during the debates leading up to the Food Security Act of 1985 that were strongly motivated by the budgetary impacts of various features of the new act.

This is hardly a new phenomenon, however. Clearly farm program costs cannot escalate upward indefinitely without hard questions being asked. During the 1950s and 1960s, for example, when agricultural surpluses became such a great burden to the American public, much concern was expressed about the magnitude of farm program costs, and considerable effort was expended by farm interests to justify these expenditures. In fact, a rather lengthy report was prepared by Congress (Joint Economic Committee 1960) in an effort to do so. In this report, much was made of the fact that agriculture is only one of the recipients of federal subsidies—other recipients include the maritime industry, air carriers, motor carriers, railroads, water carriers, mineral producers, and businesses in receipt of postal subsidies.

In any event we must continuously assess farm program costs and attempt to put them in their proper perspective for program evaluation. But determining the costs of food and agricultural programs is not as straightforward as we may like it to be. An obvious approach is to tally up the actual budget outlays or taxpayer costs associated with administering the programs. While this process sounds easy enough, it is not without its difficulties. A major problem here is that one must make some assumptions

119

as to what constitutes farm programs. Is the Food for Peace Act (Public Law 480), for example, a program designed to aid farmers by increasing the demand for food, or is it strictly a food aid program that would be in place even without support for domestic agriculture? The same question must be asked of domestic food aid programs.

The first section of this chapter will present the budgetary costs of federal farm programs according to our definitions and assumptions about what constitutes farm programs. Budgetary costs, however, do not include all costs associated with farm programs. Thus, in the remaining sections of this chapter I examine the source and nature of additional costs and indicate how they arise and how they may be estimated.

Budgetary Outlays

As we saw in Chapter 2, the federal government is required by law to support the prices of cotton, wheat, rice, peanuts, tobacco, wool, mohair, honey, milk, corn, barley, oats, rye, grain sorghum, soybeans, and sugar. The secretary of agriculture also has discretionary authority to support other agricultural commodities as needed. Nonrecourse loans, open-market purchases, deficiency payments, and a number of supply management tools are used in this support activity. This activity is carried out by the Commodity Credit Corporation (CCC), which is a wholly owned government corporation operating within the U.S. Department of Agriculture under authority of the Commodity Credit Corporation Charter established by the Agricultural Adjustment Act of 1933. The CCC has an authorized capital stock of $100 million and authority to borrow up to $25 billion from the U.S. Treasury. Borrowed funds are used to carry out the activities of the CCC and appropriations are requested from Congress to offset any losses incurred.

The CCC programs authorized under federal legislation have "entitlement" status so that there is no preestablished ceiling on the total amount that can be spent in a given year. The CCC has no choice but to make available its loans, payments, and related benefits to all farmers who apply and qualify. The benefit levels and eligibility requirements are set in advance by Congress and the administration, and once in place the CCC cannot decide that some farmers should receive them and others not.

Estimated budgetary outlays for the 1950–88 fiscal years and for those programs directly affecting commodity prices, farmers' incomes, and the supply of farm products are shown in Table 5.1. The data for 1950–73 were taken from Cochrane and Ryan 1976, and the data for 1974–88 were calculated by the author using the procedure and data sources described in Cochrane and Ryan 1976.[1] Program costs shown here consist of total CCC

Table 5.1. Government cost of farm and related programs by commodity group, 1950–88

Year	Feed grains	Wheat	Cotton	Soybeans	Tobacco	Rice	Dairy	Other	Total
					($1,000,000)				
1950	**764**	**519**	**301**	**−31**	**68**	**10**	**188**	**773**	**2,592**
1951	−263	32	−571	−1	46	3	−86	440	−400
1952	−284	203	148	0	107	0	7	500	681
1953	336	1,013	352	10	36	−1	275	821	2,842
1954	590	1,042	939	−9	42	12	393	620	3,629
1955	**481**	**742**	**360**	**41**	**158**	**106**	**58**	**571**	**2,517**
1956	701	476	975	−9	155	106	118	721	3,243
1957	585	532	−99	111	97	19	208	1,145	2,58
1958	837	747	−450	78	45	36	172	1,284	2,749
1959	756	1,286	951	120	33	26	84	1,338	4,594
1960	**911**	**877**	**270**	**−49**	**−111**	**68**	**171**	**760**	**2,897**
1961	1,230	1,069	−159	−4	13	70	244	1,086	3,549
1962	897	513	809	202	−49	67	474	1,087	4,000
1963	1,193	1,231	960	−59	181	101	556	688	4,851
1964	1,348	666	794	70	300	110	262	1,062	4,612
1965	**1,106**	**1,077**	**944**	**−41**	**227**	**96**	**217**	**883**	**4,509**
1966	1,154	1,328	1,068	24	−9	84	45	811	4,505
1967	1,000	790	378	188	16	97	291	833	3,593
1968	1,429	1,392	295	260	176	80	290	836	4,758
1969	1,966	1,497	1,033	561	115	136	255	947	6,510
1970	**2,025**	**1,553**	**996**	**−76**	**219**	**128**	**194**	**844**	**5,883**
1971	1,247	1,164	681	−575	176	140	290	975	4,098
1972	1,873	1,552	804	−57	−112	130	205	1,266	5,661
1973	1,328	707	872	−18	−106	159	103	1,046	4,091
1974	750	563	740	30	−141	173	45	667	2,827
1975	**259**	**457**	**253**	**23**	**−135**	**151**	**485**	**1,227**	**2,721**
1976	177	695	15	−6	381	180	140	1,862	3,443
1977	808	2,343	114	54	195	247	809	1,921	6,490
1978	2,293	1,310	236	81	151	50	516	3,570	8,207
1979	1,635	1,082	163	43	257	158	333	2,552	6,223
1980	**1,827**	**1,550**	**71**	**71**	**21**	**90**	**1,352**	**1,600**	**6,581**
1981	−142	2,376	356	131	27	180	3,181	1,702	7,810
1982	5,542	2,925	1,207	163	107	182	5,104	1,587	16,817
1983	13,575	7,301	2,513	217	881	1,217	5,828	703	32,235
1984	787	3,664	647	−501	351	542	3,529	2,758	11,778
1985	**5,127**	**3,064**	**1,519**	**739**	**364**	**373**	**6,259**	**−409**	**20,726**
1986	20,956	7,153	4,313	879	−80	1,481	5,252	3,215	43,171
1987	18,369	4,977	2,370	−496	−14	1,019	4,225	5,358	35,808
1988	10,308	3,151	2,066	−595	−455	475	3,586	11,319	29,856

Source: For 1950–73, Cochrane and Ryan 1976; for 1974–88, calculated by author.

expenditures on all farm programs less a credit for nonfarm benefits. Total CCC expenditures consist of

1. CCC inventory and loan transactions for price support programs estimated as
   a. purchases plus outlays for nonrecourse loans on farm products plus the cost of acquiring inventories under purchase programs, plus
   b. storage, handling, transportation, and reseal expenses, less
   c. sales of commodities acquired through price support operations plus nonrecourse loan repayments on loan redemptions (included here are foreign sales for dollars and for barter and commodities paid for by other government agencies or by special congressional appropriations).
2. Direct payments made to producers participating in commodity programs. This item includes deficiency payments, payments in kind to farmers, disaster payments, annual land retirement payments not included elsewhere, payments made under the National Wool Act of 1954, and dairy diversion and dairy buyout payments.
3. CCC interest charges allocated to commodities on the basis of average value of inventories and loans.
4. Operating and administrative expenses of allotment programs, quota programs, the sugar program, Agricultural Stabilization and Conservation Service, etc.
5. Long-term land retirement and soil conservation program expenditures—Soil Bank (1956–59), Agricultural Conservation Program (1949–70), Rural Environmental Assistance Program (REAP) (1971–74), Rural Environmental Conservation Program (RECP) (1974–88), and Conservation Reserve Program (1986–88).
6. Expenditures on export subsidy and export credit sales programs,[2] including differential payments for exporting private and CCC stocks as well as the more recent Targeted Export Assistance Program (TEAP) and Export Enhancement Program (EEP), and 1950–66 expenditures under the International Wheat Agreement Program.
7. Public Law 480 program costs less sales for foreign currency and dollar repayments, and including credit sales for dollars, Title II donations, and costs of ocean transportation on commodities donated through voluntary agencies and other costs of exporting CCC inventories. This item also includes foreign aid expenditures for agricultural products specifically identified as made in the United States. Sales of foreign currencies include proceeds to the CCC from the sale and use of foreign currencies. Dollar repayments include recovery of costs through dollar repayments to the CCC by foreign governments and private trade entities.

8. Removal of surplus agricultural commodities under section 32 of the Agricultural Adjustment Act of 1933 as amended in 1935, and under section 416 of the Agricultural Act of 1949.[3]
9. Special School Milk Program (1955–88) and School Lunch/Child Nutrition Program (1950–88). The School Lunch Program was incorporated into the Child Nutrition Program in 1967.[4]

Credits for nonfarm benefits include one-half of the following: CCC donations from inventories under section 32 of the Agricultural Adjustment Act of 1933 as amended and under section 416 of the Agricultural Act of 1949, removals of surplus agricultural commodities, Special School Milk Program, School Lunch/Child Nutrition Program, Public Law 480, and foreign aid.

Expenses included in the Total category of Table 5.1 that could not be allocated to commodity groups include (1) CCC operating expenses prior to 1981, (2) long-term land retirement and soil conservation program expenses, (3) removals of surplus agricultural commodities under section 32, and (4) School Lunch/Child Nutrition Program expenditures. Total program costs are negative for some commodities and for some years because CCC revenues from the disposition of commodities plus total credits exceed total expenditures in those years.

Although the pace of change in total program costs has been quite uneven, they have trended upward at an average rate of about $620 million per year over the period 1950–88. Even if we omit the exceptionally expensive years of 1982–87, the trend is still decidedly up, increasing at an average rate of about $148 million per year. Through the 1950s and mid-1960s they were relatively stable; during the shortages of the 1970s they trended downward as one would expect; through the 1980s they trended steeply upward.

An initial perspective on the magnitude of these expenditures can be obtained by expressing them on a per capita basis as has been done in Table 5.2. For comparison purposes we note that in 1988 per capita expenditures on domestically produced meat products were $463, on domestically produced fruits and vegetables $360, on domestically produced dairy products $230, and on domestically produced poultry and eggs $128 (see Dunham 1989)!

Another assessment of the magnitude of these expenditures can be obtained by expressing them as a percentage of cash receipts from farm marketings for the respective commodity groups (Table 5.3). Here we see that with the exception of the early 1950s and the second half of the 1970s, program expenditures relative to cash receipts were exceptionally high for feed grains, wheat, cotton, and rice. For the twenty-six years since 1953 and exclusive of the strong demand years of 1973–81, expenditures as a proportion of cash receipts for feed grains exceeded 25 percent in nineteen years; for wheat, they exceeded 25 percent in twenty-four years and exceeded 50 percent in twelve years; for cotton they exceeded 25 percent in nineteen years; and for rice, they exceeded 25 percent in sixteen years. Over the period 1968–70 program expenditures for wheat averaged 80.6 percent of cash receipts from the sale of wheat, and over the period 1983–88, 76.7 percent. Over the period 1983–88 program expenditures on rice averaged nearly 107 percent of cash receipts from the sale of rice, and in one of those years, 205 percent of cash receipts from the sale of rice! Program expenditures for dairy were kept at what appear to be fairly reasonable levels (i.e., generally well below 10 percent of cash receipts) until 1981. Thereafter program expenditures in dairy have been in the range of 18–35 percent of cash receipts.

The distribution of total farm program costs across commodity groups remained remarkably stable over the period 1950–74 (Table 5.4). During this period feed grains generally accounted for one fourth to one third of the total, wheat one fourth or slightly more of the total, cotton one fifth or slightly more of the total, and dairy generally between 2–8 percent but never more than 12 percent of the total. Beginning in 1975, however, some interesting changes began to occur in the distribution of program costs.

Table 5.2. Cost of farm and related programs per capita, 1950–88

Year	Cost per capita	Year	Cost per capita
1950	**$17.02**	**1970**	**28.69**
1951	−2.58	1971	19.73
1952	4.32	1972	26.97
1953	17.74	1973	19.31
1954	22.26	1974	13.22
1955	**15.17**	**1975**	**12.60**
1956	19.20	1976	15.79
1957	15.11	1977	29.46
1958	15.72	1978	36.87
1959	25.84	1979	27.65
1960	**16.03**	**1980**	**28.89**
1961	19.32	1981	33.94
1962	21.44	1982	72.32
1963	25.63	1983	137.29
1964	24.03	1984	49.70
1965	**23.21**	**1985**	**86.62**
1966	22.92	1986	178.67
1967	18.08	1987	146.79
1968	23.70	1988	121.17
1969	32.11		

Source: See source for Table 5.1.

Table 5.3. Government cost of farm and related programs as a percentage of cash receipts by commodity group, 1950–88

Year	Feed grains	Wheat	Cotton	Soybeans[a]	Tobacco	Rice	Dairy	Total
1950	**35.7**	**25.5**	**12.4**	**−4.2**	**6.4**	**5.1**	**5.1**	**9.1**
1951	−12.6	1.5	−20.0	−0.1	3.9	1.4	−2.0	−1.2
1952	−12.5	7.4	5.0	0.0	9.8	0.0	0.2	2.1
1953	14.0	42.3	11.1	1.4	3.3	−0.4	6.3	9.2
1954	23.1	50.0	40.8	−1.1	3.6	4.1	9.6	12.2
1955	**18.8**	**40.0**	**14.0**	**4.9**	**12.9**	**39.4**	**1.4**	**8.5**
1956	26.5	24.0	39.0	−0.9	13.3	44.2	2.6	10.7
1957	24.4	28.8	−5.6	11.1	10.0	8.6	4.5	8.7
1958	28.8	29.3	−21.0	6.7	4.4	17.2	3.8	8.2
1959	27.3	65.2	35.4	11.5	3.1	10.6	1.8	13.7
1960	**30.5**	**37.1**	**11.4**	**−4.1**	**−9.6**	**27.4**	**3.6**	**8.5**
1961	44.3	47.3	−6.4	−0.3	1.0	25.1	4.9	10.1
1962	30.3	23.0	31.7	12.9	−3.7	20.2	9.8	11.0
1963	34.9	58.0	33.8	−3.4	14.3	28.7	11.4	12.9
1964	39.1	37.9	31.5	3.8	21.2	30.7	5.2	12.4
1965	**29.9**	**60.6**	**40.5**	**−1.9**	**19.1**	**25.5**	**4.3**	**11.5**
1966	26.6	62.4	67.3	0.9	−0.7	20.0	0.8	10.4
1967	22.8	37.7	34.5	7.7	1.2	21.8	5.1	8.4
1968	33.1	72.1	22.4	9.7	15.0	15.4	4.9	10.8
1969	43.0	83.7	75.7	21.1	8.9	29.9	4.1	13.5
1970	**39.6**	**86.1**	**79.4**	**−2.4**	**15.8**	**29.6**	**3.0**	**11.6**
1971	22.6	53.7	45.8	−16.1	13.3	30.6	4.3	7.8
1972	32.0	57.0	43.6	−1.0	−7.8	22.6	2.9	9.3
1973	12.5	10.5	31.2	−0.2	−6.8	12.4	1.3	4.7
1974	5.4	7.7	25.6	0.4	−6.7	13.7	0.5	3.1
1975	**2.1**	**6.0**	**10.9**	**0.3**	**−6.2**	**14.1**	**4.9**	**3.1**
1976	1.3	10.3	0.4	−0.1	16.5	22.2	1.2	3.6
1977	6.8	50.5	3.3	0.4	8.4	26.2	6.9	6.7
1978	20.1	26.2	6.7	0.7	5.8	4.5	4.1	7.3
1979	11.6	14.4	3.8	0.3	11.3	13.3	2.3	4.7
1980	**10.0**	**17.5**	**1.6**	**0.5**	**0.8**	**5.9**	**8.3**	**4.7**
1981	−0.8	24.1	8.8	1.1	0.8	10.4	17.6	5.5
1982	31.8	29.7	27.1	1.3	3.2	12.0	28.0	11.8
1983	87.4	83.2	67.8	1.8	32.0	137.9	31.1	23.6
1984	5.0	42.7	17.6	−4.2	12.5	49.4	19.7	8.3
1985	**22.8**	**38.7**	**41.2**	**6.6**	**13.4**	**35.7**	**34.7**	**14.4**
1986	122.1	146.4	119.6	9.6	−4.1	205.1	29.6	31.9
1987	140.2	100.1	58.0	−5.0	−0.7	174.7	23.9	25.7
1988	67.4	49.0	44.3	−4.8	−22.3	38.2	20.3	19.7

Source: See source for Table 5.1.

[a]Percentage of cash receipts from oil crops.

Table 5.4. Government cost of programs by commodity group as a percentage of total cost of farm and related programs, 1950–88

Year	Feed grains	Wheat	Cotton	Soybeans	Tobacco	Rice	Dairy
1950	**29.5**	**20.0**	**11.6**	a	**2.6**	**0.4**	**7.3**
1951[b]							
1952	a	29.8	21.7	0.0	15.7	0.0	1.0
1953	11.8	35.6	12.4	0.4	1.3	a	9.7
1954	16.3	28.7	25.9	a	1.2	0.3	10.8
1955	**19.1**	**29.5**	**14.3**	**1.6**	**6.3**	**4.2**	**2.3**
1956	21.6	14.7	30.1	a	4.8	3.3	3.6
1957	22.5	20.5	a	4.3	3.7	0.7	8.0
1958	30.4	27.2	a	2.8	1.6	1.3	6.3
1959	16.5	28.0	20.7	2.6	0.7	0.6	1.8
1960	**31.4**	**30.3**	**9.3**	a	a	**2.3**	**5.9**
1961	34.7	30.1	a	a	0.4	2.0	6.9
1962	22.4	12.8	20.2	5.1	a	1.7	11.9
1963	24.6	25.4	19.8	a	3.7	2.1	11.5
1964	29.2	14.4	17.2	1.5	6.5	2.4	5.7
1965	**24.5**	**23.9**	**20.9**	a	**5.0**	**2.1**	**4.8**
1966	25.6	29.5	23.7	0.5	a	1.9	1.0
1967	27.8	22.0	10.5	5.2	0.4	2.7	8.1
1968	30.0	29.3	6.2	5.5	3.7	1.7	6.1
1969	30.2	23.0	15.9	8.6	1.8	2.1	3.9
1970	**34.4**	**26.4**	**16.9**	a	**3.7**	**2.2**	**3.3**
1971	30.4	28.4	16.6	a	4.3	3.4	7.1
1972	33.1	27.4	14.2	a	a	2.3	3.6
1973	32.5	17.3	21.3	a	a	3.9	2.5
1974	26.5	19.9	26.2	1.1	a	6.1	1.6
1975	**9.5**	**16.8**	**9.3**	**0.9**	a	**5.6**	**17.8**
1976	5.1	20.2	0.4	a	11.1	5.2	4.1
1977	12.5	36.1	1.8	0.8	3.0	3.8	12.5
1978	27.9	16.0	2.9	1.0	1.8	0.6	6.3
1979	26.3	17.4	2.6	0.7	4.1	2.5	5.4
1980	**27.8**	**23.6**	**1.1**	**1.1**	**0.3**	**1.4**	**20.5**
1981	a	30.4	4.6	1.7	0.3	2.3	40.7
1982	33.0	17.4	7.2	1.0	0.6	1.1	30.3
1983	42.1	22.7	7.8	0.7	2.7	3.8	18.1
1984	6.7	31.1	5.5	a	3.0	4.6	30.0
1985	**24.7**	**14.8**	**7.3**	**3.6**	**1.8**	**1.8**	**30.2**
1986	48.5	16.6	10.0	2.0	a	3.4	12.2
1987	51.3	13.9	6.6	a	a	2.8	11.8
1988	34.5	10.6	6.9	a	a	1.6	12.0

Source: See source for Table 5.1.
[a]Cost for this commodity group was negative.
[b]Total cost of the government programs was negative in this year.

First, cotton now accounts for a much smaller proportion of the total. This is clearly the result of a new policy for cotton that emphasizes stricter production controls as a means to support income rather than direct payments. Second, feed grains and wheat expenditures, while still accounting for the bulk of total program expenditures, have become a much more variable proportion of the total, reflecting increased variability in prices and export demand. Third, beginning in 1981, dairy has become the third most expensive commodity group, with expenditures soaring to 30 percent or more of the total. The reason for this can be traced directly to the reluctance of Congress and/or the administration to lower support prices for milk in the face of declining feed grain prices in the late 1970s and early 1980s. Low feed grain prices mean low dairy ration prices, and since feed costs amount to about 60 percent of the total cost of producing milk, low dairy ration prices in the face of stable milk prices mean good times for dairy farmers. Good times for dairy farmers, in turn, mean record milk production levels. And finally, record milk production levels (in the face of at best stable milk demand) mean record CCC expenditures needed to maintain the milk price support level!

The distribution of total CCC expenditures across cost categories has changed considerably (Table 5.5), reflecting the changing structure of the programs over this thirty-nine-year period. The greatest changes to be noted here are (1) the declining importance of Public Law 480 in more recent years, (2) the increasing importance of direct payments as a means of supporting farm incomes beginning in the early 1960s and continuing through the mid 1970s and to a lesser extent until the present time, and (3) the declining relative importance of CCC inventory and loan operations since the heyday of Public Law 480 and until the 1980s. Table 5.5 also highlights the declining relative importance of long-term land retirement and land conservation in the total farm program package since the 1950s and 1960s. When price supports for the major crops were lowered to world levels in the mid-1960s, and the surplus-producing capacity of agriculture was held in check by a system of voluntary acreage controls, the shift from government commodity operations to farm income support through deficiency payments and annual land retirement rental payments became more prominent (Table 5.5).

Costs to Consumers

Consumers suffer in two ways from farm programs that keep market prices above free-market equilibrium levels: (1) the price of food is higher then it otherwise would be, and (2) as a consequence of higher food prices, per capita consumption of these foods

Table 5.5. Expenditures by category as a percentage of total cost of farm and related programs, 1950–88

Year	CCC inventory and loan operations	P.L. 480 expenses[a]	Direct payments to farmers	Payments for land retirement	Payments for land conservation[b]
1950	**53.6**	...	**2.3**	...	**8.2**
1951	c	...	c	...	c
1952	c	...	8.9	...	38.4
1953	71.5	12.2	2.2	...	9.6
1954	77.4	9.5	1.8	...	4.7
1955	**65.2**	**26.5**	**2.8**	...	**9.3**
1956	59.2	32.5	3.8	0.1	6.6
1957	6.3	69.4	4.6	21.1	10.1
1958	18.2	48.6	3.1	23.7	7.8
1959	48.9	28.3	3.3	17.0	5.2
1960	**17.8**	**50.1**	**4.4**	**11.2**	**8.2**
1961	16.9	50.3	13.1	10.2	7.0
1962	20.6	44.5	23.0	8.6	6.2
1963	32.1	36.6	25.7	6.4	4.3
1964	22.1	37.0	27.4	6.4	4.6
1965	**4.5**	**37.1**	**49.0**	**4.5**	**4.8**
1966	c	36.3	66.3	3.5	4.7
1967	c	40.3	80.7	5.5	6.0
1968	12.8	25.7	54.6	4.4	4.6
1969	29.8	15.1	46.0	2.9	3.1
1970	**10.1**	**16.1**	**59.1**	**2.0**	**3.1**
1971	c	23.7	84.2	1.9	4.1
1972	16.0	18.7	52.2	1.2	3.3
1973	c	20.4	89.7	1.3	4.0
1974	c	25.2	83.2	1.7	0.1
1975	**19.7**	**38.7**	**20.6**	**1.5**	**9.0**
1976	23.4	38.5	10.8	1.1	6.9
1977	50.9	19.6	9.2	0.3	2.8
1978	25.9	15.6	26.2	...	3.2
1979	15.6	20.6	31.0	...	3.7
1980	**48.3**	**19.3**	**6.4**	...	**2.8**
1981	26.6	19.1	13.2	...	2.5
1982	55.3	8.3	8.9	...	1.0
1983	34.9	3.5	43.9	...	0.6
1984	c	10.1	41.2	...	1.5
1985	**28.3**	**8.7**	**35.2**	...	**0.9**
1986	21.0	2.8	54.6	0.3	0.3
1987	c	2.9	42.7	3.8	0.4
1988	c	0.9	40.6	1.0	0.7

Source: See source for Table 5.1.
[a]Includes foreign food aid assistance.
[b]Excludes payments for long-term land retirement.
cThe item was negative for the indicated year.

declines. Up until the mid-1960s, there is general consensus that food prices would have been lower had the farm programs been eliminated. Farm prices were supported above equilibrium levels and little or no attempt was made to restrict production.

Following 1965, however, when direct payments were used to support farm incomes and loan rates were reduced to near equilibrium levels, the results are less clear. Cochrane and Ryan (1976) suggest that food prices were not greatly inflated by farm policies after 1965 and up to 1973. Gardner (1981) suggests otherwise for 1978–79 as we saw in Chapter 4.

The issue is controversial because our estimation methods are less than perfect and because it is impossible to know what the structure of the agricultural sector would have been in the absence of government programs. It is clear, however, that even though the market price is not forced up by the loan rate, it is forced up to a certain extent by any supply control programs in effect. That is, when land is idled as a result of the farm programs, we saw in the previous chapter that the supply curve moves to the left because farmers in the aggregate can no longer produce as efficiently as they could with no restrictions on land use. Gardner (1981) estimates, for example, that in 1978–79 the cost to consumers of the farm. policies was as follows:

Wheat	$ 400 million
Feed grains	$ 700 million
Tobacco	$ 425 million
Sugar	$1,440 million
Peanuts	$ 200 million

Even more insidious, in many ways, are the higher food costs associated with import controls. Here no direct transfers of any kind take place—no payments to producers, no costs to taxpayers, no border taxes on imports, etc. But restrictions on imports reduce the supply available to domestic consumers, and the rationing of lower supplies causes consumer prices to be higher then would otherwise be the case. This is the principal reason why the consumer cost for sugar was so high in 1985 as noted in Chapter 1. A historical perspective of the impact of import restrictions on sugar prices since 1950 is provided by the data in Table 5.6. U.S. sugar prices have been considerably above world prices in most of these years, but the disparity has been greatest in the 1980s. U.S. sugar policy has, in part, been justified on the basis of stabilizing sugar prices. The data in Table 5.6 suggest that on this score the policy has been reasonably successful. The cost of this policy, however, has been quite high when measured in terms of higher consumer prices and in production opportunities forgone.

Import restrictions are also responsible for the fact that beef and

Table 5.6. Wholesale prices for raw sugar in the United States and in the world, 1950–90

Year	New York spot price[a]	World market price[b]	Ratio of New York to world price
		(cents/lb.)	
1950	**5.93**	**5.82**	**1.02**
1951	6.06	6.66	0.91
1952	6.26	5.08	1.23
1953	6.29	4.27	1.47
1954	6.09	4.14	1.47
1955	**5.95**	**4.19**	**1.42**
1956	6.09	4.47	1.36
1957	6.24	6.10	1.02
1958	6.27	4.36	1.44
1959	6.24	3.86	1.62
1960	**6.30**	**4.09**	**1.54**
1961	6.30	3.85	1.64
1962	6.45	3.87	1.67
1963	8.18	9.41	0.87
1964	6.90	6.79	1.02
1965	**6.75**	**3.07**	**2.20**
1966	6.99	2.81	2.49
1967	7.28	2.95	2.47
1968	7.52	2.96	2.54
1969	7.75	3.37	2.30
1970	**8.07**	**3.75**	**2.15**
1971	8.52	4.52	1.88
1972	9.09	7.43	1.22
1973	10.29	9.61	1.07
1974	29.50	29.99	0.98
1975	**22.50**	**20.49**	**1.10**
1976	13.30	11.58	1.15
1977	11.00	8.12	1.35
1978	13.93	7.81	1.78
1979	15.56	9.66	1.61
1980	**30.11**	**29.02**	**1.04**
1981	19.73	16.93	1.17
1982	19.72	8.42	2.34
1983	22.04	8.49	2.60
1984	21.74	5.18	4.20
1985	**20.34**	**4.04**	**5.03**
1986	20.95	6.05	3.46
1987	21.83	6.71	3.25
1988	22.12	10.18	2.17
1989	22.81	12.79	1.78
1990	**23.26**	**12.55**	**1.86**

Source: U.S. Department of Agriculture, *Agricultural Statistics,* various annual issues.
[a]Wholesale price of raw sugar in New York.
[b]Duty paid wholesale price of raw sugar, c.i.f. at New York.

perhaps also milk cost U.S. consumers more than would be the case in the absence of such restrictions (although milk is a more complex story, as we shall see later). Unfortunately costs due to import restrictions are hidden in the sense that they do not involve out-of-pocket costs as in the case of budgetary costs borne by taxpayers, nor does the consumer see them at the grocery store checkout counter since cheaper foreign products are not available. Furthermore, these costs are quite small for any individual consumer; hence they do not cause a great deal of opposition to farm programs that bring them about. Nevertheless, they are real and must be recognized as such.

Social Costs

A final class of costs that must be considered are referred to as "social costs," or costs over and above budgetary or taxpayer costs and "hidden" consumer costs.[5] These social costs are not gained by anyone—they are pure losses. They arise because the restrictions imposed by the policies implemented cause resources to be used in a nonoptimal (from a production efficiency standpoint) manner. Land is left idle that could be used to produce corn even if at a lower price. Land is taken out of soybean production and put into corn production because the government-supported returns for producing corn are higher than for producing soybeans. Farmers refuse to grow the more profitable crop (soybeans) in preference to the less profitable crop (corn) because they fear a loss of future corn program benefits will result from a reduction in their current corn base acreage should they convert some of their corn acreage to soybean production at this juncture. Land is used to produce milk instead of apples because of the support price for milk. Lobbyists park in Washington, D.C., full-time in anticipation of keeping favorable farm programs or making them even more favorable.[6] The Agricultural Soil Conservation Service (ASCS) is forced to hire many employees to help administer the farm programs. Some farmers are discouraged from taking up employment in their best alternative, which is *not* farming.

Some of the social costs of an acreage reduction program (indeed, probably most) can be illustrated with the aid of the simple diagram I used in Chapter 4 (Figure 4.1), reproduced here as Figure 5.1. In this diagram, the area labeled (A) represents losses in production efficiency that come about because farmers are required to produce the reduced output in a nonoptimal way. This area represents an efficiency loss to society as the supply curve shifts to the left. Consumers lose from the program because they must pay a higher price per unit for a reduced number of units of output. A measure of consumers' loss is provided by the area between price lines P' and P_e bounded by the demand curve on the right and the vertical axis of the diagram on the left.

Producers as a group may or may not lose depending upon the specifics of the program and the shape of the supply and demand curves. If, for example, the program is voluntary and an incentive payment in the form of an annual land retirement rental payment is offered producers to encourage them to participate in the program, then it seems clear that producers as a group will gain from the program. But taxpayers must foot the bill for these incentive payments. And since farmers are both consumers and taxpayers, some or all farmers may in the end also be net losers from this program.

Yet there is more. Idling of the land input literally means that it cannot be used to produce any other commodity. If it is put to absolutely no productive use, then there is an additional social loss represented by the area labeled (B) in Figure 5.1. In fact, this land is generally put to soil-conserving uses, which may have positive social benefits—if not for the current generation, then for future generations. It may also be planted to trees so that it is indeed put to productive current use. It may also have benefit as "open spaces" or as hunting land. Thus, all of area (B) may not appropriately be considered a social loss.

The reader should by now be able to demonstrate that any program aimed at supporting farm incomes by raising commodity prices and/or restricting production distorts the market in such a way as to cause social losses. Is there a better way to support farm incomes than with, say, price supports and/or nonrecourse loans? Most economists argue in the

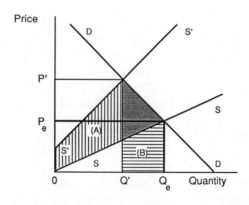

Figure 5.1. Social costs of an acreage reduction program

affirmative, suggesting that a more appropriate alternative is direct income payments. Direct income payments or transfers accomplish the objective of enhancing farm incomes without interfering with the price-equilibrating function of the marketplace. Further, by letting commodity prices reach their free-market levels, we are more likely to remain competitive with other producing nations. It is largely for this reason that the 1973 farm bill initiated the system of target prices and direct deficiency payments. Whether target prices and deficiency payments are less market distorting than price supports is debatable and is a subject to which I will return in Chapter 7.

Estimating social costs is grist for the applied welfare analyst's mill. It is a subject that generates much controversy among economic scientists because our tools of analysis are not exact and because applying these tools involves the use of value judgments on the part of the analyst that not everyone can accept. The implication that political lobbying is not a productive or useful activity, for example, is a value judgment. The welfare analyst assumes that the competitive or free-market solution is the "best" market solution and, therefore, the standard with which adopted or contemplated policies should be compared. This too is a value judgment.[7]

Some critics contend that the welfare analyst does not account for all the benefits of policy options (or, alternatively, of the costs of the free-market solution). Such benefits might include (1) the market information provided by the program used to implement the chosen policy, (2) market stability brought about by the program adopted, (3) benefits associated with maintaining an environment in which smaller farm and/or processing units can survive, etc. On the estimation side, there is a real question as to whether or not one can ever know (or estimate) with any degree of certainty (or reliability) what the agricultural sector will look like under free-market conditions. After all we have no recent history of free-market conditions in agriculture. How then, one may ask, can our estimation methods, which are historically based, lead to results that can simulate a free-market structure that we have never observed in the past?

In sum, estimation and use of social costs is problematic and subject to controversy. Not all costs or benefits are quantifiable. There may even be disagreement over what should be considered benefits. Typically, social costs are estimated relative to a free-market situation. Unfortunately, we can never be sure that we know what value prices or production or structural variables would have taken on under free-market conditions. At best, social costs as typically estimated should be used as one of several guides to policy choices rather than the sole guide.

Summary

Budgetary costs of the farm programs were quite high for the major commodities (wheat, feed grains, and cotton) during the 1950s and 1960s, fairly low for all commodities during the late 1970s and early 1980s when demand was high and surpluses had diminished, and on an upward path again following 1982 for the major commodities and dairy. In more recent years, cash receipts from the major crops have dropped substantially more than government costs of the crop programs have risen. Hence the large percentages shown in Table 5.3 for 1986–87 are a function of market conditions as well as of changes in government programs.

Domestic food programs have become more costly in the 1980s, but the foreign food aid program (Public Law 480) has diminished in relative importance. Both food aid programs are currently supported on their own merits rather than as mechanisms to support farm incomes even though they were originally designed to help support farm incomes by expanding food demand (see chapter 6).

The distribution of expenditures across cost categories has changed considerably, reflecting the changing structure of the programs over the years. Public Law 480 expenditures have diminished as a proportion of the total. Direct payments to farmers account for an increasing proportion of the total, reflecting the fact that direct payments as opposed to price supports are now the major tool with which to support farm incomes. Consistent with the increasing importance of direct payments, CCC inventory and loan operations have become a smaller proportion of total budgetary costs of the farm programs.

A full accounting of the costs of farm programs, however, must go beyond budgetary costs. Consumers face higher prices for food then they would without the farm programs because of the price support and/or output control features of the farm programs. Restrictions on imports also lead to higher consumer costs of food—costs that are "hidden" because consumers do not pay any direct transfers to effect the import restriction policies nor do they see the less expensive imports in the grocery store, so shoppers are prevented from making a comparison. Further, higher food costs mean reduced consumption since the demand curve for food is downward sloping. Thus, consumers are willing to take less of the agricultural produce off the market at artificially elevated prices.

Finally, there are social costs that arise to the extent that restrictions imposed by the farm programs cause resources to be used in a nonoptimal manner. Nonland inputs are used more intensively on the restricted land input by farmers in an attempt to increase their output. Land is left idle that could otherwise be used to produce agricultural products. Land is used to produce a supported commodity when pure economics would dictate that it should be used to produce some other commodity.

Notes

1. Basic data sources for these calculations include (1) Commodity Credit Corporation, *Report of Financial Condition and Operations of the Commodity Credit Corporation* (U.S. Department of Agriculture) for the fiscal years 1974–88, and (2) *Budget of the U.S. Government.* 1974-88. More recent data is not yet available.

2. See Glossary for a description of the Export Credit Guarantee and Export Credit Sales programs and for blended credit.

3. See Glossary for a description of section 32 of the 1933 act and for section 416 of the 1949 act.

4. The Food Stamp Program (1950–88) and the Women, Infant, and Children (WIC) Program (1965–88) were not included as farm program costs. This is at variance with Cochrane and Ryan, but was predicated on the assumption that these expenditures would exist even if the farm programs did not and hence should not be charged as a cost of the farm programs.

5. See the Glossary for a formal definition of *social costs* and for a description of how these costs can be determined.

6. As we saw in Chapter 1, whenever the potential exists for some to benefit and some to lose from pending or prospective legislation, considerable lobbying activity is likely to result. This activity requires resources that could productively be employed elsewhere and hence entails social costs. Unfortunately there are no estimates of such costs available and it is beyond the scope of this book to make such estimates. Part of these costs, though, include political contributions through the relevant political action committees (PACs). The magnitudes of these contributions are readily available starting with those made prior to the 1978 elections (see Greevy, Gore and Weinberger 1984).

7. See the discussion in Chapter 1 concerning the role of value judgments in policy analysis. See also Just, Hueth, and Schmitz (1982) for a thorough treatment of modern welfare analysis.

Suggested Readings and References

Cochrane, W. W., and M. E. Ryan. 1976. *American Farm Policy, 1948–73.* Minneapolis: University of Minnesota Press.

Dunham, Denis. 1989. "Food Cost Review, 1988." U.S. Department of Agriculture. Economic Research Service. Agricultural Economic Report no. 615. July.

Gardner, Bruce L. 1981. *The Governing of Agriculture.* Lawrence: Kansas Regents Press of Kansas.

Greevy, David U., Chadwick R. Gore, and Marvin I. Weinberger, eds. 1984. *The PAC Directory: Book II, The Federal Committees.* PAC Research, Inc. Cambridge, Mass.: Ballinger Publishing Co.

Infanger, Craig L., William C. Bailey, and David R. Dyer. 1983. "Agricultural Policy in Austerity: The Making of the 1981 Farm Bill." *American Journal of Agriculture Economics* 65 (1) :1–9.

Joint Economic Committee. 1960. "Subsidy and Subsidylike Programs of the U.S. Government." Washington, D.C.: U.S. Government Printing Office.

Advanced Readings

Harberger, A. C. 1971. "Three Basic Postulates for Applied Welfare Economics: An Interpretive Essay." *Journal of Economic Literature* 9 (3): 785–97.

Just, Richard E., Darrell L. Hueth, and Andrew Schmitz. 1982. *Applied Welfare Analysis and Public Policy.* Englewood Cliffs, N.J.: Prentice-Hall.

Wallace, T. D. 1962. "Measures of Social Costs of Agricultural Programs." *Journal of Farm Economics* 44 (1): 580–94.

III

Policy Analysis

6 Demand Expansion Programs

A variety of programs aimed at increasing the demand for U.S. farm products and thereby increasing U.S. farm prices and incomes have been pursued. Promotional programs targeted at both the domestic and foreign markets have been used extensively. Programs designed to distribute surplus commodities to the needy at home have been implemented. Efforts to dispose of surplus agricultural products abroad for both humanitarian and developmental reasons have been pursued. Finally, programs designed to subsidize sales to foreign buyers at prices below domestic market prices have been proposed and implemented. In this chapter, we will study these programs in some detail. First I examine such programs generically in an effort to provide a basis for understanding the consequences of these types of programs for U.S. farmers in particular and for U.S. society in general. Next I identify the major domestic and foreign "demand expansion" programs currently in force.

Promotional Programs

Domestic Level. Efforts to promote agricultural products concentrate on providing consumers with increased information about the product, providing consumers with information about and access to new forms of the product, or otherwise trying to convince consumers to buy more of the product. The objective of such promotion efforts is to shift the demand curve to the right. If the demand curve shifts to the right, equilibrium price and quantity sold will increase from P_e to P' and from Q_e to Q', respectively, as indicated in Figure 6.1. The net result will obviously be an increase in farm revenues.

Food manufacturers have long engaged in promoting their product

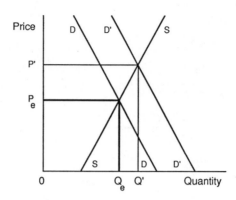

Figure 6.1. Economic impact of demand promotion policy

through advertising. To the extent that this causes a shift in the demand curve for the retail product, it also causes a shift in the (derived) demand curve for the farm product from which this processed product is made.[1] In recent years, food advertising expenditures have increased dramatically, in large part as a result of food companies' attempts to differentiate their products. Connor (1981) notes that "media advertising by food manufacturers alone averages well over 3 percent of sales if one excludes highly perishable products; food retailers spend an additional 1 percent of their sales on advertising. Other forms of sales promotion such as coupons, incentives, samples, and some direct sales force activity would probably raise the intensity of advertising to 8 percent of sales" (p. 20).

More recently farm groups have become involved in promotional programs for farm commodities. Federal legislation designed to provide agricultural commodity groups with the authority to collectively develop and support research and promotion of their product has been enacted for several commodities—cotton, potatoes, eggs, wool and lambs, wheat, and milk (see Morrison 1984). Although these acts differ on specifics, they all provide for the collection of funds from all producers by means of a "checkoff" on the sale price of the farm product. Provisions are generally included so that producers not in favor of supporting the program can obtain a refund of their checkoff, or assessment. Some states also have programs of this type for commodities not covered by federal legislation. In addition, most marketing orders (to be discussed in more detail in Chapter 8) provide for some type of assessment with which to support research and promotion programs.

The Dairy and Tobacco Adjustment Act of 1983, for example,

authorized a mandatory \$0.15 per hundredweight assessment on all milk produced in the contiguous forty-eight states and marketed commercially by dairy farmers. The act provides that dairy farmers can receive credit of up to \$0.10 per hundredweight for contributions to ongoing regional, state, or local dairy product promotion or nutrition education programs. The program is administered by the National Dairy Promotion and Research Board. Expenditures are made for advertising and promotion campaigns, nutrition education and research projects, and product development research.

The Beef Research and Information Act of 1976 authorized a similar promotion scheme for the beef industry, but the plan failed to receive the necessary producer support in a national referendum. Finally, the Food Security Act of 1985 authorized checkoff programs for beef, pork, honey, and watermelons, and the Food, Agriculture, Conservation, and Trade Act of 1990 authorized checkoff programs for soybeans, pecans, mushrooms, and limes as well as a processor-funded program for milk.

Evidence concerning the impact of advertising on product demand is mixed. Ward, Thompson, and Armbruster (1983) cite a few studies that suggest a positive payoff from advertising in dairy, grapefruit, and oranges. Kinnucan and Forker (1986) estimate that optimal seasonal allocation of generic advertising in 1979–81 would have resulted in a 9 percent increase in returns to producers supplying milk to the New York City market. Early studies commissioned by the National Dairy Promotion and Research Board failed to show consistent payoffs to promotion of dairy products (National Dairy Promotion Board 1987). A more recent study, however, shows a positive response to fluid milk advertising since the act providing for milk promotion assessments was enacted (see Ward and Dixon 1989).

These mixed results concerning consumer responses to generic advertising should lead one to be cautious if not skeptical about the efficacy of checkoff programs designed solely for the purpose of increasing generic advertising of agricultural products. The more likely scenario is that advertising pays when focused on a new product or newly discovered attribute of an old product up to the point at which the majority of consumers become well informed. Beyond this saturation point, any more than "maintenance" levels of advertising expenditures are likely to sell little additional product. If one subscribes to this thesis, then it makes little sense for the nation's dairy farmers, say, to attempt to match the soft drink industry's level of advertising expenditures. A much more effective use of the funds collected from the milk checkoff program would likely accrue from developing new product forms and new merchandising methods as have the other beverage industries.

It is instructive to ask, By how much will prices and quantities change *if* there is a positive response to these kinds of promotional expenditures? A full answer to this question requires a knowledge of the numerical values of the parameters in the retail demand function, the farm supply function, and the marketing margin relation. Changes in retail and farm prices and in farm revenues as a result of changes in the checkoff for promotion depend not only on the response of consumers to generic advertising but also on the elasticity of both supply and demand, on farmers' output response to the checkoff, and on marketers' propensity to transmit price changes through the marketing system.[2] These relationships will all vary from industry to industry.

A crucial parameter here is consumers' response to generic advertising. Regardless of the industry, there is a value for this parameter below which there will be a negative payoff (in terms of farm revenue) from a checkoff and above which there will be a positive payoff. Thus much research has been and continues to be devoted to determining the numerical value of this parameter.

Foreign Level. In order to encourage private investment in export market development, the Foreign Agricultural Service (FAS) of the U.S. Department of Agriculture initiates programs to augment the individual investments made by various U.S. agricultural organizations. The two major activities of FAS are the Market Development Cooperator Program and the Export Incentive Program. The former is jointly sponsored with private industry and focuses on generic promotion. The latter assists in the export of branded products. Both programs are designed to seek out and develop new export markets for existing and newly developed products, to reduce trade barriers, and to improve market access in various countries throughout the world. In 1980, FAS spent almost $20 million on this effort. United States and foreign cooperators raised the total spent under the market development program effort to $67.3 million (U.S. Department of Agriculture 1980). In 1986 these expenditure levels were slightly over $40 million and $120 million, respectively.

Subsidized Consumption

Domestic Subsidies. The demand curve can also be shifted to the right by means of a subsidy of a specified amount per unit of the product, which is designed to lower the effective price of food to consumers. As with promotion efforts, farmers look with favor on pure subsidy programs (as opposed to policies that keep both retail and farm prices below equilibrium) because consumption subsidies cause farm prices and revenues to rise. The problem with subsidy programs of this nature, however, is that although they do involve a transfer of benefits from taxpayers to both consumers and producers, they also involve a net social loss.

To see this, consider Figure 6.2. Here a per unit subsidy of an amount $(P' - P'')$ will shift the effective demand curve to $D'D'$. The subsidy will require a total government expenditure of area $P'ABP''$. Producer price will rise from P_e to P' so that the portion of government expenditure represented by the area between P_e and P' bounded on the right by the supply curve will be transferred to producers. Effective consumer price, on the other hand, will fall from P_e to P'' so that the portion of government expenditure represented by the area between P_e and P'' bounded on the right by demand curve DD will be transferred to consumers in the form of added utility. Unfortunately, the shaded area in the diagram represents a net loss to society since it represents a portion of government expenditure that is recovered by no one. Although this loss might be small relative to the transfers effected, it is a loss nevertheless and one that governments would ordinarily wish to avoid.

We should note that the social loss shown in Figure 6.2 is brought about because something has been introduced to distort the market—namely, the consumption subsidy. A better solution, most would argue, is to give producers and consumers a direct income or lump-sum transfer so that no (or minimal) market distortion takes place in the product market.

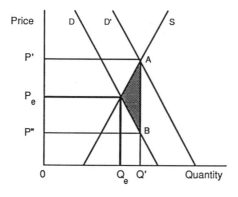

Figure 6.2. Economic impact of a food subsidy program

Export Subsidies. Food aid to foreign countries has long been a policy objective of the government of the United States. In part this objective has been pursued to assist foreign countries in times of need. In part also this objective has been pursued to assist American farmers in disposing of surplus agricultural production. Further, for several years prior to 1972, direct export subsidies were used to encourage increased grain exports. Under this program, a private U.S. grain exporter could obtain a special payment for each bushel of grain exported. This payment was designed to make up the difference between the U.S. market price and a lower price on the world market.

Export subsidies of this type lost favor with the public in 1972 when subsidies based on low world prices were being paid on exports to the USSR while an extraordinary surge in demand for grain exports was pushing world-market prices above U.S. loan rate levels. Thus export subsidies of this nature were terminated in 1972. Subsequently, however, weaker world demand plus the propensity of the European Economic Community[3] to offer surplus grain and dairy products to foreign buyers at subsidized prices has caused a reversal of U.S. actions. The Food Security Act of 1985 instituted new programs enabling export subsidies on major agricultural commodities: the Export Enhancement Program (EEP) and the Targeted Export Assistance Program (TEAP). Both were continued by the 1990 farm bill.

Whatever name these programs are given, they constitute export subsidization and amount to export "dumping." They are expensive in terms of U.S. taxpayer dollars. Furthermore, they involve a transfer of U.S. dollars to *foreign* buyers. Figure 6.3 helps to make this clear. The panel on the left in Figure 6.3 represents the supply-demand situation for the United States. The panel on the right represents the supply-demand situation for the rest of the world (ROW). The middle panel collapses the information about the United States and the ROW into the two curves labeled ES and ED, respectively. ES represents the "excess supply" schedule of the United States. It is measured as the horizontal difference between supply and demand in the United States at the various levels of price above the no-trade equilibrium price, P_u. ED represents the "excess demand" schedule of the ROW. It is measured as the horizontal difference between demand and supply in the ROW at various levels of price below the no-trade equilibrium price, P_r. The equilibrium "world price," P_w, assuming there are no barriers to trade between nations and no domestic protection, will be determined by the intersection of ES and ED. If price in the United States and the ROW is at the "world price" level, the excess of supply in the United States over demand in the United States will be exactly equal to the excess of demand in the ROW over supply in the ROW. Thus the United

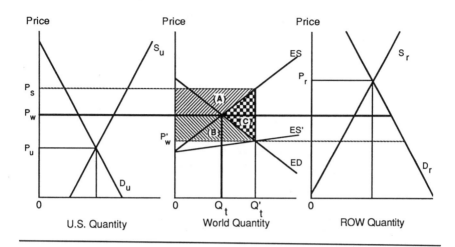

Figure 6.3. Economic impact of an export subsidy

States will export its excess supply to the ROW. (I am, of course, simplifying a great deal here by assuming there are no transportation costs. But this assumption is not crucial to my argument.)

Now suppose the United States offers an ad valorem export subsidy in order to maintain the domestic price above the world price level, that is, a price of P_s. The quantities offered to the ROW by the United States are less expensive to the ROW by the amount of the subsidy. This is reflected in Figure 6.3 by a rotation of the "effective" excess supply schedule to ES'. Equilibrium world price is now P'_w and the United States has driven a wedge between the domestic (support) price, P_s, and the resulting world price. The quantity traded, though, increases from Q_t to Q'_t.

The total subsidy required is $(P_s - P'_w) \times Q'_t$. That portion of this subsidy labeled (A) is transferred to U.S. producers—it is in fact the amount by which U.S. producers' surplus is increased over and above the amount by which U.S. consumers' surplus is diminished. That portion of the subsidy labeled (B) is transferred to *foreign* consumers—it is in fact the amount by which foreign consumers' surplus is increased over and above the amount by which foreign producers' surplus is diminished. Note that as in the case of domestic subsidies, the portion of the subsidy labeled (C) is transferred to no one. Area (C) is, then, a pure social loss. Thus again we have a case of a market distortion generating a social loss. This time the distortion is caused by an export subsidy. Presumably everyone would have been better off had we made lump-sum transfers to both U.S. producers and foreign consumers equal to areas (A) and (B), respectively, but otherwise leaving the market undistorted.

Subsidization of exports is by no means a new phenomenon. Fornari (1973) notes that in 1725, Francis Rawle urged the legislative authority in the colonies to implement a subsidy on wheat exported to Europe so as to bring the farmer back from the "brink of ruin." Rawle suggested that the subsidy be financed by a land tax or a duty on rum. Section 32 of the Agricultural Adjustment Act amendments of 1935 authorized export subsidies for several agricultural commodities and the Agricultural Acts of 1948 and 1949 authorized export subsidies when U.S. domestic prices were "noncompetitive" with those of other nations. As Morgan (1979, 122–123) notes, export subsidies are officially justified on the basis of "making American agricultural products more competitive abroad." In fact, however, it is difficult to disguise a pure subsidy program under a false name. Wheat exports have historically received the greatest amount of assistance under the various subsidy programs. The International Wheat Agreement of 1949 and subsequent extensions set price ranges for wheat trade below the U.S. domestic price, thus implicitly sanctioning export subsidies in an effort to avoid criticism for "export dumping."

Food Donation Programs

Domestic Programs. The federal government operates several food donation programs designed to improve the nutritional intake of needy families and schoolchildren and, for the most part, to do so with surplus farm commodities. Under these programs, food is provided at no or modest cost to qualified needy persons and schoolchildren and at reduced cost to low-income consumers. Federal expenditures since 1950 under each of these programs are given in Table 6.1.

COMMODITY DISTRIBUTION PROGRAM. The Commodity Distribution program was inaugurated under section 32 of the Agricultural Adjustment Act amendments of 1935 and continued under section 416 of the Agricultural Act of 1949. Its main objective is to find constructive uses for a portion of basic agricultural commodities that do not move through commercial channels at current domestic prices. Under this program, Commodity Credit Corporation-acquired surpluses can be donated to school lunch programs, charitable and nonprofit institutions serving needy persons and children, and low-income families.

The Commodity Credit Corporation (CCC) arranges for and finances the necessary processing and packaging of these surplus commodities and

Table 6.1. Expenditures under various food donation programs, 1950–88

Year	Commodity distribution[a]	Food stamp	School lunch	School breakfast	Special milk	Special food	WIC
				($1,000,000)			
1950	**25**	...	**65**	...	...	...	...
1951	25	...	68	...	...	...	...
1952	7	...	66	...	...	...	...
1953	17	...	67	...	...	...	...
1954	61	...	67	...	...	...	...
1955	**97**	...	**69**	...	**17**	...	...
1956	135	...	67	...	46	...	...
1957	104	...	84	...	61	...	...
1958	110	...	84	...	66	...	...
1959	137	...	94	...	74	...	...
1960	**75**	...	**94**	...	**81**	...	...
1961	174	...	94	...	84	...	...
1962	253	13	99	...	89	...	...
1963	233	19	109	...	94	...	...
1964	235	29	121	...	99	...	...
1965	**257**	**33**	**130**	...	**97**	...	...
1966	151	65	141	...	97	...	...
1967	116	106	150	1	99	...	...
1968	147	173	160	2	102	...	...
1969	250	229	204	5	101	1	...
1970	**312**	**550**	**300**	**11**	**101**	**8**	...
1971	346	1,522	532	19	91	21	...
1972	338	1,793	739	25	90	37	...
1973	282	2,212	882	35	91	45	...
1974	218	2,845	1,085	59	49	61	10
1975	**79**	**4,599**	**1,289**	**86**	**123**	**97**	**89**
1976	49	5,632	1,492	114	139	148	143
1977	62	5,399	1,684	155	157	218	256
1978	96	5,499	1,831	184	135	228	380
1979	150	6,822	2,006	236	137	265	525
1980	**194**	**9,117**	**2,279**	**288**	**145**	**317**	**725**
1981	237	12,287	2,381	332	101	415	869
1982	467	12,065	2,185	317	18	384	948
1983	1,357	13,136	2,402	344	17	422	1,123
1984	1,490	13,131	2,508	364	16	471	1,386
1985	**1,439**	**11,701**	**2,578**	**379**	**16**	**535**	**1,488**
1986	1,381	11,435	2,714	406	16	579	1,581
1987	1,313	10,509	2,797	447	16	640	1,680
1988	1,073	12,032	2,916	474	19	648	1,798

Source: U.S. Department of Agriculture. *Agricultural Statistics,* various annual issues.
[a]Section 32 and Section 416 expenditures.

pays transportation charges to designated receiving points. Local governments then distribute the food to qualified recipients—hospitals, orphanages, homes for the aged, nonprofit summer camps for children, child-care centers, public welfare agencies serving needy persons, needy Indians, public schools with lunch programs, and victims of natural disasters. From time to time, and when surpluses are low or the CCC does not have sufficient supplies to meet the needs, the CCC has been authorized to purchase supplies on the open market.

In the early days of the program, the list of commodities that could be distributed was quite limited. In the 1960s, the list was expanded and an effort was made to emphasize high-protein items and generally those items that would improve the nutritional value and variety of foods offered. Thus by the early 1970s, the emphasis of the commodity distribution program had shifted from a surplus disposal program to a program designed to meet the nutritional needs of recipients.

CHILD NUTRITION PROGRAMS. The National School Lunch Act was passed in 1946 "to safeguard the health and well-being of the Nation's children and to encourage the domestic consumption of nutritious agricultural commodities and other food." Under the school lunch program, the federal government provides cash, commodities, and technical assistance to schools operating the program. Both public and nonprofit private schools for grades 1 through 12 are eligible to participate in the program. Schools are required to serve lunches meeting minimum nutritional requirements, to serve meals without cost or at reduced cost to poor children, to operate the program on a nonprofit basis, and to utilize surplus and donated agricultural commodities as far as is practicable.

Federal assistance consists of cash payments for local food purchases, federal donations of food, and cash for acquiring food service equipment. Commodities donated to the schools come from stocks acquired by the CCC under its price support activities, from purchases by the CCC of agricultural commodities in surplus supply, and from special purchases by the U.S. Department of Agriculture needed to meet the needs of the schools.

In addition to the school lunch program and operated in conjunction with it are the Special School Milk Program added in 1954 designed to encourage the consumption of fluid whole milk by children, and the school breakfast program added in 1966 designed to provide additional nutritional content to the diet of needy pupils and children traveling long distances to school. Some nonschool food programs were added in 1966, including those to provide food to nonprofit day-care centers in low-income areas and areas with many working mothers and in summer camps for needy children. The program for expectant mothers and infants (the Women, Infants, and Children Program) was added in 1974.

FOOD STAMP PROGRAM. The most celebrated of the domestic food donation programs is the Food Stamp Program made part of permanent legislation in 1964 and initiated on a pilot basis by executive order in 1961. This program was designed "to strengthen the agricultural economy; to help achieve a fuller and more effective use of food abundances; [and] to provide for improved levels of nutrition among economically needy households." The program provides stamps to eligible households with which these households can buy food. The stamps are worth more in food than they cost the needy household to obtain.

A household's food stamp allotment is based on the cost of the Thrifty Food Plan, household size, and the household's net monthly income. The Thrifty Food Plan is designed to serve as a basic guide for the selection of low-cost meals—those with less red meat, poultry, and fish and more dry beans and grain products than the average household consumes. The maximum food stamp allotment any household may receive is equal to the cost of the Thrifty Food Plan for that household's size. The allotment actually received is equal in value to the cost of the Thrifty Food Plan less an amount equal to 30 percent of the household's net monthly income.

To be eligible for food stamps, households must either receive public assistance or be below certain minimum income and resource levels, and be living as an economic unit, excluding roomers, boarders, and live-in attendants. Except for the disabled elderly, drug addicts, and alcoholics, households must cook their food at home. All household members who are able-bodied and over 18 years of age must register for employment and accept it when offered. College or university students who are claimed as a dependent by a taxpayer who is not a member of an eligible household are not eligible to receive stamps.

Food stamps cannot be sold, given away, or used to pay creditors. In 1975, participants with monthly incomes in excess of the stated minimums were required to pay a certain amount, called the purchase requirement, in order to obtain their entire monthly food stamp allotment. Food stamps which the household was required to purchase were referred to as bonus stamps. The amount of bonus stamps received was dependent entirely upon the household's monthly income level—the higher the income level the greater the purchase requirement. In 1977, under the Food and Agriculture Act, eligibility standards were tightened to reduce program costs and the purchase requirement was eliminated on bonus stamps. That is, a household now receives food stamps to ensure that only 30 percent of its net cash income need be allocated to food. Under the current scheme, food stamps may free up income that would have been spent on food. Some of the income freed up may be spent on additional food. Recipients are not required to use all of the stamps they receive.

Current eligibility standards for food stamp recipients are (1) net household income must be below the nonfarm poverty level, (2) financial assets of the household must not exceed $3,000, and (3) the head of household must register for and accept when offered suitable employment unless he or she is responsible for dependents or is a student attending school more than half-time.

There is a tendency to view the Food Stamp Program as one in which recipients should be *required* to use the stamps to purchase food since the objective of the program is to improve the nutritional well-being of low-income families. Many program analysts, however, argue this position is not so clear. To see the latter argument, consider Figure 6.4. In this diagram, I_1, I_2, and I_3 are household indifference curves for food and nonfood purchases. The curve labeled I_1 indicates various combinations of food and nonfood purchases which give the household the same level of satisfaction or utility. It is thus termed an indifference curve. Indifference curves I_2 and I_3 yield successively higher levels of utility. The straight lines labeled B and B′ are termed budget lines. They indicate various combinations of food and nonfood purchases the household can buy at the lower and the higher income level, respectively. The household will maximize its utility by purchasing that combination of food and nonfood indicated at the tangency of the budget line with an indifference curve.

Budget line B represents the household's food-nonfood purchase possibilities in the absence of a food subsidy program. The point of

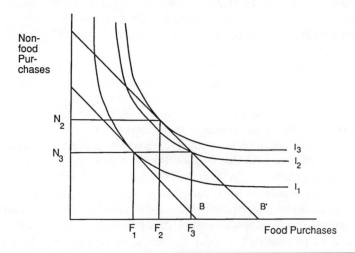

Figure 6.4. Welfare effects of eliminating the food purchase requirement

tangency between line B and the household's indifference curve, I_1, gives the optimal (equilibrium) level of consumption of food, F_1, and of nonfood, N_3. We can view B′ as representing the food-nonfood purchase possibilities under the assumption that the household has received an income subsidy. Let us assume that the household receives such a subsidy in the form of food stamps it has purchased (at below market value) that must be redeemed for food and nothing else. These stamps are of a sufficient value to enable the household to increase its food purchases from F_1 to F_3. None of the household's income has been freed up for increasing its purchases of nonfood items. The same amount of household income will be spent on food as before the receipt of food stamps. With food stamps, however, the household can obtain more food than before because in effect the price of food to this household has been reduced. The household is clearly better off than without the food stamps because it can now purchase more food while maintaining its previous level of nonfood purchases. Hence the household is able to attain the higher utility level represented by indifference curve I_2.

Note, however, that if the consumer is given a cash subsidy and instructed that it may do with the cash as it wants, it will purchase somewhat less food (F_2) than under the food stamp approach with a purchase requirement, but it will be able to purchase a greater quantity of nonfood items (N_2) *and at the same time* reach a higher utility level, as represented by indifference curve I_3. This same point can be reached with an equivalent subsidy given in the form of food stamps with no food purchase requirement. Under the latter scheme, some of the household income previously allocated to food would be freed up for the purchase of additional nonfood items.

CONTRIBUTION TO FARM INCOME. It should be noted that the economic consequences of food subsidy programs are the same whether the food subsidy is given in cash or in kind. Thus while some of these programs distribute food and some distribute cash with which to purchase food, the result is the same, as indicated in the previous section on domestic food subsidies. Nevertheless, to subsidize food consumption for the purpose of increasing the nutritional level of some or all consumers is a vastly different objective than subsidizing food consumption for the purpose of increasing farm income. While the effects are the same, the costs society is willing to accept to meet these two different objectives will likely be different.

In any event, some straightforward calculations indicate that food donation programs do not have a large impact on farm incomes. Consumer expenditures on domestically produced food in the United States in 1988 amounted to \$396.6 billion (Dunham 1989). Federal expenditures on the programs itemized in Table 6.1 sum to $19.0 billion for 1988. If we assume that 50 percent of the federal expenditures on the Food Stamp Program is actually diverted to nonfood expenditures by food stamp recipients,[4] this brings the sum down to $12.9 billion, or 3.2 percent of total consumer expenditures on domestically produced food in the United States. Since U.S. farmers' share of the consumer's food dollar averages about 30 percent, this sum represents about $3.9 billion in farm value. Cash receipts from farm marketings were $151.4 billion in 1988. Hence the programs itemized in Table 6.1 contributed only about 2.6 percent to gross farm income in 1988.

Foreign Donations. The Agricultural Trade Development and Assistance Act—more commonly known as the Food for Peace Act or simply Public Law 480—was passed in 1954 in an effort to draw together under one authority all existing foreign aid programs. This act made it U.S. policy to

> expand international trade among the United States and friendly nations, to facilitate the convertibility of currency, to promote the economic stability of American agriculture and the national welfare, to make maximum efficient use of surplus agricultural commodities in furtherance of the foreign policy of the United States, and to stimulate and facilitate the expansion of foreign trade in agricultural commodities produced in the United States by providing a means whereby surplus agricultural commodities in excess of the usual marketings of such commodities may be sold through private trade channels, and foreign currencies accepted in payment thereof. It is further the policy to use foreign currencies which accrue to the United States under this Act to expand international trade, to encourage economic development, to purchase strategic materials, to pay United States obligations abroad, to promote collective strength, and to foster in other ways the foreign policy of the United States. (Cochrane and Ryan 1976, 144)

Three important conditions existed at the time of enactment of Public Law 480. First, public sentiment to continue aid to Western Europe and other developing countries for reconstruction after the war was still quite strong. Second, the costs of dealing with the continuing growth of farm surpluses was mounting. Third, there was a severe shortage of foreign exchange among the trading partners of the United States which reduced their demand for U.S. agricultural products.

A major impetus, therefore, for the enactment of the Food for Peace Act was to provide a means of disposing of agricultural surpluses. It was fairly successful judging by the expenditure levels shown in Table 6.2! In

Table 6.2. Donations under Public Law 480 and foreign aid food assistance, 1955–88

| Year | P.L. 480 Title I[a] | P.L. 480 Title II | | Foreign Aid Food assistance[b] |
		Government to government	Voluntary agencies	
		($1,000,000)		
1955	**73**	**52**	**135**	**170**
1956	439	63	185	28
1957	908	52	165	8
1958	658	51	173	9
1959	724	30	131	0
1960	**824**	**38**	**105**	**0**
1961	951	75	144	0
1962	1,050	88	169	7
1963	1,147	89	170	14
1964	1,112	81	189	24
1965	**1,301**	**57**	**183**	**26**
1966	1,079	87	180	42
1967	975	110	157	37
1968	1,032	100	152	17
1969	773	111	154	11
1970	**815**	**113**	**128**	**12**
1971	743	138	142	56
1972	678	228	152	66
1973	659	159	128	84
1974	575	147	145	76
1975	**762**	**148**	**191**	**123**
1976	650	65	192	216
1977	760	92	250	419
1978	739	112	223	475
1979	793	128	265	304
1980	**859**	**222**	**254**	**194**
1981	790	241	302	237
1982	722	158	227	467
1983	810	178	207	130
1984	775	291	311	104
1985	**928**	**395**	**304**	**90**
1986	766	179	241	131
1987	660	125	123	60
1988	703	236	227	75

Source: U.S. Department of Agriculture, *Agricultural Statistics,* various annual issues.

[a]Foreign currency sales plus long-term dollar credit sales. Includes Title III sales authorized under the 1977 act and used for local development projects.

[b]Donations, grants, and loans under the Mutual Security Act of 1953, the Agency for International Development (established in 1962), and Section 416 food aid begun in 1984.

1963, for example, Public Law 480 and other foreign food aid expenditures represented 27.7 percent of the value of U.S. agricultural exports. Over the years, however, the focus of Public Law 480 has changed significantly. It is now viewed more as a program aimed at facilitating economic development and transforming developing countries into regular trading partners than as a surplus disposal program. In fact, in recent years foreign food aid expenditures have stabilized at around 3 percent of the value of U.S. agricultural exports.

Title I of the Food for Peace Act authorized the sale of surplus agricultural commodities for foreign currencies to "friendly nations." Sales under Title I are identified as "concessional sales" since they are made at terms more favorable to the recipient nation than to a commercial buyer. These sales are made at prevailing world-market prices but at low interest loans extending for as long as forty years. The CCC finances the sale and export although actual sale is made by private U.S. suppliers. The commodity is then usually resold in the recipient nation and the local currency generated is used for projects specified in the sales agreement. The majority of the currency generated is used for development projects in the recipient nation, a lesser amount is allocated to replace U.S. dollar expenditures on U.S. projects, and the remainder is allocated to miscellaneous projects, including military aid.

By 1966, large surpluses of U.S.-owned foreign currencies were accumulating under Title I concessional sales. In large part this prompted the Food for Peace Act of 1966, which stipulated that foreign currency sales be phased out and replaced by credit sales for dollars and convertible currencies. This act also formally shifted the focus of foreign aid from a surplus disposal program to an economic development program. More specifically the act was designed to shift recipient nations to regular dollar customers as soon as possible. To facilitate this transition the act authorized government-to-government dollar credit sales in which the importing governments were required to guarantee convertibility into dollars of all local currency repayments over a period of time as long as forty years. Such long-term dollar credit sales expanded rapidly at the expense of foreign currency sales.

Donations to foreign countries for disaster relief were authorized under Title II of the Food for Peace Act. Both government-to-government grants and donations to needy persons through voluntary relief agencies were supported under the act. In both cases, commodities were to come from CCC stocks of surplus agricultural commodities. The Food for Peace Act removed the requirement that commodities for donation come from CCC stocks and paved the way for Title II donations to be used in projects with longer range economic development significance.

In 1977, Title III sales were authorized under which local currency proceeds utilized for development purposes could be credited against dollar repayment obligations incurred by the Title I agreement. Title III aid must be used for new or complementary development efforts—it cannot be used to replace existing development efforts.

Export Subsidies. The Export Enhancement Program (EEP) authorized by the Food Security Act of 1985 is the most important of our current export subsidy programs. This program permits the U.S. Department of Agriculture to use CCC-owned commodities as export bonuses to private U.S. exporters so as to make U.S. agricultural commodities more competitive in the world market—that is, to enable U.S. exporters to sell specified agricultural commodities to foreign buyers at prices below domestic U.S. prices. CCC expenditures on the EEP in fiscal years 1985, 1986, 1987, and 1988 were $58.8 million, $32.0 million, $731.0 million, and $1,302.9 million, respectively (U.S. Department of Agriculture 1986–89). Wheat is by far the largest beneficiary of this program. In these same fiscal years, of the total amount of U.S. wheat exports, 19 percent, 44 percent, 62 percent, and 42 percent, respectively, were supported under the EEP. In these same fiscal years, wheat exporters were given an average bonus over the representative export price of 22 percent, 31 percent, 28 percent, and 11 percent, respectively (U.S. Department of Agriculture 1989).

The 1985 act also authorized a Targeted Export Assistance Program whereby priority assistance may be provided to producers of commodities who have been found to have suffered from unfair trade practices under section 301 of the Trade Act of 1974 or who have suffered from retaliatory actions of various nations. Up to $325 million annually was authorized for this program. This authorization was subsequently changed to $110 million for 1986-88.

The Foreign Agricultural Service (FAS) operates two programs under which it guarantees repayment of credit offered by private U.S. exporters to foreign countries for specified agricultural commodities. The Export Credit Guarantee Program (GSM-102) authorized in 1980 provides short-term (up to three years) credit guarantees.[5] Under this program $1.1 billion of wheat, $400 million of corn, $385 million of cotton, and $135 million of rice were sold in 1985. The Intermediate Credit Guarantee Program (GSM-103) authorized in 1985 provides intermediate-term (up to ten years) credit, but otherwise is operated as is GSM-102. Additional subsidy programs authorized by the Food Security Act of 1985 include a Cottonseed Oil Assistance Program, a Sunflowerseed Oil Assistance Program, mandated CCC purchases of red meat for export, a Dairy Export Incentive Program, and a Mandated Dairy Sales Program (see Ackerman and Smith 1990).

Summary

A variety of programs aimed at improving farm incomes via demand expansion have been attempted. These programs have a simple and straightforward objective—to move the demand schedule to the right and in this way cause both equilibrium farm prices and quantities, and thus also farm incomes, to increase. The more recent farmer-funded promotion programs are in general more controversial than are the food subsidy programs. A minimum level of generic as well as product-specific advertising can probably be justified on the basis of informing the consuming public about the availability of food products, about food prices, and/or about nutritional content. Massive levels of generic advertising, however, probably do more for the advertising industry and its suppliers than for agriculture, as is surely the case for much of soft drink and beer advertising.

Regardless of their contribution to farm incomes, most domestic food subsidy programs are now pursued for the additional, if not primary, purpose of satisfying humanitarian objectives. Providing subsidies to domestic consumers, though, entails social or deadweight losses as well as substantial transfers from taxpayers to farmers and targeted consumers. Subsidizing exports amounts to export dumping and leads to direct transfers from domestic taxpayers to foreign consumers as well as to social or deadweight losses. Donating surplus agricultural commodities to domestic or foreign consumers is no doubt consistent with society's overall goals but contributes only marginally to increasing farm incomes.

It could be argued that sales of surplus commodities to developing nations is a positive-sum game so long as the value of these commodities to recipient nations exceeds the cost of transportation. Basically these commodities have no value to the United States and may even have a negative value in the sense that if the CCC cannot move them into commercial channels, high storage costs will accrue. It is also tempting to extend this reasoning to the subsidization of sales through Public Law 480 of commodities purchased on the open market on the basis that otherwise these purchases would eventually become surplus commodities and therefore be carried by the CCC anyway. This is hardly consistent, however, with balancing aggregate commercial demand and aggregate U.S. supply at current market prices.

It is by no means clear that U.S. food aid programs are in the long-run interests of either the United States or recipient nations. There have been instances, for example, in which food aid has diverted the attention of recipient nations from useful development projects while planners in these countries concentrate on developing the internal infrastructure and dock facilities necessary to handle the food aid. In such cases the food aid may well have had a negative value to the recipient nations in the sense that these were not the priority development projects. Food aid may have provided the incentive to postpone hard policy decisions and implementation of programs needed for domestic agricultural development. Further, food aid may have depressed domestic agricultural prices and lowered incentives to increase production in the recipient nations.

On the other hand, food aid has been of considerable benefit to those nations with hard-currency shortages, has fostered increased savings and investment, and has been important in increasing food consumption levels overall as well as at times of severe local food shortages. From the U.S. perspective, food aid has been instrumental in increasing rates of economic growth and development in some countries that have subsequently become important commercial markets for U.S. agricultural exports. Outstanding examples here would include Western Europe, Japan, Taiwan, South Korea, and India. Thus Public Law 480 exports not only contribute to meeting the humanitarian food needs of the world, they also provide a catalyst for agricultural and economic development in less developed countries and in turn lay the foundation for future expanded U.S. exports.

Notes

1. See Tomek and Robinson 1972 for a discussion of derived demand.

2. This dependence can be demonstrated analytically by solving a system of equations—a demand equation, a supply equation, and a marketing margin relation—simultaneously for equilibrium price and quantity. Here the demand equation would have retail price and generic advertising as two of its arguments, the supply equation would have farm price and generic advertising as two of its arguments, and the marketing margin relation would also have generic advertising as one of its arguments.

3. See the Glossary for a description of the European Economic Community.

4. At first glance it would seem that food stamp benefits issued as food coupons would increase food expenditures by the face value of the coupons since coupons can only be used to buy food items. In fact, however, food coupons are used as a substitute for cash when purchasing food so that the cash expenditures substituted for are freed up for use on other consumer items as well as for additional food (see Boehm and Nelson 1978).

5. See the Glossary for a description of these programs.

Suggested Readings and References

Ackerman, Karen Z., and Mark E. Smith. 1980. "Agricultural Export Programs: Background for 1990 Farm Legislation." U.S. Department of Agriculture. Economic Research Service. Staff Report no. AGES 9033. May.

Boehm, William T., and Paul E. Nelson. 1978. "Food Expenditure Consequences of Welfare Reform." *Agricultural-Food Policy Review* (U.S. Department of Agriculture, Economic Research Service) AFPR-2, pp. 46–50.

Cochrane, Willard W. and Mary E. Ryan. 1976. *American Farm Policy, 1948–1973*. Minneapolis: University of Minnesota Press.

Connor, John M. 1981. "Advertising, Promotion, and Competition: A Survey with Special References to Food." *Agricultural Economics Research* 33 (1): 19–28.

Dunham, Denis. 1989. "Food Cost Review, 1988." U.S. Department of Agriculture. Economic Research Service. Agricultural Economic Report no. 615. July.

Fornari, Harry. 1973. *Bread upon the Waters*. New York: Aurora Publishers.

Kinnucan, Henry, and Olan D. Forker. 1986. "Seasonality in the Consumer Response to Milk Advertising with Implications for Milk Promotion Policy." *American Journal of Agricultural Economics* 68 (3): 562–71.

Morgan, Dan. 1979. *Merchants of Grain*. New York: Viking Press.

Morrison, Rosanna Mentzer. 1984. "Generic Advertising of Farm Products." U.S. Department of Agriculture. Economic Research Service. Agricultural Information Bulletin no. 481. September.

National Dairy Promotion Board. 1987. "Report to Congress on the Dairy Promotion Program." U.S. Department of Agriculture. July 1.

Robinson, Kenneth L. 1989. *Farm and Food Policies and Their Consequences*. Englewood Cliffs, N.J.: Prentice-Hall.

Tomek, William G., and Kenneth L. Robinson. 1972. *Agricultural Product Prices*. Ithaca: Cornell University Press.

U.S. Department of Agriculture. 1980. "Foreign Market Development." Foreign Agricultural Service. December.

_____. 1986–89. "Report of Financial Condition and Operations of the Commodity Credit Corporation." Commodity Credit Corporation.

_____. 1989. "Wheat Situation and Outlook Report." Economic Research Service. WS-287. November.

_____. 1990. "Food Consumption, Prices, and Expenditures, 1967–88." Economic Research Service. Statistical Bulletin no. 804.

Ward, Ronald W., and Bruce L. Dixon. 1989. "Effectiveness of Fluid Milk Advertising since the Dairy and Tobacco Adjustment Act of 1983." *American Journal of Agricultural Economics*. 71 (3): 731–40.

Ward, Ronald W., Stanley R. Thompson, and Walter J. Armbruster. 1983. "Advertising, Promotion and Research." In *Federal Marketing Programs in Agriculture*, edited by Walter J. Armbruster, Dennis R. Henderson, and Ronald D. Knutson. Danville, Ill.: Interstate.

7 Income Support through Price and Output Control

We have seen that a cornerstone of agricultural policy in the 1930s and since has been income support through the marketplace—prices supported above equilibrium levels with or without government purchases of surplus production, and/or output control for the program commodities. Chapter 5 provided an overview of the Treasury costs and the social costs of these policy instruments. In this chapter we will study the economic consequences of these instruments in more detail. In particular, I will examine the consequences of price and income support on the output market as well as on the input markets. The reader may wish to review the concept of *social costs* discussed in Chapter 5 and defined in the Glossary before proceeding with this chapter. Our concern here will be with the domestic impacts only, leaving the international consequences for Chapter 9.

Price Supports with No Supply Control

Policies initiated in the 1930s supported prices of program commodities above equilibrium levels by explicit price guarantees or nonrecourse loans and with government purchases of surpluses. Supply control was a secondary aim of this early legislation. High price supports (at near 90 percent of parity) were set during World War II and continued after the war as price supports became entrenched in the thinking of farmers and of policymakers. Figure 7.1 clarifies why such a policy could be expected to lead to surpluses. Here P_s is the desired support level. At this price producers, responding along their supply curve, produce output level Q_s. But at price P_s consumers are only willing to purchase Q_d of the quantity supplied. The government must act as residual buyer in order to maintain the support price at P_s; i.e., the government will be required to purchase the surplus of $(Q_s - Q_d)$ at price P_s.

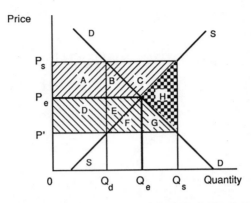

Figure 7.1. Impact of price support with no supply control

Producers will be very happy with this program because collectively their income will have been increased by the areas labeled A, B, and C. Part of this increase in producer income is financed through government purchases of the surplus. The remainder is financed by transfers from consumers, who have been forced to sacrifice utility as a result of the policy. That is, consumers are forced to pay a higher price and are, therefore, willing to consume less than they would under a free-market situation. In fact, consumers lose utility in the amount of areas A plus B. All of consumers' utility loss, then, is transferred to producers.

This program, in contrast to a free-market equilibrium, encourages the employment of "too many" resources in the production of the program commodity as output increases from Q_e to Q_s. These are resources that presumably could have been more productively employed elsewhere in the economy. To this loss must be added the costs of storage, which again we can consider costs associated with using resources that could presumably have been productively employed elsewhere in the economy.

To be fair to this type of program, we must also admit to some potential benefits. The surpluses can help to contribute to a national reserve that will come in handy in times of droughts or wars. Further, these surpluses enable the U.S. government to pursue its policies of helping the needy both at home and abroad, even though, as we saw in Chapter 6, subsidized sales of government surpluses also lead to social losses. Indeed there may be some revenue to the federal government from the sale of these surpluses even if the sales are made at distressed prices. Thus it may well be that some of the out-of-pocket taxpayer costs will be recovered.

Note that the consequences are the same whether the government uses an explicit price guarantee or a nonrecourse loan program with the price guarantee or the loan rate pegged at a level above equilibrium—that is, at P_s. In either case, surpluses will develop and the government will need to enter the market as residual buyer in order to maintain the price support or loan rate program.

As we saw in Chapter 2, the Agriculture and Consumer Protection Act of 1973 authorized the establishment of target prices and the use of deficiency payments as a means of supporting farm incomes for the major crops. The consequences of this policy can also be examined with the aid of Figure 7.1. Assume that the target price is established at P_s. As before, this will encourage output level Q_s. But in this case, the government is assumed not to purchase any surpluses. Rather, consumers responding along their demand function will drive the market price down to P' as output level Q_s is placed on the market. Direct deficiency payments paid to farmers of amount $(P_s - P')$ will be required. A net social loss is again incurred (area H in this case) as producers are encouraged to produce well beyond the equilibrium output level. Here producers gain areas A, B, and C while consumers gain areas D, E, F, and G. Government costs in the form of deficiency payments include *all* of the areas of Figure 7.1 labeled with a capital letter. Thus net social loss in this case is area H. It represents an expenditure by the government (taxpayers) that is recovered by no one in society. Again, too many resources are devoted to production of the program commodity as a result of the government program adopted.

It is sometimes suggested that direct government payments in the form of deficiency payments are a more attractive means of giving farmers income support than are price supports and government purchases. Such an argument is based on the notion that the former cause no market distortions and no social losses. But as the above analysis makes clear, this is not the case if the direct income payments are accompanied by a market-distorting target price maintained at a level above competitive equilibrium.

It is important to note also that the extent of market distortions, producers' gains or losses, consumers' gains or losses, and government costs will depend on the slope of the demand and supply curves. For example, when price is supported above equilibrium and the government maintains this support price via an open-market purchase program, then for a given price support level and a given demand elasticity, government costs will be higher the more elastic is the supply curve. This is true because for a given price support level, a more elastic (i.e., flatter) supply curve will call forth a greater increase in production when price increases than would a more inelastic (i.e., steeper) supply curve. Thus, the excess of supply over demand at the support price will be greater when supply is elastic than when supply is inelastic, and the government will be required to spend a greater amount to maintain the support level.

The relative costs of a price support/government purchase program (with no supply control) and of a target price/deficiency payment program (again with no supply control) will, on the other hand, depend primarily on the elasticity of the demand curve. As can be verified by graphing these two policy situations with alternatively sloped demand curves, when the target price and support price are maintained at the same level, if the demand curve is inelastic, a price support/government purchase program will be less costly to the government (excluding storage charges) than will a target price/deficiency payment program, regardless of the elasticity of supply.

This last result is an important point often overlooked by policymakers. It is all the more important in view of the fact that (1) domestic demand for most U.S. agricultural products is highly inelastic and (2) a major instrument of U.S. farm policy is currently the target price/deficiency payment program. It is interesting to note that this last conclusion still holds when the commodity under consideration is a major export commodity like corn or wheat and when export demand is *elastic* (say, in the vicinity of −2.0).

The above analysis captures the essential economic impacts of price support and target price/deficiency payment programs although it oversimplifies actual operations. While some might consider price supports for U.S. agriculture to be quite high, they are substantially lower than they were twenty or thirty years ago and the market is permitted to play a greater role now in price determination. Nevertheless, Commodity Credit Corporation (CCC) loan rates still exist for the major crops so that the market price may not be determined by commercial demand but rather by the loan rate. If this is the case, then deficiency payments are based on the difference between the target price and the loan rate. Further, there are complications like Findley payments, marketing loans, use of generic certificates, and payment limitations (see Glossary)that must be considered in a full analysis. Also payments are made on only a portion of a farmer's production and are based on "normal yield" rather than on actual yield. Finally, not all producers of a given program commodity participate in the program in force; therefore, not every producer of the program commodity receives deficiency payments.

Impact on Input Markets. As we saw in Chapter 4, we can normally expect a policy that supports the price of the program commodity above the equilibrium price with no supply control to cause a rightward shift of the demand curve for inputs. This will result in an increase in the equilibrium price (and quantity marketed) of farm inputs.

The extent to which the demand curve for an input shifts is clearly a function of whether or not this input is used in the production of other commodities and, if so, the relative importance of the program commodity in the total mix of products that use this input.

When there are several inputs used in the production of the program commodity, it is of interest to know whether the different input markets will be impacted differently. In Chapter 4, we saw that the input with the more elastic supply curve would experience a smaller price increase but a greater quantity increase than would the input with a more inelastic supply curve.

Normally we expect the supply curve for land to be more inelastic than that for labor or fertilizer or pesticides or machinery. Again as we saw in Chapter 4, this is important in determining the distribution of benefits from price supports. In Chapter 4, for example, we discovered that a majority of the benefits will get "capitalized" into the value of the land resource.

Price Supports with Supply Control

The Soil Conservation and Domestic Allotment Act of 1936 and the Agricultural Adjustment Act of 1938 relied on a *voluntary* conservation reserve program and a *voluntary* acreage allotment program in an effort to control the accumulation of price-depressing surpluses in agriculture. Those farmers who complied with these programs received rental payments for acreage diverted and had access to nonrecourse loans. Noncompliers were severely penalized by not having access to nonrecourse loans.

The 1938 act also authorized a *mandatory* marketing quota program when supplies exceeded a "normal supply" level. The quota was not applied at the individual-farmer level. Rather, individual farmers were given an acreage allotment that the secretary of agriculture estimated would produce the amount of the national quota.

By 1941 quotas of this type were in effect for tobacco, cotton, wheat, and peanuts, and voluntary acreage allotments had been used for corn, potatoes, and rice. Following the Korean War when surpluses again became excessive, quotas were reinstated for wheat, cotton, and rice and maintained for tobacco and peanuts. Corn allotments were in force until 1959 although they were not very effective in reducing corn acreage. Authority for mandatory quotas for corn was repealed by the Agricultural Act of 1954.

The mandatory programs of the 1950s were continued into the 1960s. Neither they nor the lower price supports also enacted, though, were very successful in reducing production. Productivity grew rapidly as nonland inputs were substituted for land. Furthermore, these programs were not applied as strictly as they could have been. Cross-compliance features were not implemented or not enforced so diverted acres in one crop became surplus acres in another. Small acreages were exempted from acreage allotments. In general, "slippage" was quite high.

In fact, a principal disadvantage of voluntary acreage reduction programs is the slippage involved. That is, the government seeks a certain acreage reduction to accomplish a targeted reduction in aggregate production based on an average yield for the nation, the region, or even the individual farm. At a uniform payment rate, the individual farmer will always enroll his lowest yielding acreage first. Thus there is slippage in the sense that crop yields on the enrolled acres are lower than yields expected (by policymakers) on these acres. Another source of slippage is cropland enrolled in the program that would normally be left idle after planting anyway—for example, fallow land, flooded land, or land on which there was a crop failure due to a drought. A third source of slippage is overplanting by nonparticipants in response to the higher price expected as a result of the acreage reduction program. Finally, participants may use nonland inputs more intensively on the land remaining in production so that crop yields on program land may be higher than expected. The slippage factor has been estimated to be as high as 0.4–0.5 for the land reserve programs of 1956–73 (Erickson 1977). Thus these programs were only 50–60 percent effective in reducing production. Recent research suggests that under current provisions, a 5 percent acreage reduction for feed grains is only 40 percent effective in controlling corn production and a 40 percent acreage reduction is only 25 percent effective (Harrington and Price 1990).

In 1956 the Soil Bank Program was initiated. This program was intended to retire more land than was retired under the acreage allotment or conservation schemes of earlier years. The Soil Bank Program had two components. First was the short-term acreage reserve program for wheat, corn, tobacco, peanuts, cotton, and rice under which producers received a rental payment for converting land normally planted to program crops to soil-conserving uses. Second was a conservation reserve scheme designed to retire land for between three and ten years in return for an annual cash rental payment. The Soil Bank Program, however, was not much more successful than were the earlier acreage reduction schemes. It did remove considerable acreage from production, but at politically unacceptable costs. It was terminated in 1959. The conservation reserve component of this program removed mostly marginal land, so it had little impact on production. About 70 percent of the land removed under the conservation reserve scheme was in the form of whole farms. This had a serious impact on communities and local businesses in areas with heavy participation.

As we saw in Chapter 2, the conservation reserve program introduced by the Food Security Act of 1985 emphasizes taking highly erodible cropland out of production. Thus, it has a stronger soil conservation objective than a price enhancement or supply control objective. I will return to a more complete assessment of this program in Chapter 10.

Mandatory marketing quotas applied to individual farms were actively sought by Secretary of Agriculture Freeman in the Kennedy administration. After much political haggling, approval was granted to try this program on a test basis in the wheat industry. The program failed to win the support of wheat farmers in a national referendum. No further effort at implementing marketing quotas for the major crops has been attempted.

After the unsuccessful bid for mandatory marketing controls, emphasis shifted back to voluntary controls for the major program commodities. Now, however, acres diverted were required to be planted to soil-conserving uses rather than to competing program commodities. This is currently pursued via the acreage reduction provisions for the major field crops. Cotton was, for several years, under the dual control of marketing quotas for the regular allotment and voluntary acreage controls as for the other crops for further acreage reductions. Mandatory quotas remained for peanuts, rice, and tobacco. By 1971 a poundage quota, or true marketing quota applicable to individual farms, was in effect for all tobacco, and in 1977 poundage quotas were added for peanuts. In 1975 the rice quota program was eliminated and replaced with a voluntary set-aside program.

Consequences of an Acreage Reduction Program.
The consequences of an acreage reduction program on prices and output were outlined in Chapter 4 (see, in particular, Figure 4.1). In summary, it will be recalled that restricting acreage causes the supply curve for the program commodity to shift to the left because nonland inputs must now be used in a nonoptimal way with the restricted land input. This supply shift results in a higher market price for the program commodity and reduced output relative to the situation that would be expected under a free-market equilibrium. The social costs of such a program were outlined in Chapter 5 (see Figure 5.1 and accompanying text). The repercussions of this program on the input markets are somewhat more complicated to unravel than for the price support programs considered earlier.

IMPACT ON NONLAND INPUTS. The immediate effect of an acreage restriction on the market for nonland inputs is a leftward shift in the demand curve for these inputs (i.e., a shift to D_sD_s in Figure 7.2) causing a reduction in both price and quantity used. This occurs in response to the reduction in the land input and before farmers have time to make significant substitutions of nonland inputs for the land input. The 1983

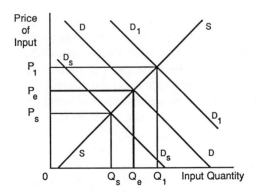

Figure 7.2. Impact of acreage restriction program on nonland input market

payment-in-kind (PIK) program, for example, resulted in dramatic reductions in nonland inputs with significant repercussions on input suppliers and local communities as demand for these inputs waned significantly and as agricultural input manufacturers and dealers laid off significant numbers of workers.

Over time, however, the increased program commodity price will induce farmers to make output-increasing substitutions of nonland inputs for the land input to the extent that this is possible and/or to adopt new output-increasing technology to apply to the restricted acreage. This will cause the demand curve for nonland inputs to shift back to the right and could even cause it to shift further right than its original position (e.g., to D_1 in Figure 7.2) as farmers respond to the higher output price. The long-run impact then could well be to cause an increase in nonland input prices and quantities utilized. This is precisely the situation that was observed in the mid-to-late 1950s and the 1960s as farmers confounded policymakers with output-increasing input substitutions and with adoption of output-increasing technology.

The long-run impact on wage rates is more difficult to ascertain. It is conceivable that in the long run both the wage rate and the quantity of labor utilized will decline. This will depend on the substitution possibilities for labor. If, for example, the opportunities for substituting labor for land in production are quite limited, the increase in the price of the program commodity may not be sufficient to compensate the labor input. This will be particularly true if the other nonland inputs used and the new technology adopted are labor saving as well as output increasing. In such a situation, we could expect labor to move out of agriculture. This again is precisely what has happened over the years since World War II and is the reason why the farm labor input has been dramatically reduced.

IMPACT ON THE LAND INPUT. The land market impacts are more complex still. The land supply curve will immediately shift to the left because of the restriction on land use. That is, at any given price of land, less will be offered for use in the program commodity because of the restriction. The basic structure of demand for land, however, will not have been altered. Consequently, the value of land in production will rise (Figure 7.3). If the land removed from production is idled, its rental value will be reduced to zero and if this reduction in value offsets the rise in value of land in production, the average price of all land will fall. I assume, however, that the program is voluntary, which means the government must offer farmers an incentive to participate in the program. If the incentive offered is equal to the value of land in production, as is usually the case, then clearly the value of all land will increase. This is the result we have observed in the past.

DETERMINATION OF RENTAL PAYMENT RATE. One of the difficulties for policymakers is how to determine the rental value (i.e., incentive payment) for land to be idled in order to encourage the desired acreage reduction. The usual procedure is to base this rental value on the average market value of land that is used to produce the program commodity. Currently, deficiency payments are presumed to be sufficient to entice the desired program participation. Judging by the number of acres actually idled

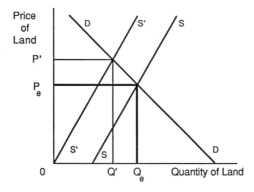

Figure 7.3. Impact of acreage restriction program on the land market

in the last few years (see Tables 2.7–2.9), we must conclude that deficiency payments are indeed high enough to encourage idling of land.

In general, per acre net returns over variable costs will not be the same for all acres of a given crop. Assume the distribution of net returns is normal, or bell-shaped. Assume also that the proposed acreage reduction program is voluntary. Then rational producers will participate in the program if the per acre payment rate equals or exceeds their expected net returns from the acres they would idle. At relatively low payment rates, only that cropland with low expected returns will be enrolled. The payment rate will need to be quite high to attract cropland with high expected net returns.

The S-shaped curve labeled 0ABC in Figure 7.4 shows the expected number of acres that would be enrolled in the program at various payment rates. The shape of this curve is dependent upon the distribution of per acre net returns across all acres of the program commodity.

Assume the payment rate is initially established at $0P_1$. This rate will only encourage $0E_1$ acres to be enrolled—far short of the goal of $0E_3$. Assume the payment rate is increased to $0P_2$. This will encourage $0E_2$ acres to be enrolled—closer to the goal but still short. But note the effect of this higher payment rate on farmers with an expected per acre net return of $0P_1$. At payment rate $0P_2$ these farmers will receive a windfall of amount AA'! That is, they would have been willing to participate for a payment rate of $0P_1$. Any amount above this rate is simply icing on the cake!

If the government establishes a *uniform* payment rate of $0P_3$ so as to encourage E_3 acres to be enrolled in the program, the government expenditures required to meet its acreage reduction goal will equal the area

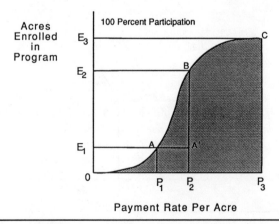

Figure 7.4. Alternative payment rates for an acreage reduction program

$0P_3CE_3$. Of this amount, only the area *above* the curve 0ABC (the unshaded area) is necessary to encourage the desired participation. Miller (1974) calls this a "production adjustment" payment. The area *below* curve 0ABC (the shaded area) Miller calls an "income supplement" payment. The income supplement is a payment in excess of that necessary to encourage the desired participation.

How might each producer be paid exactly what he or she requires to participate in the program and no more? The answer is to operate the program on a bid basis. That is, individual farmers would make an offer to the government to enroll the requisite number of acres at a specified payment rate. The government would accept the lowest such bids until the desired total acreage is enrolled. This is the system currently being used to attract crop acreage into the conservation reserve program authorized under the Food Security Act of 1985. It is also the system used for implementing the Milk Production Termination Program authorized by the same act.

Prices Supported with Marketing Quotas

Under a marketing quota program, output can be reduced by any method the farmer chooses. If the farmer is economically rational, he or she will attempt to minimize cost subject to this output restriction. Hence, the farmer will merely move down the existing supply curve. As we saw in Chapter 4, this leads to reduced output and reduced social losses in the output market and to the creation of a new asset. The value of this new asset at the beginning of the first year of the quota program can be estimated from the discounting formula:

$$V = \sum_{t=1}^{T} [N_t(1 + r_t)^{-t}] \qquad (7.1)$$

where T is the number of years the program is expected to be in effect, r_t is the interest rate expected to be relevant for discounting purposes in year t of the program, and N_t is the annual returns to the quota expected in year t. N_t is a direct function of the difference between the price at which farmers would be willing to produce under the quota program and the market price that would prevail under the quota program (see Figure 4.2). Hence by equation (7.1) the value of the asset is a direct function of this price difference and, therefore, of the quota supported price. If the program is assumed to be in effect indefinitely, and if N_t and r_t are assumed to be equal to N and r, respectively, for all t, then equation (7.1) becomes

$$V = \sum_{t=1}^{\infty} [N_t(1 + r_t)^{-t}] = N/r \qquad (7.2)$$

A marketing quota program rewards the initial recipient of the quota with a windfall of amount V. If the quota is assigned to the *landowner* rather than to the *farm operator*, the windfall accrues to the landowner. If the quota cannot be sold or rented, it remains with the individual to whom it was originally assigned. The quota not only accumulates value to the quota owner, it also gives the quota owner the right to produce (or, more properly, market) the restricted output. If this quota owner is *not* the most efficient producer available, then there will be an efficiency loss associated with prohibiting the sale or lease of the quota.

Furthermore, if the quota is not tradable, the existing regional production pattern *and* size structure of farms producing the program commodity will be maintained. Under such a scheme, regional production patterns will not be permitted to change as regional competitive advantages change. The losses in efficiency that might result could be substantial. Similarly, by prohibiting the transfer of quota rights, and unless the quota amount is revised upward over time, future reorganizations of the farm may be discouraged. Thus, for example, if new technology becomes available that would bring about cost economies associated with a larger scale of operation, the farmer would be prevented from taking advantage of these economies, and society would be a net loser.

Impact on Inputs. If the quota program calls for nontransferable quotas, the quota value accumulates into the value of the fixed resource to which the quota applies. More to the point, the quota value goes to the owner of the right to produce the restricted quantity. If this right can be sold with or without another fixed resource such as land, then we have merely created a new factor of production—the quota—that must be used in fixed proportions to the output sold. The quota value then accrues to the new factor of production.

What then is the impact of a tradable quota program on the market for other inputs used to produce the program commodity? Clearly here we have the reverse of the results under price supports with government purchases. Since output of the program commodity is restricted, the demand for all inputs will fall. That is, there will be a leftward shift in the demand curve for all inputs so that input prices and quantities will both fall. Those inputs with the most elastic supply curve will experience the greatest decline in quantity and the least decline in price.

Controlling Output by Limiting Nonland Inputs

Land is, of course, not the only input that could be restricted. It is certainly conceivable that an alternative policy could be to restrict by legislation the amount of chemicals farmers can apply. Although there does not at the moment appear to be a strong argument for doing so, current emphases on limited-input agricultural production systems could conceivably lead to some such limitation. It is just as conceivable that some limit on machinery usage could be imposed in the future, although again the basis for doing so is not at present evident.

At one time "controlling" the use of the labor input was given some consideration. But by now much of the surplus labor has been moved out of agriculture. What is left is undoubtedly concentrated on the 72 percent of the farms that produce less than 15 percent of the farm output (see Tables 3.9 and 4.2). Transferring labor away from these farms would contribute very little to reducing output. In fact, if some of the small holdings on which there may now be surplus labor are consolidated into larger farm units, reducing the number of farm workers might actually increase agricultural output as the larger units capture some scale economies!

Capital may be a another candidate for input reduction. Capital has over the past substituted for labor and land. The cost of capital has been declining relative to the cost of labor and land. A substantial share of the increase in capital use has been for new technology which is labor saving, output increasing, and profitable to early adopters. Thus limiting the capital input would appear to be a step backward.

Summary

High price supports with no supply control or with supply control of limited effectiveness will inevitably lead to production in excess of what the market will take at the supported price. Government must then assume the role of residual buyer. Production will exceed the equilibrium output level, and social losses will be incurred in that more resources are committed to production of the supported commodity than would be expected in the absence of government intervention. This market distortion will lead to increased demand for the factors of production and a bid up of prices in the input markets, and especially in the land market since the supply of land is so inelastic. Further, pressures will develop to dispose of the surpluses by domestic or export subsidy plans, which would add to the distortion (see Chapter 6).

Prices supported with acreage reduction programs lead to more complex results. First, the output supply curve shifts to the left because nonland inputs must now be used in a nonoptimal way with the restricted land input. This not only results in a higher "equilibrium" price and a lower "equilibrium" output but also causes a significant social loss in the sense that an inefficient combination of inputs is encouraged. The short-run impact of such a program will be reduced demand for all inputs and lower prices for nonland inputs. The price of land remaining in production, however, will increase as the land supply curve shifts to the left. In the long run, the demand for most nonland inputs can be expected to increase as can the prices of these inputs as nonland inputs are substituted for land. Farm wage rates may also increase, although this will depend on the substitution possibilities that exist in production. In the short run, there may be serious and negative impacts on rural communities as the demand for inputs and other services falls off sharply.

Over the years there has been much discussion about whether supply control programs should be voluntary or mandatory. Voluntary programs are usually implemented as acreage reduction programs with farmers being offered incentives to participate in the program or subject to a penalty for not participating in the program. Mandatory supply control programs have generally been implemented with acreage allotments and marketing quotas.

Voluntary programs, of course, preserve the right of farmers to make a free choice as to whether or not they will participate. Also voluntary programs give the farmer the opportunity to select the specific acreage to be removed from production. Furthermore, voluntary programs *can* be designed, via a bid system, so that program incentive payments encourage the desired acreage reduction with no excess payments. Finally, voluntary programs can fairly easily be designed to remove the more marginal land, which might subsequently be used for, say, recreation purposes, where it may have greater social value.

Marketing quotas with high price supports lead to underproduction and to the opposite effects on the farm input markets. A marketing quota program leads to the creation of a new asset—the quota—which accumulates value depending upon the level of the price support chosen. If the quota is marketable, no other distortions are introduced. If, however, the quota is prohibited from being sold or leased, then additional social costs may accrue in the sense that society cannot take advantage of shifting regional advantages, emerging technologies enabling scale economies, and/or the fact that nonquota holders may be more efficient producers than the quota recipients.

Suggested Readings and References

Bowers, Douglas E. 1987. "USDA Acreage Reduction Programs: 1933–1987." *Policy Research Notes* (U.S. Department of Agriculture, Economic Research Service) 23:34–49.

Erickson, Milton. 1977. "Use of Land Reserves in Agricultural Production Adjustment." *Agricultural-Food Policy Review.* U.S. Department of Agriculture. Economic Research Service. AFPR-1, pp. 76–84.

Erickson, Milton, and Keith Collins. 1985. "Effectiveness of Acreage Reduction Programs." *Agricultural-Food Policy Review: Commodity Program Perspectives.* U.S. Department of Agriculture. Economic Research Service. Agricultural Economic Report no. 530. pp. 166–84.

Harrington, David H., and J. Michael Price. 1990. "How Effective Is the ARP?" *Agricultural Outlook.* U.S. Department of Agriculture. Economic Research Service. AO-161. pp. 22–25. March.

Miller, Thomas A. 1974. "Estimating the Income Supplement in Farm Program Payments." U.S. Department of Agriculture. Economic Research Service. Technical Bulletin no. 1492. March.

Advanced Readings

Floyd, John E. 1965. "The Effects of Farm Price Supports on the Returns to Land and Labor in Agriculture." *Journal of Political Economy* 73 (2): 148–58.

Wallace, T. D. 1962. "Measures of Social Costs of Agricultural Programs." *Journal of Farm Economics* 44 (1): 580–94. May 1962.

8 Income Support through Marketing Control

There are substantial economies of size in farm product handling and processing. Consequently the number of handlers and processors buying farm products in any given area is typically quite small relative to the number of farmers producing the product. Because of this inequality of numbers, handlers and processors are generally believed to have the ability to exert a degree of monopsony control over the terms of trade with farmers. The resulting disparity in market power puts farmers at a considerable disadvantage, as was particularly evident during the 1920s and 1930s, when demand for farm products was low and/or supplies were excessive. During this period, for example, dairy farmers were often denied access to an outlet for their milk on a day's notice.

Various mechanisms were used in an effort to alleviate the problem. Such mechanisms included agricultural cooperatives spurred on by the Sapiro movement in the 1920s (see Chapter 1), labor union–type farmer organizations in the case of milk (Manchester 1983), and "clearinghouses" of voluntarily participating handlers in the case of fruits and vegetables (Farrell 1966). Some of these efforts were successful in achieving gains in the form of improved marketing conditions and in enhanced farmer prices. All eventually failed, however, in sustaining initial successes. A common limitation of these mechanisms was the "free-rider" problem. Nonparticipants enjoyed the same price benefits as did participants but were not bound by any of the marketing limitations imposed by the agency of collective action. Furthermore, nonparticipants did not have to pay any of the costs of maintaining the agency of collective action.

Farmers thus sought and gained legislation which built on the principles of collective action they had been trying to achieve on their own and enabled them to overcome the limitations experienced with the previous voluntary organizations. The end result was the Agricultural Marketing

Agreement Act of 1937, which constitutes the statutory basis for federal marketing orders as we know them today. In this chapter I will examine the role and function of marketing orders as a mechanism of market control, and we will study the economic consequences of these instruments of support for farmers both in terms of what theory would suggest and in terms of what the evidence indicates.

Marketing Orders Defined

A marketing order provides for the regulation of the quantity, price, and/or timing of the product marketed. It is a legal instrument which sets the limits within which an agricultural industry can operate a program of self-regulation. It defines the terms of trade for handler and producer, the commodity to be regulated, and the area to be covered by the order. Its regulations are incumbent on all producers *and* on all handlers in the industry once approved by two thirds of the producers in the specified area. The Agricultural Marketing Agreement Act of 1937 provides for marketing orders that are administered at the federal level. Since that time, however, individual states have enacted their own marketing order legislation following the general pattern of the federal statutes. In addition there are several "freestanding" orders provided by specific legislation or as part of general farm legislation. The latter are for the most part limited to providing for generic promotion activities. It is beyond the scope of this chapter to examine the state and freestanding orders. In general, however, the character and consequences of all marketing orders are similar regardless of legislative origin.

Provisions of the 1937 act were heavily influenced by the views and activities of agricultural cooperatives at that time, as well as by the economic climate. The clear intent of the act was to improve the economic well-being of producers of farm products. Marketing orders were initiated as mechanisms to accomplish this task through *orderly marketing* procedures.

The 1937 act provided for marketing orders for milk, fresh fruits and vegetables, nuts, tobacco, hops, honeybees, and naval stores. Specifically excluded were honey, cotton, food grains, feed grains, sugarcane and sugar beets, wool and mohair, livestock, soybeans, poultry except turkeys, and eggs except turkey-hatching eggs. In the case of milk marketing orders, the act went to considerable lengths to spell out the details that were to be included in the orders. Here minimum-pricing regulations were specified. For the remaining commodities, fewer specifics were given, although product-pricing regulation was explicitly ruled out while authority to "manage" marketed supply was granted. All orders, however, were to contain one or more of the following terms and conditions (U.S. Department of Agriculture 1971):

1. Prohibitions against unfair methods of competition and unfair trade practices in the handling of the agricultural commodity
2. Provisions specifying that except for milk and cream to be sold in fluid form, the commodity or its products shall be sold only at prices filed by the handlers in a manner specified by the order
3. Specifications on how the order is to be administered by an agency selected by the secretary of agriculture and on the payroll of the U.S. government
4. Consistent with any other specific provisions of the act

The number of orders currently in place by commodity is shown in Table 8.1. This table also shows the estimated 1986–87 volume of total U.S. production of these commodities covered by the orders indicated.

Marketing Order Provisions

Quality Control. Quality control provisions of federal orders permit the setting of minimum grades, sizes, and maturity standards normally enforced through mandatory federal inspection. These provisions are justified on the basis of improving producer returns. Discarding off-grade product increases the average quality of the product moving to market. A higher quality product should command a higher price and, therefore, larger producer returns. Furthermore, by imposing a degree of uniformity on a commodity, an industry can add to buyer confidence in the quality of its product and thereby create a more rational and attractive basis for trade. By restricting lower quality products during periods of large supply, the order can prevent the price of higher quality products from being reduced as much as might otherwise occur.

All but three fruit and vegetable marketing orders provide for some form of quality control. The cranberry order authorizes grade and size standards only for the restricted portion of the crop. The Florida grapefruit orders do not authorize size and grade standards, but fruit sold under these orders is subject to such standards under the Florida citrus order.

Quantity Control. Quantity controls are used to divert excess supplies from primary market channels into alternative outlets. These provisions represent the "strongest" form of regulation permitted under marketing orders because the direct regulation of quantity has the greatest

Table 8.1. Number of marketing orders by commodity and percentage of total U.S. production regulated by orders, 1986–87

Commodity	Number of orders	Percentage of production regulated
Oranges	4	100.0
Grapefruit	2	100.0
Lemons	1	100.0
Limes	1	100.0
Tangelos	1	100.0
Tangerines	1	65.2[a]
Avocados	1	9.2
Nectarines	1	100.0
Peaches	4	32.5[a]
Apricots	1	5.4
Sweet cherries	1	36.2
Fresh prunes and plums	2	96.0[a]
Tokay grapes	1	3.1[a]
Desert grapes	1	91.7
Pears	3	89.6[a]
Papayas	1	100.0
Cranberries	1	100.0
Olives	1	85.9
Kiwi fruit	1	100.0
Fall potatoes	5	68.5[a]
Summer potatoes	1	17.0[a]
Onions	2	21.1
Tomatoes	2	34.5
Celery	1	19.0
Lettuce	1	3.4
Almonds	1	100.0
Filberts	1	100.0
Walnuts	1	100.0
Peanuts[b]	1	94.2
Dates	1	89.6
Raisins	1	100.0
Dried prunes	1	97.0
Spearmint oil	1	28.6
Milk	44	90.0

Source: Jesse and Johnson 1981; Glaser 1986; and U.S. Department of Agriculture 1987.
[a]Percentage of volume regulated in 1977–79 as estimated by Jesse and Johnson 1981.
[b]A marketing agreement is in effect for peanuts.

potential for affecting producer revenue. Such controls appear in the form of direct volume management and market flow programs.

VOLUME MANAGEMENT. One method of managing the volume marketed is by use of *producer allotments* under which a producer is restricted in the amount that can be sold in the marketplace in a given marketing season. This mechanism is authorized for cranberries, Florida celery, and spearmint oil, although it has never been used in the cranberry order. Producers of commodities for which this provision is invoked are given an allotment base which is set on the basis of sales during a specified period of time. The sales allotment is then simply a specified percentage of this allotment base. A producer allotment scheme is the most direct and effective way to raise prices received by farmers, if, of course, the market demand curve is inelastic.

A second method of managing volume marketed is by use of a *market allocation* program. Under this program, the order specifies the maximum sales permitted in a primary market outlet for the farm commodity. The remainder of the commodity must be diverted to one or more of the nonprimary markets. This program will only work if arbitrage among the different markets is not possible—that is, if it is impossible to buy the product in one market for immediate resale in a second market. It is only advantageous in the sense of increasing producer returns if the elasticity of demand in the primary market is less in absolute value than is the elasticity of demand in the secondary markets.

Marketing orders for cranberries, almonds, walnuts, filberts, California dates, and raisins authorize this form of volume control. For each such commodity, prior to harvest a salable percentage is determined based on crop size and other market conditions. Each handler then applies this percentage to the total quantity handled in order to determine the quantity that may be marketed without restriction. Sales in excess of this percentage must be allocated to the export market, manufactured food products, or livestock feed.

Milk marketing orders also implement this form of volume control, although with a different mechanism: with minimum class prices. Here all regulated milk is permitted to flow into the primary, or class I, market at the established minimum class I price. Any milk remaining will automatically be diverted into the less lucrative manufacturing, or class II, market.

The third method of controlling volume marketed is with the aid of *reserve pools*. This method is similar in principle to a market allocation program except that in this case the restricted portion of production is set aside into a reserve pool rather than diverted immediately to secondary markets. Sales out of the reserve pool can subsequently be made on the primary market if demand conditions improve or supplies fall short of initial expectations. Reserve pools are authorized for spearmint oil, almonds, walnuts, raisins, and prunes.

MARKET-FLOW PROGRAMS. These types of quantity control programs are designed to enhance producer returns by regulating the amount sold each day, week, or month during the shipping season to avoid seasonal gluts with their corresponding low prices and seasonal shortages with their corresponding high prices and lost sales. *Handler prorates* specify the maximum quantity a handler may ship over a stated period of time. Receipts in excess of this amount must be held for shipment in subsequent time periods or diverted to secondary markets. Prorates are used extensively in the citrus industry, because citrus fruit may be stored on the tree for periods of time without significant loss of quality. Prorates are also used for Tokay grapes, Florida celery, and South Texas lettuce. *Shipping holidays* are periods during which all commercial shipping is prohibited and are used for some of the citrus crops, onions, Tokay grapes, celery, and lettuce. Shipping holidays are used primarily to avoid a buildup of supplies in terminal markets during periods of restricted trade activity. In the milk marketing orders, *class I base plans* are used in an effort to encourage producers to manage their seasonal production of milk so that it is more in line with the seasonal pattern of consumption. A system of incentives and/or penalties is incorporated into these plans in an effort to encourage a more optimal pattern of production.

Market Support Activities. Market support activities do not directly affect the quantity or quality of the product sold. Rather they contribute to achieving other goals relating to the marketing of regulated products. Included here are standardization of containers and packs to promote more uniformity in packaging; assessments on producers (in the case of milk) or handlers (in the case of fruits, nuts, and vegetables) to raise funds to support new-product development and/or promotion efforts; price posting by handlers; prohibition of unfair trade practices by handlers; and provision of shipping information necessary for administering the orders.

Classified Pricing. Producer milk that meets U.S. Public Health Service standards for drinking-milk purposes—that is, grade A milk—is regulated by the federal milk marketing order system. Manufacturing-grade (grade B) milk is unregulated except insofar as its price is buttressed by the dairy price support program (see Chapter 2) and is sold directly by farmers or farmer cooperatives to unregulated butter or cheese manufacturers.

Under the federal order system, milk delivered to the order is classified according to its final use, and handlers must pay a minimum price for each use class. For our purposes it will be sufficient to consider only two use classes—fluid (class I) milk and manufacturing (class II) milk—although specific orders may establish a third use class in order to price separately milk going into such soft products as yogurt and ice cream.

Prices are established for each of the forty-four federal milk marketing orders on the basis of a specified relationship to the competitive pay price for grade B milk in the Minnesota-Wisconsin area. With minor exceptions, federal order class II prices are set at or near the Minnesota-Wisconsin price. Federal order class I prices are higher than class II prices by fixed differentials. The geographic structure of class I prices has historically been designed to approximate what economic location theory would lead us to anticipate for a competitive market given that the surplus production area is in the Minnesota-Wisconsin area. The differential between class I and class II prices in the Upper Midwest order (the approximate locale of the price basing point for all federal orders) was set at \$1.20 per hundred weight by the Food Security Act of 1985. This differential reflects (1) the differences in the costs of producing grade A and grade B milk and (2) the discriminatory pricing aspects of the federal order system.

Economic Impacts of Marketing Orders

Price Enhancement. The nut, date, raisin, and cran-berry orders allocate supplies among the primary and secondary markets directly with market allocation procedures. The citrus orders allocate supplies directly with prorates. The milk orders allocate supplies among the primary and secondary markets indirectly with minimum pricing rules. In each case, the opportunity exists to enhance producer returns by exploiting the differences in demand elasticities in the two markets. To see this consider Figure 8.1, where SS is the producer supply curve, D_pD_p is the (derived) demand curve for the primary market, D_sD_s is the (derived) demand curve for the secondary market, and D_pD_t (the heavy line) is the aggregate demand curve over both markets.

The competitive solution (i.e., the solution with no price discrimination) would yield a producer price in both markets of P_e, total quantity produced of Q_e, and Q_p allocated to the primary market and Q_s allocated to the secondary market. Restricting sales (directly or indirectly) in the primary market to Q_r leads to a price increase in that market to P_p (relative to the competitive price) and causes more of the product (also Q_r in this case) to be allocated to the secondary market. Price in the secondary market will then fall (relative to the competitive price) to P_s. If the demand curve in the secondary market is more elastic than is the demand curve in the primary

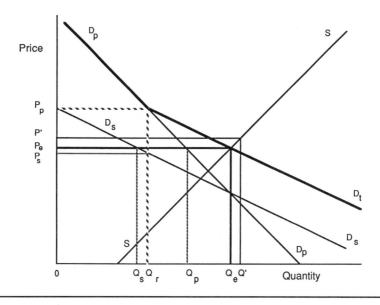

Figure 8.1. Economic effects of price discrimination

market, as is the case in Figure 8.1, the price decline in the secondary market will be less than the price rise in the primary market. Furthermore, if this elasticity condition holds, producer revenue will increase as a result of the reallocation of supply among the two markets since the revenue gain from less sales at a higher price in the primary market will exceed the revenue loss from greater sales at a lower price in the secondary market. This increase in producer revenue will be reflected in a higher (relative to P_e) average price farmers receive for the commodity they produce. The higher producer price (P') will call forth a greater volume (Q') of farm output.

Since price rises and quantity falls in the primary market, consumers of the primary product suffer a loss in utility. Conversely, since price falls and quantity increases in the secondary market, consumers of the secondary product gain utility. Because of the difference in elasticities in the two markets, however, the losses exceed the gains so that in the aggregate consumers lose. The primary market in the dairy industry is the fluid milk market. Low-income consumers may have more need for fluid milk than they do for the less necessary products of the manufactured dairy sector (e.g., butter and cheese). Thus in the case of dairy, it may well be that low-income consumers suffer the greater loss as a result of price discrimination—a fact that most of society would consider undesirable on equity grounds.

Producers, on the other hand, are clearly net gainers from price discrimination because the average price they receive increases. It can be shown, however, that the net gains to producers do not offset the net losses to consumers. Consequently, there is a net social loss from price discrimination relative to the competitive solution, and an excess of resources will be devoted to the production of the commodity under consideration.

It must be pointed out that just because two markets for a given farm commodity are separable and that demand elasticities differ in these two markets, price discrimination does not *necessarily* follow. Some element of monopoly control must also exist in the form of a producer cartel or in the form of favorable legislation. In U.S. agriculture, this favorable legislation is given in the form of marketing order legislation. Further, the monopolistic agent *must do those things necessary to effect price discrimination*!

Additionally, it must be remembered that marketing orders were instituted to provide for *orderly* marketing. In part, this means to provide for more price stability than would exist in the absence of the order. Market allocation schemes can contribute to this type of market order or price stability. This would be desirable not only to producers but to all of society if it results in lower producer costs and thus in a rightward shift of the supply curve. Price risk or uncertainty can increase production costs in three different ways. First, if price risks cannot be covered by insurance or other means, then producers will require a higher rate of return than they would on a less risky enterprise. Second, when a great deal of uncertainty exists in production of a farm commodity, farm lenders may charge higher rates of interest or limit the amount they will loan on this enterprise compared with less risky enterprise. Third, with price uncertainty the allocation of resources by both growers and handlers is subject to larger errors than when price is more certain. Reducing or eliminating any of these costs can cause the supply curve to shift to the right. The resulting efficiency gains may well more than offset any welfare losses society sustains from the price discrimination. Thus a little price discrimination may not be all that bad when weighed against the potential benefits.

In a comprehensive study of marketing orders for fruits, vegetables, and specialty crops, a team of researchers (U.S. Department of Agriculture 1981) studying those commodities for which marketing orders and market conditions permit price discrimination to happen found the following:

1. The almond, date, and cranberry orders did not regulate shipments for the crop years 1973–80 (and thus did not practice any price discrimination although they had the necessary options to do so).

2. During the last decade, the use of market allocation in raisins appears to reflect a goal of stabilizing price in the presence of unusual fluctuations in supply rather than long-run price enhancement.

3. It appears that market allocation under the walnut order as been used mainly to stabilize prices in recent years, but that some vestiges of a price-enhancing goal remain.

4. Resource misallocation attributable to market allocation is distinctly possible in the case of filberts.

5. There is no question that considerably more navel oranges, Valencia oranges, and lemons are being sold to processors than would be (the case) without order-imposed restrictions on fresh sales.

Price discrimination also occurs in the nation's milk markets as a result of the classified pricing scheme of milk marketing orders. The extent to which price discrimination occurs in this industry, however, is less clear. Song and Hallberg (1982) calculated a set of prices for the major dairy products which would have maximized consumer welfare and a second set which would have maximized producer revenue in each of the years 1960–79. Given the relatively inelastic demand for fluid milk products compared with manufactured dairy products, one would expect that prices which would have maximized producer returns from a given quantity of milk would involve considerable price discrimination. Their analysis showed that actual prices were closer to consumer-welfare-maximizing levels than to producer-revenue-maximizing levels. Further, the analysis showed that prices of dairy products moved progressively closer to the consumer-welfare-maximizing levels over the period of study. Hence if the gross class I differentials were due entirely to price discrimination, there was less discrimination in the dairy industry in 1979 than in 1960. It should be noted, however, that class I differentials also reflect transportation charges, added costs of producing milk meeting the higher sanitary standards for fluid use, and other costs associated with marketing fluid, as opposed to manufacturing, milk. Thus they are not simply a reflection of monopolistic behavior of farmer cooperatives.

OVER-ORDER PAYMENTS. Federal order class prices for milk are minimum prices to be paid by handlers. There is nothing to prevent milk producers (or their cooperative representatives) from negotiating prices above the federal order minimums (overorder payments), however. This has quite commonly been done over the past several years. Farmers often justify these payments on the basis that they are a payment for market services provided by farmer cooperatives (see Manchester 1983). Critics of the dairy industry argue that overorder payments result from the superior market power of dairy cooperatives, buttressed in part by marketing orders. Babb, Bessler, and Pheasant (1979) examined data for the late 1960s in an effort to determine what factors account for these overorder payments. A principal concern was whether market power as measured by cooperative concentration is a major or sole factor determining the size of overorder payments. They found this not to be the case. Rather it appeared that those factors that would normally be expected (on the basis of economic theory) to explain the geographic structure of prices accounted for much of the variation in overorder payments across markets.

Producer Allotments. The economic consequences of allotments and quotas were examined in the previous chapter. There we found that allotments not only have the potential of enhancing producer returns but also create windfall gains to the allotment recipients and cause social losses. The detrimental effects of allotment provisions in marketing orders are not as clear as in the case of tobacco or peanuts, but appear to exist nevertheless. The Economic Research Service team concluded the following (U.S. Department of Agriculture 1981, 39):

> While data series on the market value of transferred allotments do not exist, there is evidence that for both hops and spearmint oil, the cost of purchasing or leasing quota for new entry or expansion is considerable. Combined with the limited allocation of new allotment, this is persuasive evidence that these orders are being used to restrict supply. More specifically, the cost of marginal production is artificially elevated by the market value of allotment transfers. This potentially impedes growth and entry, even though all existing producers may not sell as many hops as their full allotment permits. The same tendency exists under the celery order. But entry and expansion are apparently not as economically attractive as in the hops and spearmint oil industries, since allotment transfers have carried a zero or negligible price, at least in recent years.
>
> Balanced against the resource misallocation cost of producer allotments are possible benefits of more stable prices and incomes to producers. However, for two producer allotment commodities, hops and spearmint oil, storage is a feasible option and could provide an alternative to stabilization by allotments.

(The hops order was terminated after this study was completed.)

Shipping Holidays, Prorates, and Base Plans. As we have seen, it is possible to enhance producer revenue via price discrimination among markets in the same time period so long as these two markets are separable and so long as the elasticities of demand in the two markets differ. By similar reasoning it will be seen possible to enhance producer revenue via price discrimination in the same market but in *different time periods* if demand elasticities differ in the two time periods. In competitive markets, we would expect price in different time periods to differ by the cost of storing the product from the first period to the second. If prices differ by more than the cost of storage, then we must conclude that some market imperfection such as price discrimination across time has occurred.

Shipping holidays, season prorates, and base plans that the various marketing orders authorize provide the opportunity for price discrimination in this sense. The Economic Research Service team (U.S. Department of Agriculture 1981, 47) saw little evidence to support the claim of excessive price discrimination in this sense in the case of fruit, nut, and vegetable marketing orders. Rather the research team concluded that at best such provisions serve to smooth out product flow over the season and thereby contribute to more efficient marketing. The base plans implemented in federal milk marketing orders also appear to serve more to contribute to efficiency than to enhancing prices (see Manchester 1983).

Marketing Boards

In most other countries of the world where marketing control of the type considered here is exercised, *marketing boards* are used rather than marketing orders. England, Canada, Australia, New Zealand, West Africa, South Africa, Israel, and the Netherlands all use marketing boards. Before leaving the subject of this chapter it would seem useful, then, to distinguish between these two institutional mechanisms. This is particularly true in view of the fact that the National Commission on Food Marketing appointed by the president in 1966 recommended that marketing boards be established in the United States as a means of protecting and improving the economic position of American farmers. Apparently the commission's view was that marketing boards would give farmers not only greater representation but also greater bargaining power.

Basically the two institutions differ very little. Marketing boards generally have the same objectives as marketing orders. Marketing board powers are stipulated in enabling legislation. At one extreme are promotional and advisory boards, just as in the case of marketing orders, whose power derives from an ability to assess producers to finance these types of activities. At the other extreme are marketing boards empowered with complete monopsony-monopoly power over domestic production and sales of farm produce. These boards may be empowered to buy, process, store, grade, and sell domestically as well as on the export market all of the farm produce. An example here would be the Australian Wheat Marketing Board. These boards may also exercise their power by licensing producers, determining to whom licensed producers may sell their produce, and even specifying the quantities that can be marketed and the price at which the transaction takes place.

The major difference appears to be that marketing boards may be given legislative authority to regulate production as well as marketing, whereas marketing orders cannot regulate production directly. Hence marketing board powers are more complete than are the powers of marketing orders. In general, marketing boards do appear to give their producers more bargaining power than do marketing orders. This, however, is probably due more to how the powers granted are exercised rather than to any inherent differences between the two institutional forms.

Why Marketing Orders and Not Price Supports?

Marketing order legislation is the product of a compromise between those who favored direct price and production control (e.g., wheat, feed grain, and cotton producers) and those who stressed orderly marketing of available supply through voluntary, cooperative means under the aegis of federal legislation. Thus we might argue that a basic philosophical difference explains in large part the different policy approaches to field crops and to horticultural crops and milk. Nevertheless, some fundamental differences in the production and/or product attributes of these two groups of commodities played an important role in the policy choices as well.

First of all, the major field crops are grown over wide areas of the country, involve many producers and markets, and have relatively long growing seasons. In contrast, the horticultural crops are grown by relatively few producers in specialized production regions and have a short growing season. It is much easier to isolate markets (necessary for the effective price discrimination operation of marketing orders) for crops grown in specialized regions and over short marketing seasons. Further, both voluntary and mandatory production control programs are less attractive when there are few producers in specialized production regions.

Second, field crops are grown for an international market; thus price and production policies must be formed in recognition of international interdependencies. Most horticultural crops, on the other hand, are (or were until recently) grown for the domestic market only and pricing policies could be established without regard to imports from other countries. Market coordination of the type exercised by marketing orders is much easier to accomplish when international consequences can be ignored.

Finally, most horticultural crops and fluid milk cannot be stored for long periods of time in their original form, as can field crops. This means that a price support/government purchase program is not a viable option in the case of horticultural crops and fluid milk.

Summary

Marketing orders were implemented to add order and stability to the agricultural product markets that, because of an imbalance of power between farmers and handlers and processors, were thought to be inherently unstable. They do this through quality control provisions, quantity control provisions, various market support activities, and, in the case of dairy, pricing regulations.

Marketing orders have come under strong attack at various times because of provisions which in effect permit what is deemed to be excessive price discrimination. The evidence suggests that there are probably excessive levels of price discrimination in the citrus orders, significant but perhaps not excessive levels of price discrimination in the milk orders, and little price discrimination of any consequence in the remaining orders. Producer allotments appear to have led to significant price enhancement and too few resources being devoted to the production of hops and spearmint oil.

On the positive side, however, marketing orders appear to have contributed to seasonal and interseasonal stability of producer prices and incomes in several commodities, have promoted higher quality assurance and standards which, together with pack and container standards, have led to reduced marketing costs, and through money collected from producer assessments have sponsored research that has led to technical efficiency in both production and marketing. Marketing orders have further provided marketing information that would not likely have been available in their absence. Not only has this information been of value to both producers and handlers, it has been useful to those charged with monitoring the marketing systems for evidence of poor marketing performance.

Suggested Readings and References

American Agricultural Economic Association. 1986. "Federal Milk Marketing Orders: A Review of Research on Their Economic Consequences." Task Force on Dairy Marketing Orders. Occasional Paper no. 3.

Babb, Emerson M., D. E. Bessler, and J. W. Pheasant. 1979. "Analysis of Over-Order Payments in Federal Milk Marketing Orders." Purdue University Agricultural Experiment Station Bulletin no. 235. August.

Farrell, K. R. 1966. "Marketing Orders and Agreements in the U.S. Fruit and Vegetable Industries." *Organization and Competition in the Fruit and Vegetable Industry.* National Commission on Food Marketing. Technical Study no. 4. June.

Glaser, Lewrene K. 1986. "Provisions of the Food Security Act of 1985." U.S. Department of Agriculture. Economic Research Service. Agriculture Information Bulletin no. 498. April.

Hoos, Sidney. 1979. *Agricultural Marketing Boards—An International Perspective.* Cambridge, Mass.: Ballinger Publishing Co.

Jesse, Edward V., and Aaron C. Johnson, Jr. 1981. "Effectiveness of Federal Marketing Orders for Fruits and Vegetables." U.S. Department of Agriculture. Economics and Statistics Service. Agricultural Economic Report no. 471. June.

Manchester, Alden C. 1983. *The Public Role in the Dairy Economy.* Boulder: Westview Press.

Song, Dae Hee, and M. C. Hallberg. 1982. "Measuring Producers' Advantage from Classified Pricing of Milk." *American Journal of Agricultural Economics* 64 (1): 1–7. February.

Thor, Peter K., and Edward V. Jesse. 1981. "Economic Effects of Terminating Federal Marketing Orders for California-Arizona Oranges." Economic Research Service. Technical Bulletin no. 1664.

U.S. Department of Agriculture. 1971. "Compilation of Agricultural Marketing Agreement Act of 1937 with Amendments as of January 20, 1971." Consumer and Marketing Service. Agricultural Handbook no. 421. October.

_____. 1981. "A Review of Federal Marketing Orders for Fruits, Vegetables, and Specialty Crops." Agricultural Marketing Service. Agricultural Economic Report no. 477. November.

_____. 1987. "Marketing Agreements and Orders for Fruits and Vegetables." Agricultural Marketing Service. Program Aid no. 1095. September.

9

Farm Policy and International Trade

Foreign countries are an important market for several U.S. agricultural products—particularly for wheat, feed grains, rice, soybeans, tobacco, and cotton. In 1950 the value of U.S. agricultural exports represented only about 10 percent of cash receipts from farm marketings. But by 1980–81, this percentage had tripled and U.S. farm incomes were at all-time highs. Clearly it is in U.S. farmers' interests to keep this percentage high. In the context of Chapter 7, increasing exports is one way to cause the aggregate demand curve for agricultural products to shift to the right, thus supporting higher market prices as well as increased output. Unfortunately, since 1981 U.S. agricultural exports have diminished sharply—down to 24 percent of cash receipts in 1990 from 30 percent in 1981. Why has this happened? Have domestic agricultural policies contributed to the problem or have they prevented a more unfavorable situation? Have international exchange rate fluctuations worsened or improved the situation? Have agricultural policies of other nations been at fault? Have permanent changes in comparative advantages and trade patterns occurred?

In this chapter I provide the background with which to address issues involving international trade. I begin with a brief review of ideas about trade—why it occurs, who gains and who loses from trade, and why trade may be discouraged by certain countries at certain times. I also review the level of protectionism at home and abroad for some of the principal agricultural commodities. Next we study the economic consequences on trade and world prices of various protectionist policies of the type used in the United States and elsewhere. Using dairy as a case study, I examine the expected U.S. and worldwide impacts of removing protectionist barriers to trade. I also briefly explore the consequences of removing protectionist barriers to trade throughout all of agriculture. Finally, I examine the impact of changes in exchange rates on trade in agricultural products.

Why Trade Occurs

Early economists adopted the principle of "absolute advantage" to explain interregional and international trade.[1] According to this principle, a country will export those commodities it can produce more cheaply than any other country and will import those commodities other countries can produce more cheaply. But what if country X has an absolute advantage in all commodities because of cheap labor, abundant resources, availability of superior technology and know-how, and/or because of a unique climate? What would this nation do with the revenue it would receive from all the other nations who purchase its commodities? Where would all the other countries get the revenue with which to purchase country X's commodities?

Questions of this sort led to the theory of "comparative advantage" as a more appropriate principle with which to explain trade. This theory postulates that it is "opportunity" costs not "absolute" costs that matter. That is, to produce one more unit of commodity A, country X must make some adjustments in the way it uses its productive resources, and these adjustments lead it to give up the opportunity to produce commodity B (and maybe other commodities as well). Country X is said to have a comparative advantage in the production of commodity A if when it expands production of commodity A, its reduction in output of commodity B is *less* then would be the case in country Y if country Y expanded production of commodity A.

The theory of comparative advantage suggests that under perfect competition and no artificial barriers to trade, country X will import those goods for which the price on the international market is *lower* than is its opportunity cost of producing these goods at home. Similarly, country Y will export those goods for which it can get a price on the international market that is *higher* than opportunities (revenues) it would forgo by producing fewer of these products at home and more of other products that it could import.

According to the theory of comparative advantage, then, comparative differences in production costs determine which goods are traded. Trade will stimulate investment in and expansion of those industries producing export goods and force contraction of those industries producing import goods. As export industries expand, they will demand inputs (including labor) and intermediate products from other industries. Hence the benefits from trade-related expansion will not be limited to the export industries but will be felt throughout the economy.

In Chapter 2 we saw that wholesale sugar prices have in recent years been much higher in the United States than in other countries. The reason this is so is that in order to protect its price support policy, the United

States effectively limits the importation of sugar from countries that have a much greater comparative advantage for sugar production. It would not be easy convincing U.S. sugar producers that removal of our protectionist policy is beneficial to the United States. After all, their jobs would be at stake. Clearly, though, this would be beneficial to U.S. consumers because with lower sugar prices, consumers could afford to buy not only more sugar but, more important, more of other products. Some of these other products would be foreign-produced goods. When U.S. consumers buy more goods for which foreign countries have a comparative advantage, consumers in these foreign countries will have more income with which to buy goods and services for which the United States has a comparative advantage. There are, then, mutual advantages from specialization and exchange.

If some or all U.S. sugar producers cannot compete with foreign producers of sugar, then expanded trade will provide U.S. sugar producers with other employment opportunities. These opportunities will permit U.S. sugar producers (1) to exploit *their* comparative advantage and in this way contribute more to the social product of their country and (2) to earn an unsubsidized income comparable to that of other workers so they too will be able to purchase more goods and services. In the end, everyone will have a better life, even the sugar producers who have been displaced!

These ideas can be demonstrated graphically as in Figure 9.1. Here I assume a two-good, two-country world: food and clothing, and the United States (US) and the Rest-of-the-World (ROW). The curve labeled P_{us} represents the production possibilities open in the United States. It shows the various combinations of food and clothing that the United States can produce given its resource endowments. If the United States produces more units of food, then it must give up some production of clothing. If it produces more units of clothing, it must give up some production of food. The curve labeled P_{row} is the production possibility curve for ROW. It is to be interpreted similarly. These two production possibility curves are drawn to reflect the fact that given the resource endowments and technical know-how in the two countries, the United States is capable of producing more food at a given (feasible) level of clothing output than is ROW, and ROW is capable of producing more clothing at a given (feasible) level of food output than is the United States.

The curve labeled U_{us} is an indifference curve for the United States which shows various combinations of quantities of food and clothing among which U.S. consumers are indifferent or, alternatively, which yield the same total consumer utility. Curve U'_{us} has the same meaning. But since it is positioned above and to the right of U_{us}, it yields a greater total consumer utility. Curves U_{row} and U'_{row} have similar interpretations, but for ROW.

If trade between the United States and ROW is not permitted, perfect

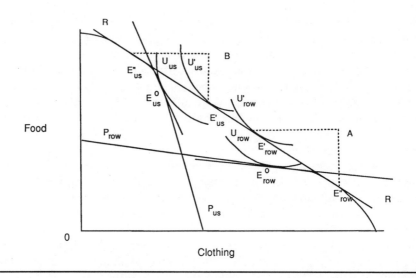

Figure 9.1.   Illustration of the gains from specialization and trade

competition in each country requires that the ratio of prices of the two commodities be equal to the ratio of the marginal rate of product transformation of the two goods in production and to the ratio of the marginal rate of substitution of the two goods in consumption. Thus the no-trade equilibrium points are E_{us}^0 and E_{row}^0 with prices defined by the slopes of straight lines tangent to the respective production possibility and indifference curves at these equilibrium points. In Figure 9.1, these price lines are the line segments going through points E_{us}^0 and E_{row}^0. Notice that the slopes of these two lines differ, implying that the prices of food and clothing in the two countries differ—food is relatively more cheap than clothing in the United States, whereas clothing is relatively more cheap in ROW—consistent with the technical possibilities in the two countries.

When trade is permitted, prices of the two goods will be equal in both countries, so they now face a common price line labeled R in Figure 9.1. In equilibrium, the ratio of prices of the two goods must still be equal to the ratio of marginal rate of product transformation in production. Hence the new equilibrium production points are E_{us}' and E_{row}'. But since the ratio of prices must also equal the ratio of marginal rates of substitution in consumption, consumers in both countries are now able to reach a higher indifference curve. Hence consumers in the United States as well as in ROW are better off with trade. ROW will specialize by producing relatively more clothing and import AE'_{row} of its food needs from the United States. The United States, on the other hand, will specialize by producing relatively more food and import BE''_{us} of its clothing needs from ROW.

The moral of this story is that both countries are better off when trade is permitted because aggregate utility in each country will have increased. Obviously, the world is better off when trade is permitted, since aggregate world utility will have increased. With trade, consumers in both countries will have available a larger and more diverse bundle of goods and services at lower overall prices than they would under total or partial isolation. Prices of imported goods and their domestic equivalents will be lower. Prices of exported goods will be higher. Nevertheless, consumers will be able to reallocate their incomes in such a way as to achieve greater satisfaction from consuming less of the now more expensive goods and more of the now less expensive goods.

Trade will also have an impact on the structure of the economy in that investment and expansion will be stimulated in the export industries and discouraged in the import industries. This in turn means there will be a change in the demand structure for inputs and in the value of resources devoted to the respective industries. Of course, some clothing producers in the United States and some food producers in ROW will need to shift into production of the alternative good where their prospects are better. They may not be happy about having to make the switch, but they will be forced to do so in order to achieve a comparable level of income.

To this point I have made the simplifying assumption that there are no other costs involved in trade. This, of course, is not true. It is costly to transport goods from one country to another. Further, there may be border costs in the form of export taxes or import duties. We cannot ignore these costs, for they do have an impact on the level of trade that will occur. For the time being, however, it is sufficient to keep in mind that trade will still occur in the face of transportation and other trade costs so long as the unit value of these costs does not exceed the equilibrium, between-country unit price differences for the goods under consideration.

Another issue that must be dealt with is the money exchange rate between trading partners. I will address the exchange rate question in a later section. For the moment, however, I assume that currency exchange rates among the various countries are in balance and do not vary with the imposition of the policy instruments to be examined.

Shifting Comparative Advantages

Production costs and hence comparative advantages differ among countries for several reasons. First, all countries are not endowed with the same number and quality of productive and natural resources. Because of its climate, Scandinavia has considerable difficulty growing bananas, oranges, and lemons no matter how strong is Scandinavia's desire to do so. Up to recent times at least, the United States had greater amounts of capital and productive technology available than did the developing and centrally planned economies. Workers in some countries have different skill levels than do workers in other countries.

Second, not all countries have the same amount of productive resources available. In general, I expect the costs associated with using plentiful resources to be lower than the costs associated with using scarce resources. Japan could produce many of the crop products produced in the United States. The scarcity of land in Japan relative to her total population, however, would make the cost of doing so quite high. The United States, Canada, Australia, and the USSR, on the other hand, have great expanses of land capable of producing cereal crops, and there is little competition for this land for other uses since population density in these countries is so low.

Finally, basic resources have differing degrees of mobility among countries. Thus a country may have higher production costs than another because it cannot export surplus resources or it cannot import resources that are in short supply locally.

In recent years we have observed a considerable shifting of comparative advantages in agricultural production among countries of the world. This is due in large part to a greater international mobility of certain key productive resources. There has, for example, been a significant increase in the transfer of technology throughout the world since World War II—to Western Europe, Japan, South Korea, Taiwan, India, etc. New crop varieties have been developed and shared with other countries. The resulting so-called Green Revolution has played an important role in permitting India to become an occasional net exporter of wheat and rice in recent years. The boundaries for corn production have been pushed further north as a result of the development of more tolerant and shorter growing-season corn varieties. Agricultural reform along with the application of new technology has contributed to make China a net exporter of corn, rice, cotton, and soybeans, and to be much less dependent on imports of wheat than in years past. Brazil and Argentina are now very competitive in cereal grains. The European Economic Community (EEC), traditionally a net importer of wheat and coarse grains, is now a major exporter of both.

One of the major achievements of the EEC has been to effect the relatively free movement of labor from one country in the Community to another. This will likely accelerate now that Spain and Portugal have joined the Community and that a new initiative to integrate the economies of the Community by 1992 is well under way.[2] This labor mobility has had an impact on comparative advantages in both agriculture and nonagriculture in the Community. Finally, capital is now much more readily available to the international community so that investment in needed industries and development efforts can more readily take place.

Clearly some resources are less easily transferred from country to country so that comparative advantages in some commodities will be more difficult to change. The advantages New Zealand, Argentina, and Ireland enjoy in the form of grassland feeding of dairy cows, for example, are not likely to be duplicated elsewhere. Similarly, the tropical regions of the world enjoy advantages in agricultural production that cannot easily or cheaply be duplicated in the temperate zones.

Protectionism

A general conclusion of trade theory is that citizens of the world would be better off in the aggregate under free trade. Trade allows consumers access to goods that would otherwise be prohibitively expensive. In the absence of trade, for example, U.S. consumers would not be able to enjoy the many tropical fruits and condiments we now enjoy at such low cost, if at all. Similarly foreign consumers would not have access to the many products manufactured in America that are cheaper because of the lower cost resources and the technical know-how that exist in America.

Nevertheless, and for a variety of reasons, countries do erect trade barriers to protect their local industries. As I discussed in Chapter 2, the United States originally levied import duties on sugar and other commodities in an effort to *generate revenue* for the federal treasury. This is probably still an objective of some countries but is certainly no longer a primary objective of the United States and, in general, is not a very widespread reason for protection.

National security interests are often used to justify protection. Japan and the nations of Western Europe, for example, give their farmers substantial protection from foreign competition in pursuit of a *food security* objective. Vivid memories of food shortages caused by trade disruption and the ravages of armed conflict during World War II still persist in these regions of the world. These memories underpin a general philosophy of generous support for local farmers which, among other things, translates into high price supports and protection against imports of domestically produced commodities.

If imports in country X have been exceeding exports for long periods of time, other nations may lose confidence in the economic strength of this country to meet its international obligations. One solution is for country X

to devalue its currency relative to other currencies (see the section to follow on exchange rates and trade). Devaluation would tend to make imports more expensive to domestic consumers and thus reduce import demand and, in turn, reduce the pressure on the *international payments* account of country *X*. A second solution is for country *X* to stave off devaluation by placing restrictive quotas or high tariffs on imports and thereby reduce domestic consumers' access to or effective demand for imported goods.

Import duties and quotas are often used to protect newly emerging, or infant, industries. The *infant industry* argument for protection is an idea that was promulgated during the early development of the United States. The infant industry argument was essentially the rationale for protecting the U.S. sugar industry against competition from Europe in the late 1800s. Unfortunately, as is well illustrated by the U.S. sugar industry, the infant quite often fails to develop into an adult! This is particularly true if the industry in question is capable of retaining for long periods of time the political power necessary to maintain protective legislation.

One form of protection often leads to a second. Dairy policy in many of the developed countries of the world is a case in point. Here domestic price supports are the rule. In general, these supports are so high that without tight import controls, imports of dairy products from other countries would undermine the domestic price support legislation. Hence these countries use import quotas and import duties to limit imports and in this way *protect domestic price policy*. In the United States, for example, imports of dairy products are limited primarily to specialty cheeses from other countries and in total are maintained at about the equivalent of 2 percent of annual U.S. milk production.

Many countries engage in various retaliatory actions when they feel that their competitors or trading partners are engaging in *unfair trading practices*. The United States has most recently been highly critical of the trading practices of the EEC, Japan, and South Korea in both agricultural and industrial goods. Much protectionist rhetoric has resulted in the halls of Congress and elsewhere. The Food Security Act of 1985, in fact, authorized a Targeted Export Assistance Program designed to enable the United States to retaliate against certain countries accused of unfair trade practices.

Rates of Protection in World Agriculture. The effect of protectionist policies is to raise domestic as well as world prices and to keep them higher than would be the case in a perfectly competitive, free-trade world. This, in turn, leads to overproduction in the protectionist countries and to reduced world trade. Recent estimates of the level of protection of selected agricultural products in the principal industrial countries of the world are given by the "nominal protection" coefficients shown in Table 9.1.

Nominal protection coefficients are calculated as the ratio of domestic price to world market price at a country's border. In the case of dairy, for example, the border price for a specific country is the New Zealand price plus the cost of transportation from New Zealand to the specific country of concern. The New Zealand price is used here since the cost of producing milk in New Zealand is estimated to be the lowest of any country in the world. The minimum value these coefficients can take on is 1, indicating an absence of domestic protection. The higher the value of these coefficients, the greater the degree of domestic protection afforded the respective commodity.

These estimates are, at best, approximations of the level of protection. Several additional factors must be weighed to get "true" measures of rates of protection. For instance, if world prices vary while domestic prices are stable, as would be expected if domestic prices were administered, nominal protection coefficients can vary widely over time. Also, domestic prices can be measured at different levels in the marketing chain: farmgate, the intervention level, or the wholesale level. Since different countries report prices at different levels, country comparisons can be somewhat misleading.

Furthermore, nominal protection coefficients by themselves are simply measures of the degree of protection in individual countries: they provide little information about the international consequences of the level of protection so measured. It is clear that agricultural policies do affect world

Table 9.1. Nominal protection coefficients for selected commodities in the industrialized countries, 1980–82

Country or region	Wheat	Coarse grains	Rice	Beef and lamb	Pork and poultry	Dairy products	Sugar
Australia	1.04	1.00	1.15	1.00	1.00	1.30	1.00
Canada	1.15	1.00	1.00	1.00	1.10	1.95	1.30
EEC[a]	1.25	1.40	1.40	1.90	1.25	1.75	1.50
EFTA[b]	1.70	1.45	1.00	2.10	1.35	2.40	1.80
Japan	3.80	4.30	3.30	4.00	1.50	2.90	3.00
New Zealand	1.00	1.00	1.00	1.00	1.00	1.00	1.00
United States	1.15	1.00	1.30	1.00	1.00	2.00	1.40
Weighted average	1.19	1.11	2.49	1.47	1.17	1.88	1.49

Source: World Bank 1986.

[a]European Economic Community. In 1980–82 the EEC consisted of Great Britain, Ireland, France, West Germany, Italy, the Netherlands, Belgium, Luxembourg, Denmark, and Greece.

[b]European Free Trade Association. In 1980–82 EFTA consisted of Austria, Norway, Sweden, Iceland, and Switzerland.

prices and trade. A good deal more information is needed to assess what would happen to world prices if protectionist policies were abandoned, however. The relative importance of the different countries as well as their internal policies must be considered. Switzerland, for example, has the highest rate of nominal protection for dairy of any country in the world. Yet Switzerland has such a small percentage of the world's milk production and engages in such a small percentage of world trade in dairy products that this high level of protection is of little consequence to the international dairy market. A large exporting country, on the other hand, with a high rate of nominal protection may also have little impact on the international market if the remainder of its internal policies act to remove the surplus production that would otherwise occur—policies such as production quotas (or acreage set-asides in the case of grains), domestic food distribution programs, stock accumulation programs, or price distortions in competing or complementary commodities.

Despite these limitations, some general conclusions are apparent from the coefficients shown in Table 9.1. First, Japanese farmers are more highly protected than are farmers almost anywhere else in the developed world. EEC farmers are also very highly protected. Second, dairy farmers everywhere except in New Zealand are heavily protected. Third, relative rates of protection between commodities vary from country to country. This suggests that even within countries there are distortions as farmers react to prices that have been set by policy rather than by comparative advantages.

Some isolated examples of the incidence of and consequences of protection as pointed out by Miller (1986) will serve to further dramatize the unattractiveness of protectionism:

1.   In some instances, EEC farmers paid more for imported feedstuff for their dairy cows than they could have received on the world market for the milk produced from it.

2.   EEC butter stocks have in the past been so large that they eventually deteriorated to the point that they had to be sold as butteroil. Butteroil brings $450 per ton, which is only 14 percent of the price paid to EEC farmers to produce it in the first place.

3.   Japanese rice producers are paid over eight times the world price for rice. This generosity has encouraged the production of large domestic rice surpluses, which are sold as animal feed or exported at discount prices, involving losses of billions of dollars per year.

4.   The EEC has sold high-quality beef to Brazil for $490 per ton despite paying its own beef producers $1,200 per ton to produce it.

5.   Since 1973, the EEC's expenditure on its price supports has jumped by nearly 80 percent in real terms but the EEC's value-added in agricultural

production has declined.

6.   Consumer transfers to U.S. beef producers under the Meat Import Law over the 1964–84 period averaged $180 million a year, but the total loss to U.S. consumers was $260 million a year. Third-country importers of beef were the net beneficiaries of the program.

7.   In 1985 EEC farmers received 18 cents per pound for sugar that was then sold on the world market for 5 cents per pound. At the same time the EEC imported sugar at the higher world price.

8.   Canadian farmers pay up to eight times the price of a cow for the right to sell that cow's milk at the government's support price.

9.   The United States subsidizes irrigation and land-clearing projects that result in increased agricultural capacity, and at the same time pays farmers not to use the land for growing crops.

The cassava-soybean story reported by McCalla and Josling (1985) is also worth repeating. In the 1970s, rapid economic growth and rising per capita incomes in Europe resulted in an increased demand for meat. This caused meat prices to rise, which, in turn, stimulated expansion of the hog and poultry industries and increased the demand for high-protein feed. But rising nonfarm incomes was a signal to EEC farmers to bring pressure for increased price supports for such commodities as feed grains. As feed grain prices rose, a mixture of 80 percent cassava and 20 percent soybean meal came to be used as a price-competitive substitute for corn since both cassava and soybean meal entered the EEC duty free. The effect of increased price supports in the EEC then was to create a huge market for Thailand cassava and to choke off the demand for EEC-grown feed grains. In the end, the EEC was forced to negotiate with Thailand and Indonesia a voluntary export restraint agreement (see Glossary) in order to protect the EEC feed grain industry.

The corn gluten caper is also instructive (McCalla and Josling 1985). Corn gluten is a by-product from wet milling corn for the production of high-fructose corn syrup (HFCS) and grain alcohol. HFCS is a substitute for high-priced sugar. Grain alcohol is a substitute for high-priced OPEC oil. Corn gluten is a good source of protein for animal feed and thus is a substitute for both corn and high-priced cassava-soybean meal feed. An expanded demand for HFCS and grain alcohol due to price distortions and protection increased the supply of gluten, reducing its price and making it more competitive with corn, soybean meal, and cassava.

Impacts of Removal of Protection: Case of Dairy. In a recent study commissioned by the World Bank, Tyers and Anderson

(1986) used a multicommodity (wheat, rice, coarse grains, meats, dairy, and sugar) simulation model of world agriculture to project expected 1985 consequences of free trade in dairy for thirty countries and country groups. The first projection assumed 1980–82 domestic-to-border price ratios would remain unchanged to 1985. This projection then assumed a continuation of 1980–82 protectionist dairy policies everywhere and was used as the basis of comparison for subsequent simulations. This simulation is referred to as the reference scenario. A second projection assumed removal of all forms of dairy market intervention—domestically as well as across borders. In the latter projection, 1980–82 domestic-to-border price ratios in all nondairy markets were assumed unchanged from their actual levels through 1985. Thus, whatever protection existed in 1980–82 in the nondairy industries was assumed to be continued through the projection period.

In both of these simulations it was assumed that the border price for milk in every country is the New Zealand producer price for milk plus an allowance for processing milk into exportable form as well as an allowance for transportation from New Zealand to the border. All milk product quantities were converted into fluid milk equivalents so all dairy products could be treated, for analytical purposes, as a single commodity.

The study found that under removal of all protectionist policies for dairy, world prices for milk and world trade in dairy products would nearly double! Imports in Japan, the USSR, China, Mexico, South Korea, and Taiwan would increase substantially. Exports from New Zealand, Australia, Argentina, Brazil, India, and the United States would also increase substantially. The European Free Trade Association's (EFTA) share of world exports would fall from 13 to 2 percent (see Glossary for countries included in EFTA). Argentina and Brazil would shift from positions of net importer to positions of net exporters, accounting for 13 percent of world exports. Exports from the EEC would also increase slightly although the EEC's share of world exports would fall from 54 percent in the reference scenario to 47 percent under trade liberalization in dairy. The U.S. share of world exports would increase from 8 percent to 14 percent as the U.S. captures some of the market freed up by trade liberalization.

Because the world price of dairy products is projected to be so high under trade liberalization, milk prices in Canada, the United States, and several of the developing countries would change very little. Milk prices in Australia and South America would increase significantly but not by as much as in the low-cost countries of New Zealand and Argentina. In the EEC, milk prices would also increase slightly in spite of the current high level of protection in the EEC. Milk prices in EFTA would drop by 18 percent, and in Japan by over 30 percent. Significant price decreases would also occur in the centrally planned economies and in Egypt, Nigeria, China,

South Korea, Taiwan, and Mexico. In general, global liberalization in dairy would raise the price to producers in the major dairy countries currently having relatively low rates of protection, while for several with relatively high rates of protection at present there would be little price impact. The major exceptions would be in Japan and EFTA.

The overall welfare gains in the industrialized countries estimated by Tyers and Anderson to result from removal of dairy intervention policies everywhere amount to about $7 per capita. As in every such case there would be some gainers and some losers. By and large all the major dairy-producing countries in North America, Europe, Australia–New Zealand, and South America would be net gainers. Many countries in the developing world would be losers. New Zealand stands to gain the most from trade liberalization in dairy—an estimated $195 per capita—as both the world price and her exports increase. The next largest gainer would be EFTA ($25 per capita) as consumer prices in this area fall. Australia and Argentina also stand to gain significantly. Strangely enough the remaining countries with relatively high rates of protection—the EEC, the United States, and Canada—would also gain somewhat. This gain would come primarily from increased exports. Thus while consumers in the United States would be only slightly affected, producers in the United States would be better off. In total, gains in the United States would more than offset the losses. The big losers would be the Caribbean and African countries, and to a somewhat lesser degree, Asia.

It is inevitable that compared with the status quo, there are both gainers and losers from free trade. The distribution of gains and losses from free trade in dairy depicted here would appear to most people to be rather inequitable—the low-income countries tend to be the big losers while most of the high-income countries end up being gainers. It must be borne in mind, however, that the analysis on which these results are based tells only part of the story. That is, rates of protection and thus prices, production, and trade in all other agricultural markets were assumed to remain unchanged at their 1980–82 levels. Clearly these variables are not likely to remain unchanged if free trade were to occur in dairy. In particular, it is to be expected that in those countries where milk production is projected to increase, resources would be bid away from the production of other agricultural products, some of which are produced in the countries identified above as losers. Hence the losses sustained in a given country as a result of free trade in dairy can be expected to be compensated for (and indeed more than compensated for) by the gains from nondairy enterprises.

Removal of All Forms of Protection in Agriculture.
If trade liberalization *in all of agriculture* were to occur, we should expect somewhat less drastic milk price changes to occur as substitutions in production take place and as lower feed prices in Western Europe lower the cost of milk production in that region. Indeed it is likely that world prices for milk would drop *below* current U.S. price levels so that U.S. milk production would fall and the United States would become a net importer rather than a net exporter of dairy products. The United States would by no means get out of milk production: no other country is a large enough milk producer to supply itself plus the United States. But by the same token, the United States would not be able to compete with New Zealand or a few other countries for much, if any, of the milk needs of other countries.

This is precisely the result found by Tyers and Anderson when all forms of protection in agriculture were removed and in all countries of the world. More recent studies[3] confirm that under multilateral trade liberalization (removal of protection everywhere in all of agriculture), U.S. prices and production of wheat, coarse grains, and rice would increase over the long term, but U.S. prices and production of milk would fall. U.S. sugar prices and production are also expected to fall. The impact of multilateral trade liberalization on U.S. meat prices and production is less certain. Some researchers see an increase while others see a decline. The consensus, though, seems to be a much smaller impact of liberalization on the U.S. meat industries than on grains, dairy, and sugar.

Multilateral trade liberalization would result in higher world prices for most agricultural commodities, if not immediately, then in the longer term. The greatest increases would likely occur in dairy and wheat, but significant increases would also occur in coarse grains, rice, beef, and sugar. Significant gains would accrue to all of the developed countries, with the biggest gainers being New Zealand and Canada. The United States would also be a net gainer from multilateral trade liberalization.

Economic Consequences of Protectionist Policies

Price and Income Support and Export Subsidies.
U.S. exports as a percentage of world trade in the major crop commodities are shown in Table 9.2. Clearly the United States is the dominant supplier of most of these commodities. As the dominant supplier, its price support activity can be expected to have an enormous impact on world prices. If support prices (or loan rates) are set above equilibrium world price levels, domestic surpluses will accumulate. This, in turn, will lead policymakers to

Table 9.2. U.S. exports as a percentage of world exports for selected agricultural commodities, 1987/88

Commodity	Percentage
Wheat	37.5
Corn	67.8
Total coarse grains	54.2
Rice	18.6
Soybeans	72.7
Cotton	28.4

Source: U.S. Department of Agriculture 1990.

consider export subsidy programs in an effort to dispose of the government accumulations of surpluses. As we saw in Chapter 6, not only are export subsidies expensive to U.S. taxpayers, they involve social losses as well as income transfers from U.S. taxpayers to foreign consumers. Further, they tend to depress world prices.

Thus policies such as the Export Enhancement Program authorized under the Food Security Act of 1985 and the Food, Agriculture, Conservation, and Trade Act of 1990 are not the most attractive policy choices. Further, such policies put the United States in a very awkward position when its president goes to an economic summit meeting with the leaders of the other major economies of the world to argue for worldwide elimination of protectionist policies; when its trade representative goes to a General Agreement on Tariffs and Trade (GATT) Round to argue for the same thing; or when it berates the EEC for subsidizing exports of wheat, coarse grains, sugar, and dairy products.

The EEC has used export subsidies extensively in recent years because domestic price support policy has encouraged huge surpluses in commodities for which the EEC was previously a net importer. In the EEC these subsidies are termed export restitutions. They are changed frequently to reflect the fact that world prices and exchange rates vary daily but the domestic support price is fixed or changes much less frequently.

Since 1973, the United States has used target prices and deficiency payments as a means of supporting farm incomes. It will be recalled that deficiency payments are determined on the basis of the difference between mandated target prices and market prices or loan rates. Since these payments in effect constitute a price guarantee above the market price or loan rate, we would expect deficiency payment recipients to produce greater than equilibrium levels of output. In the absence of supply control or government purchases, the effect of this policy, then, is to increase domestic output above equilibrium levels and to drive domestic market prices below equilibrium levels, as was demonstrated in Chapter 7.

The international consequences of a target price/deficiency payment program of the type considered here can be studied with the aid of Figure 9.2. The basic format of this diagram is the same as that of Figure 6.3 in Chapter 6. The left panel reflects the supply-demand situation in the United States. The right panel represents the supply-demand situation in the rest of the world (ROW). The middle panel collapses the information from the left and right panels into a single supply-demand situation for the world, with the schedule of excess demand from the ROW measured by the ED curve and the schedule of excess supply from the United States measured by the ES curve.

A major difference in Figure 9.2 is that under a target price/deficiency program, the U.S. supply curve is perfectly inelastic at Q_u and up to the point at which price equals P_t, while for prices higher than P_t it follows the original curve. The reason for this is that effectively there can be no producer price in the United States below P_t, and hence U.S. farmers will always produce as if price is at least P_t—that is, they will respond with Q_u output.

This "kinked" supply curve in the United States also causes the excess supply curve to be kinked, as shown by the heavy line in the middle panel of Figure 9.2. The net result, then, is a reduction in the world market price relative to the equilibrium level (that is, from P_w to P'_w) and an increase in U.S. exports to the rest of the world (from Q_t to Q'_t). ROW production will fall and ROW consumption will rise in response to the reduced world price.

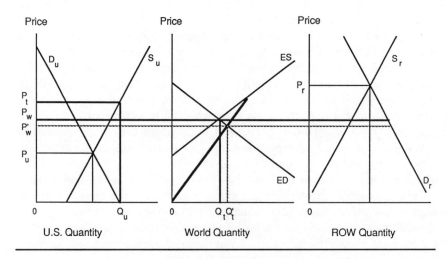

Figure 9.2. Economic impacts of target price and deficiency payment program on world markets

The welfare of ROW consumers will rise, but that of ROW producers will fall. When the social costs in the United States are combined with those of the ROW, total world welfare will be seen to be reduced.

Supply Control with Acreage Restrictions. Any program of the dominant exporting country that restricts output would be expected to lead to restricted exports and upward pressure on world prices. The conservation reserve and acreage reduction programs in the United States will clearly have these effects. As we have seen before, an acreage reduction program causes the supply curve in the United States to shift to the left. There will accordingly be a corresponding shift in the excess supply (ES) curve in the middle panel of Figure 9.3. The consequences of an acreage reduction program can then be deduced directly from the diagram—world price rises and world consumption declines relative to the no-acreage-reduction, free-trade situation. ROW production will rise slightly in response to the higher world price and to partially compensate for the reduced exports from the United States. Hence a reduction in world trade is induced and world welfare falls accordingly.

Import Quotas and Duties. Import quotas and duties are important trade-restricting devices and are frequently used throughout the world. Agitation by the National Cattlemen's Association resulted in the

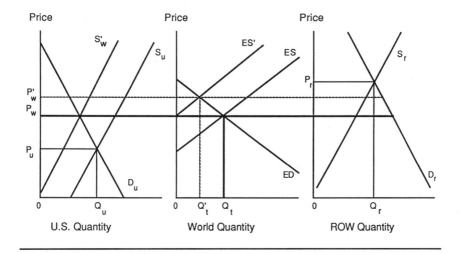

Figure 9.3. Economic impacts of acreage reduction program on world markets

Meat Import Quota Act of 1964, which authorized quotas on certain meat and meat products to limit imports to no more than 4 percent of domestic beef and veal production. The secretary of agriculture was directed to impose quotas whenever imports were projected to exceed this amount. The secretary could suspend quotas if, in the opinion of the president, it was in the overriding interest of the United States to do so because of national security reasons, domestic price interests, or superseding trade agreements. The quota for imports was programmed to increase in proportion to expansion in domestic production.

During the next few years domestic cattle prices were low enough to depress beef imports below the levels set in the quota legislation and it was not until 1970 that imports became significant enough to call for imposition of quotas again. But in 1970 the president suspended the quota in an effort to control food price inflation. In 1972 calls for imposition of beef quotas were denied on the same basis. Voluntary export restraint agreements negotiated between the United States and beef and veal importers have generally been sufficient to keep imports within the intended limits, so the quota has not been implemented in recent years.

As we have seen before, the United States also uses import quotas on dairy products, and a country-by-country quota to most favored nations on sugar (see Glossary for a definition of *most favored nations*). Import quotas are also established on raw cotton, although imports of raw cotton have not approached the quota limits in recent years. Import quotas on cotton textiles, however, are effective. The latter were negotiated through the Multifiber Arrangement under the auspices of GATT in 1974.

The consequences of import quotas can be examined with the aid of Figure 9.4. In this figure, the positions of the United States and the ROW have been reversed so that the United States is now viewed as the major importer. The effect of a binding quota is to put a wedge between the internal U.S. price, P', and the world price, P'_w, the latter being determined by the intersection of the world ES curve and the new ED curve that is now truncated by the quota (the kinked curve indicated by the heavy line in the middle panel of Figure 9.4). There is no revenue collected by the U.S. government. Importers, however, are granted rights to import, and these rights receive an economic rent equal in value to $(P' - P'_w)$. Presumably importing agencies could sell these rights to the highest bidder and in this way capture the gains for their nation. Under a restrictive quota program, U.S. consumption is reduced but production is increased. It is in this sense that an import quota protects local producers. In the ROW, production falls as world price falls, partially compensating for the reduced trade.

Another way to limit imports is via licensing requirements. Under this type of program an importer is granted an import license issued for a

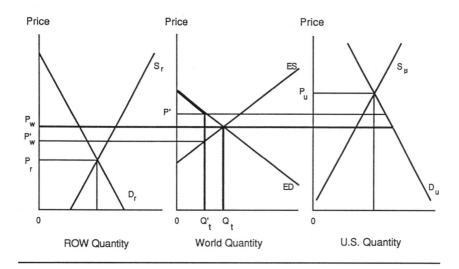

Figure 9.4. Economic impacts of an import quota or duty on world markets

limited period of time when domestic supplies are considered inadequate. The licensee is required to make a deposit on the intended transaction. If the transaction is not made during the license period, the deposit is forfeited. Canada and the EEC use import licenses in grains and oilseeds. Japan uses licensing in wheat and red meat.

A third way to limit imports is to levy a tax or duty on imports. Many nations, including the United States, utilize this policy instrument. The United States imposes import duties on beef, veal, mutton, and pork, on dairy products, on tobacco, on cotton products, and on wool. In 1984 import duties on cotton yarn, woven cotton fabrics, and cotton wearing apparel amounted to 8 percent, 11 percent, and 25 percent, respectively, of the foreign export values of these products. In 1979 import duties on wool were 25.5 cents per pound but, as a result of GATT negotiations, have since been lowered to 10 cents per pound. Long-standing U.S. policy permits import duties on any agricultural commodity (e.g., on grain imported from Canada) any time importation is judged to interfere with the price support programs of the United States.

"Variable" import levies are the cornerstone of EEC agricultural policy. Variable import levies equal the difference between the price of imports at the EEC port (e.g., Rotterdam) and an officially fixed entry price (the so-called threshold price) at which foreign goods can be sold in the EEC. The levy varies because world prices and exchange rates vary (daily) whereas domestic support or threshold prices tend to remain fixed for long periods of time.

An import duty or tax of size $(P' - P'_w)$ has precisely the same effect on producing and consuming nations and on world prices as does an import quota. The difference is that with an import duty there is an annual stream of duty revenue to the importing country and there are no quota rents. An import duty imposed on imports in the United States can be viewed as a tax on local consumption. As such, it causes the effective demand curve in the United States (D_u) to shift downward by the amount of the tax. The effective excess demand curve (ED), then, also shifts downward, with the same domestic and world price and production and trade impacts as import quotas. If the duty imposed is ad valorem, D_u (and ED as well) will shift in a nonparallel fashion to reflect the fact that the per unit tax is greater at higher prices than at lower prices.[4]

Exchange Rates and Trade

Prior to World War I currency exchange rates were maintained through the operation of the gold standard. Under this standard, national currency values were defined in terms of gold, and the central monetary institutions of the day (located principally in London) could manage credit and investment mechanisms to aid the flow of goods and capital and maintain exchange rate stability. While the gold standard existed, no great exchange rate problems occurred because no great corrections were required. As a nation's gold supply increased or decreased, internal inflationary or deflationary pressure led to shifts in the demand and supply schedules for imports and exports of goods and services and of capital. Thus all necessary economic adjustments were made internally, and in the absence of any major world disturbances, sufficient exchange rate stability was maintained to serve the needs of the day and to facilitate trade expansion.

But World War I drastically altered production and trade patterns, and postwar reconstruction permanently altered capital flows and credit balances. A variety of trade restrictions were imposed that were unrelated to long-run economic considerations. Agriculture was not excepted as import duties, import quotas, import licenses, and export subsidies were employed to protect local farm interests. World War II served only to intensify these problems.

In an effort to find a formula for creating a stable world economic order in a world of fixed exchange rates, discussion among the world's leading economic powers began even before the war was over. These discussions culminated in the Bretton Woods (New Hampshire) Conference in 1944. This conference resulted in two international monetary institutions: the International Bank for Reconstruction and Development (IBRD, or the World Bank), which initiated operations in 1945,[5] and the International Monetary Fund (the IMF), which initiated operations in 1947. The conference also led to the General Agreement on Tariffs and Trade (GATT), which was to serve as the permanent international agency responsible for implementing a code of conduct for international trade (see Glossary for a complete description of those institutions).

The IMF implemented a *fixed* exchange rate system based again on gold under which the U.S. government was to fix the value of the U.S. dollar in terms of gold by agreeing to buy and sell gold at a price of $35 per ounce. All other nations were to peg the value of their currency to the U.S. dollar. The IMF also established a revolving pool of foreign currencies and gold (financed by assessments as a requirement of membership) to facilitate international capital flows. The system worked well enough to survive for nearly a quarter of a century. Changes were made over this period in an effort to make the system compatible with changing world economic realities. The most innovative of these was the development of "special drawing rights," or SDRs ("paper gold" as they were sometimes called), created to supplement national currencies and to create liquidity and thus help members manage their balance of payments (see Glossary for a definition of SDRs).

Nevertheless, the system proved to be too inflexible to survive, and it broke down in 1971 when the United States suspended its commitment to the IMF because the U.S. dollar had gotten seriously out of line with respect to the Japanese yen and the West German mark. Subsequently the U.S. dollar was devalued by 8 percent against the mark in August of 1971 and by 10 percent in February of 1973. These adjustments proved insufficient, so in late 1973 the U.S. dollar was allowed to seek its "natural" value with respect to other currencies on the open market—that is, it was allowed to "float" rather than be tied to gold. All other Western countries followed suit.

From that point on economic adjustments throughout the world were permitted to be reflected automatically through changes in the rate at which one country's currency could be exchanged for another. And change they did. Immediately after the major currencies of the world were allowed to float, the U.S. dollar continued its downtrend against the West German mark and the Japanese yen until late 1979, by which time it had depreciated another 50 percent. But then the dollar rebounded so that by late 1984 it was back to mid-1970 levels against the mark and the yen. By 1987 it was again back down to 1979 levels against the mark and the yen (see Table 9.3).

Table 9.3. Selected exchange rates in foreign currency per U.S. dollar, 1970–90

Year	United Kingdom	West Germany	Japan	Brazil	Korea	Singapore	Australia	Canada	Mexico	SDR	ECU
1970	0.4166	3.6600	360.00	0.0045	310.56	3.0612	0.8928	1.0476	12.500	1.0000	0.9783
1971	0.4109	3.4907	349.33	0.0053	347.15	3.0478	0.8826	1.0097	12.500	0.9970	0.9544
1972	0.4003	3.1886	303.17	0.0060	392.89	2.8092	0.8387	0.9899	12.500	0.9211	0.8914
1973	0.4081	2.6726	271.70	0.0061	398.32	2.4436	0.7041	1.0000	12.500	0.8388	0.8119
1974	0.4277	2.5877	292.08	0.0068	404.47	2.4369	0.6966	0.9780	12.500	0.8315	0.8384
1975	0.4520	2.4602	296.79	0.0081	404.00	2.3713	0.7638	1.0171	12.500	0.8236	0.8059
1976	0.5565	2.5180	296.55	0.0107	484.00	2.4708	0.8182	0.9860	15.430	0.8662	0.8945
1977	0.5732	2.3221	268.51	0.0141	484.00	2.4394	0.9018	1.0634	22.570	0.8565	0.8763
1978	0.5215	2.0086	210.44	0.0181	484.00	2.2740	0.8736	1.1406	22.770	0.7987	0.7847
1979	0.4721	1.8328	219.14	0.0269	484.00	2.1746	0.8946	1.1714	22.800	0.7740	0.7296
1980	0.4302	1.8176	226.74	0.0527	607.43	2.1412	0.8782	1.1692	22.950	0.7683	0.7182
1981	0.4976	2.2600	220.54	0.0931	681.03	2.1127	0.8702	1.1989	24.510	0.8481	0.8957
1982	0.5724	2.4265	249.08	0.1795	731.08	2.1400	0.9858	1.2337	56.400	0.9058	1.0207
1983	0.6597	2.5532	237.51	0.5770	775.75	2.1131	1.1100	1.2324	120.09	0.9355	1.1233
1984	0.7518	2.8459	237.52	1.8480	805.98	2.1331	1.1395	1.2950	167.83	0.9756	1.2674
1985	0.7792	2.9439	238.54	6.2000	870.02	2.2002	1.4318	1.3654	256.87	0.9849	1.3104
1986	0.6821	2.1714	168.52	13.656	881.45	2.1774	1.4959	1.3895	611.77	0.8524	1.0166
1987	0.6119	1.7973	144.64	39.229	822.57	2.1060	1.4281	1.3259	1378.00	0.7734	0.8663
1988	0.5614	1.7562	128.15	262.38	731.47	2.0124	1.2572	1.2307	2273.00	0.7441	0.8447
1989	0.6099	1.8800	137.96	2833.9	671.46	1.9503	1.2618	1.1840	2461.30	0.7802	0.9071
1990	0.5603	1.6157	144.79	68,300	707.76	1.8125	1.2799	1.1668	2812.61	0.7371	0.7855

Source: International Monetary Fund, *International Financial Statistics*, various monthly issues.
Note: The currency units for the countries listed are the pound, mark, yen, cruzado, won, Singapore dollar, Australian dollar, Canadian dollar, and peso, respectively. SDR refers to the Special Drawing Right and ECU to the basket of Western European currencies (see Glossary).

Exchange rate variations such as these are of critical importance to trade flows around the world. To see why this is so, note the essentially equivalent phenomenon of a change in the price ratio of two agricultural commodities such as corn and sorghum. If the price of sorghum rises relative to the price of corn, we should expect (all else equal) some resources to shift out of corn production and into sorghum production. Changes in currency exchange rates have a directly analogous effect. If the U.S. dollar rises in value relative to the German mark so that a German citizen can now obtain fewer U.S. dollars for 1 mark, we should expect Germans to buy fewer American goods and services than before the dollar appreciated. The German mark will not "go as far" in the United States as it did before. Thus under a dollar appreciation, American export industries will suffer because demand from Germany will have been choked off. However, German export industries will benefit because by similar reasoning, demand for German goods and services by Americans will increase!

We can see the full implications of such currency rate changes on trade and international commodity prices with the aid of Figure 9.5. In this diagram the effect of a dollar appreciation is shown as a downward and nonparallel shift in the "effective" excess demand curve (ED' in the middle panel). When the dollar appreciates against all other currencies, at any

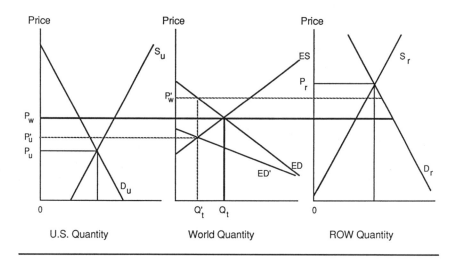

Figure 9.5. Economic impacts of an appreciation of the dollar on world markets

given U.S. dollar price of the good exported from the United States, importers will be willing to buy less because they will be required to give up more of their local currency per unit of the commodity purchased. At low (U.S.) prices, the difference in amount of local currency importers will be required to give up will be small. At high (U.S.) prices, however, the difference will be relatively high. Hence the excess demand curve will rotate rather than shift in a parallel fashion.[6]

The trade implications of an appreciation of the U.S. dollar against all other currencies is then clear: world price expressed in dollars will rise (to P'_w); price in the United States will fall (to P'_u); U.S. exports will decline (to Q'_t); consumption in the United States will rise and production in the United States will fall, thus making up for some of the reduction in exports; and the higher price of the commodity in the ROW (expressed in importers' currencies) will cause ROW consumption to fall and ROW production to rise, thus making up for the remainder of the reduction in exports from the United States.

But note that the economic consequences do not stop here. A dollar appreciation not only causes a reduction in world and hence also U.S. prices of farm products that we produce in sufficient quantities to export, but also results in reduced farm incomes and drives down the price of land used to produce these products. This was the situation by 1984. Furthermore, interest rates were quite high in the United States in 1984. The net result of all this was that farmers' ability to service their debt was reduced because incomes were low and interest payments increased. In addition, because land prices were driven down, the collateral used to support farmers' debt diminished in value before they even knew they were in trouble. It is understandable, then, that many farmers in the United States were in deep financial difficulties during 1984 and 1985.

The economic consequences to be expected from a depreciation of the U.S. dollar are a mirror image of those to be expected from an appreciation of the dollar. That is, a depreciation of the dollar against all other currencies should result in an increase in U.S. farm prices and thus production, a decrease in U.S. consumption, and an increase in U.S. exports. All this in turn should mean an increase in U.S. farm incomes and land prices. Thus the 1985 and subsequent turnaround in the value of the dollar against the major currencies of the world was welcomed by U.S. farmers. Unfortunately, the expected turnaround in farm incomes and land prices has been slow in coming, and U.S. policymakers have resorted to export subsidies to help push the process along.

One factor that must be kept in mind is that not all currency ratios change by a like amount. The U.S. dollar has fallen substantially against the German mark, the Japanese yen, and the British pound as well as the other major European currencies. The currencies of Brazil and Argentina have fallen even more than has the dollar against these same currencies. Thus to the extent possible, the Japanese and Europeans may be expected to buy grain from South America rather than North America because it appears to them even cheaper. The currencies of South Korea, Taiwan, Hong Kong, China, and Singapore, on the other hand, have remained relatively stable with respect to the U.S. dollar. These countries are collectively major importers of agricultural commodities. But since the dollar has also not changed much against the currencies of these countries, demand for U.S. products from these countries has not changed much.

Finally, there appears to be excess capacity for grain production around the world today—particularly in the United States and Western Europe. Thus when grain becomes cheaper overseas, these countries tend to encourage local production even at considerable taxpayer expense in order to protect local farmers. Hence the adjustments predicted by our diagram are not always permitted to happen.

Why Exchange Rates Vary. Currency markets can be viewed in much the same way as can a normal commodity market—the forces of supply and demand still apply. When the demands for different currencies change by different rates, for example, we should expect the rate at which one currency exchanges for another to change as well. Most analysts agree that three factors are important in bringing about exchange rate changes: inflation rates, interest rates, and expectations. If the general inflation rate in the United States falls below the general inflation rate in other countries, U.S. goods and services become cheaper relative to goods and services in other nations. As U.S. goods and services become cheaper to foreign buyers, demand for these goods and services increases, which means that demand for U.S. dollars increases. As demand for U.S. dollars increases, the price of the U.S. dollar relative to other currencies rises. In this situation the U.S. dollar is said to have appreciated. Similarly, if the U.S. interest rate rises relative to the interest rate in other countries, there will be an increase in demand for U.S. dollars as foreign citizens and businesses find investment opportunities in the United States more attractive than those in their own countries. Hence again the U.S. dollar will appreciate. It is clear then that government monetary and fiscal policy that affects inflation rates and interest rates has an important bearing on exchange rates.

The role of expectations is a bit more complex but nevertheless just as important. Assume that currency dealers *expect* the U.S. dollar to depreciate against the major currencies of the world. Their expectations might be based on anticipated actions of U.S. monetary authorities, anticipated actions of the U.S. Congress, or expected election results. In any event, based on these expectations currency dealers will unload U.S. dollars because they believe this currency will buy fewer U.S. goods and services in the future. If a sufficient quantity is unloaded, the U.S. dollar will in fact depreciate because there is now a surplus of dollars on the market.

Overvalued and Undervalued Dollar. The dollar is said to be *overvalued* when the dollar price of U.S. goods exceeds the dollar price of comparable foreign goods. Overvaluation will occur when the United States intervenes in the international money market by selling some of its holdings of foreign currencies in an effort to prop up the value of the dollar. The effect of this action is the same as if the dollar had appreciated: U.S. exports will diminish and world prices will fall. From the U.S. perspective, overvaluation is like an export tax levied on itself, which would tend to dampen export demand. If an importer's currency is overvalued, the effect is like an import subsidy. If there is no unemployment in the United States, an overvalued currency will be deflationary since demand for products of the export industries is choked off.

By contrast, the dollar is said to be *undervalued* when the dollar price of U.S. goods is less than the dollar price of comparable foreign goods. Undervaluation will occur when the United States deliberately sells dollars on the international money market in order to obtain more foreign currency or to prop up the value of foreign currency. The effect of this action is the same as if the dollar had depreciated: U.S. exports will increase as world prices increase. From the U.S. perspective, undervaluation is like an export subsidy paid for by the U.S. government. If an importer's currency is undervalued, the effect is like an import duty imposed by the importer. If there is no unemployment in the United States, an undervalued currency will be inflationary as the demand for products of export industries rises.

Our definitions of overvaluation and undervaluation follow from the concept known as purchasing-power parity. That is, when the dollar price of U.S. goods *equals* the dollar price of comparable foreign goods, purchasing power parity is said to be achieved. In reality, however, overvaluation and undervaluation are subjective ideas since *equilibrium* exchange rates are difficult, if not impossible, to determine.

Summary

Worldwide trade expansion is a goal whose achievement would have positive payoffs to all people and all nations. Adjustments required to achieve this goal may be painful to some in the short run. U.S. dairy and sugar farmers as a whole, for example, would not look with favor on free trade, because some would need to adjust out of these industries. Similarly consumers in the less developed countries would not appreciate the higher cost of some of their food items under free trade. In the longer term, however, compensating adjustments throughout these economies would lead to greater overall welfare.

Nevertheless, a variety of measures are used to distort trade flows in agricultural (as well as nonagricultural) products and thus prevent the magnitude of trade flows that would generate maximum world welfare. Some of these measures are overt or direct—import quotas, licenses, and duties, and export subsidies and taxes. Some are covert or indirect: domestic price supports and supply control measures. Some are aimed at a broader level but filter down to the agricultural sector: fiscal and monetary policies that affect interest rates, which, in turn, affect exchange rates, and direct international currency policy that dictates the degree to which exchange rates are permitted to change.

The fact that in the 1930s or even before many countries initiated agricultural policies of a highly protectionist nature is for the most part quite understandable. Dairy provides a prime example here. Milk and dairy products are highly perishable commodities and, given the technology of those days, could neither be stored for long periods of time nor be transported long distances. Furthermore, the production technology available did not permit farmers to milk large herds or specialize solely in milk production. Thus, in order to supply consumers' demand for not only processed dairy products but also fresh milk, many farmers throughout the rural landscape were involved in dairying. Most policymakers deemed it crucial to implement whatever policies were necessary to ensure the continued survival of these dairy farmers so that a steady and uninterrupted supply of fresh milk was available to consumers.

The fact that these protectionist policies have persisted through the 1980s, however, is less easily explained. Production, marketing, and transportation technology is vastly different today than it was fifty years ago. Milk can now be transported great distances without deterioration in quality sufficient to prevent its use as drinking milk at the destination.[7] Further, the technology exists to produce forms of drinking milk (e.g., sterilized milk or a product that can be reconstituted into drinking milk at the destination) that can be transported longer distances, even across international borders. Most of the more essential manufactured products can now be shipped satisfactorily anywhere in the world. Today's production technology in conjunction with new marketing and transportation technology has permitted specialization in dairy production, which, in turn, has meant fewer farmers producing milk. Herds with 1,000 or more cows are not uncommon in the United States today.

The study by Tyers and Anderson (1986) suggests that world welfare could be increased substantially by dismantling the protectionist policies in dairy and moving closer to a free-trade position. Further, there is currently a great deal of concern among the major developed countries about the mounting costs of protecting local dairy industries. Expenditures on dairy price supports alone in the EEC amounted to 5,442 million ECU (49.5 ECU per ton of milk produced) in 1984 and in the United States $1,598 million ($26 per ton of milk produced).[8] Expenditures associated with supporting the dairy industry in Austria, Finland, Norway, Sweden, and Switzerland have also been exceedingly high. There is no indication that this situation will improve until these countries take action to curb the production growth of recent years.

Why then do protectionist policies persist? Anderson and Hayami (1986, 37–38) offer the following explanation:

> Much of the explanation is to be found on the supply side of the political market for rural assistance policies. Direct price- or income-support schemes are simply much more costly politically per dollar of assistance to farmers. For a start they are more overt because they involve direct treasury payments, which are open to periodic budget scrutiny. Import controls, by contrast, do not involve government payouts, and may even add to treasury revenue through tariffs on imports; domestic consumers pay the subsidy in the form of high domestic prices for food. And ... consumers have an ever-decreasing incentive to oppose such price distortions as their incomes grow.
>
> Another major reason why agricultural protection policies are less

costly politically than more direct assistance policies is that the former can be argued to be necessary for reasons of food security. However, although protection is certainly the first-best policy instrument for boosting food self-sufficiency, food self-sufficiency is not synonymous with food security, especially when the raw materials for the crucial inputs (fertilizer, animal feedmixes) must be imported. . . .

A final reason why first-best adjustment assistance policies have not been adopted is the divergence between the real interests of the farm population and those of farm organizations. The cooperative organizations, for example, benefit from the sale of farm inputs, especially to small farms, and the marketing of their produce. (Large farms can often obtain better deals through private traders.) Cooperatives thus have an interest in ensuring that the agricultural output of small farms does not shrink. If the government were to adopt policies that assisted farm households to earn larger incomes off the farm, or strengthened the incentive for farms to increase in size, it is possible that the political and economic power of cooperatives would diminish. Cooperatives therefore do not lobby for better rural education (which would encourage more part-time and full-time off-farm employment) or more research (which might generate greater economies of scale and increase the incentives for small farms to amalgamate), even though such policies would benefit farm people more than do protectionist policies.

Notes

1. This section draws heavily on Houck and Pollak 1978.

2. This new initiative was provided with the passage of the Single European Act in 1985 (see Hallberg 1989).

3. See Baker, Hallberg, and Blandford (1989) for a review of several of these studies.

4. Just as import duties drive a wedge between domestic and world prices, so do costs of transporting the good from exporter to importer. Thus the economic consequences of transportation costs can be analyzed in the same way as import duties with a downward shifting of the ED curve. Per unit costs of transportation, however, would normally be expected to be the same regardless of the quantity exported. Hence in the case of transportation costs, the ED curve should shift in a parallel fashion.

5. The initial focus of the World Bank was on postwar reconstruction. Now it is concerned with economic development. The bank's chief function is to channel capital into investments in less developed countries. It supports projects to aid production and marketing of exportable commodities, to finance the importation of

capital equipment, and to train country specialists.

6. Technically, if all three price axes are denominated in the currency of the exporter, the supply and demand curves of the ROW should be shown as shifting downward (again in a nonparallel fashion) so as to produce the downward shift in the excess demand curve. I ignore this technicality here in the interests of keeping the diagram as uncomplicated as possible.

7. In the United States, fluid grade milk destined for the drinking-milk market regularly moves from the Upper Midwest to Florida. Technically there is no reason why it could not move from *any* two points in the United States, although legal restrictions or local sanitary regulations may present an obstacle.

8. One ECU (European currency unit, the monetary unit of account for the EEC) was equivalent to $1.05 in 1984. See the Glossary for a definition of the ECU.

Suggested Readings and References

Baker, Derek, Milton Hallberg, and David Blandford. 1989. "U.S. Agriculture under Multilateral and Unilateral Trade Liberalization—What the Models Say." Department of Agricultural Economics and Rural Sociology, Pennsylvania State University. A.E. & R.S. 200. January.

Hallberg, M. C. 1989. "Europe 1992: What Does It Mean for United States Agriculture?" *Farm Economics*. Pennsylvania State University Cooperative Extension Service. September/October.

Hallberg, M. C., and Woong-Je Cho. 1987. "The World Dairy Market." Department of Agricultural Economics and Rural Sociology, Pennsylvania State University. A.E. & R.S. 191. August.

Herendeen, James B. 1987. "Agriculture in an Unstable Economy." *Farm Economics*. Pennsylvania State University Cooperative Extension Service. September/October.

Houck, James P. 1986. *Elements of Agricultural Trade Policies*. New York: Macmillan.

Houck, James P. and Peter K. Pollak. 1978. "Basic Concepts of Trade." *Speaking of Trade: Its Effect on Agriculture*. University of Minnesota Agricultural Extension Service Special Report no. 72. November.

McCalla, Alex F. and Timothy E. Josling. 1985. *Agricultural Policies and World Markets*. New York: Macmillan.

Miller, Geoff. 1986. *The Political Economy of International Agricultural Policy Reform*. Canberra: Department of Primary Industry.

Sorenson, V. L. 1975. *International Trade Policy: Agriculture and Development*. East Lansing: Michigan State University Press.

World Bank. 1986. *World Development Report 1986*. International Bank for Reconstruction and Development. New York: Oxford University Press.

Advanced Readings

Anderson, Kym, and Yujiro Hayami. 1986. *The Political Economy of Agricultural Protection*. New York: Allen and Unwin.

Bale, M. D., and E. Lutz. 1981. "Price Distortions in Agriculture and Their Effects: An International Comparison." *American Journal of Agricultural Economics* 63 (1): 8–22.

Bale, M. D., and E. Lutz. 1979. "The Effects of Trade Intervention on International Price Instability." *American Journal of Agricultural Economics* 61 (3): 512–16.

Bredahl, M. E., W. H. Meyers, and K. J. Collins. 1979. "The Elasticity of Foreign Demand for U.S. Agricultural Products: The Importance of the Price Transmission Elasticity." *American Journal of Agricultural Economics* 61 (1): 58–63.

Kost, William E. 1976. "Effects of an Exchange Rate Change on Agricultural Trade." *Agricultural Economics Research* 28 (3): 99–106.

Schuh, G. Edward. 1974. "The Exchange Rate and U.S. Agriculture." *American Journal of Agricultural Economics* 56 (1): 1–13.

Tyers, Rodney, and Kym Anderson. 1986. "Distortions in World Food Markets: A Quantitative Assessment." Unpublished report.

U.S. Department of Agriculture. 1990. "World Agricultural Supply and Demand Estimates." Economic Research Service. WASDE-239. February 9.

10 Resource Use, Abuse, and Preservation

Through most of U.S. history, agricultural policy has focused on developing the nation's natural resources and ensuring the survival of an agrarian structure based on a large number of relatively small, family-oriented farm units. This focus has led to specific policy prescriptions. The land resource was viewed as the nation's source of material well-being. Rights to this resource were to be reserved for private individuals of the chosen agrarian structure rather than for monopolistic units more typical of the industrial sector. The private market was to be the chief instrument for effecting adjustments in land use and landownership dictated by changing economic conditions. That is, private interests should guide allocation decisions, not "public" interests as reflected through the regulatory process.

Conservation of the land resource has from time to time been on the policy agenda, but conservation has clearly been a much more important issue when incomes in the farm sector were depressed and when agricultural surpluses were excessive (see Chapter 2). Policy concerning development and use of the nation's water resource was initially, at least, governed by a commitment to the expansion of agriculture but again consistent with the adopted agrarian structure. Farmers were granted priority rights to this resource. Little concern was given to water conservation.

Science has generally been viewed as the path to human betterment. Few policymakers questioned the wisdom of the use of the technology science produced: pesticides that can be harmful to the environment, fertilizers that lead to pollution of underground and surface water, machines that compact the soil, management practices that lead to soil erosion, etc.

More recently, there have been increasing calls for new priorities in agricultural policy that would recognize conservation of natural resources, protection of the environment, and preservation of open spaces. Not only are questions being asked about the price and income policy enacted by

Congress for agriculture, but the market system itself is recognized as having shortcomings in the way it allocates resources. Some argue that the private market encourages resource exploitation rather than resource stewardship and that it favors individual interests over both the short-run (current generation) and long-run (future generation) interests of society as a whole.

In this chapter we will study these issues in some detail. Specifically, we will examine the ways in which private and social interests in agriculture diverge. We will also ask to what extent current policy exacerbates or ameliorates this situation. Finally, we will look at alternative ways of attempting to ensure that social as well as private interests are attended to.

Land Use

The United States has a land area of about 2.26 billion acres. In 1987, 20 percent of its land area was in cropland, including cropland used only for pasture; 26 percent was in other grassland pasture and range; 29 percent was forested, excluding forestland in parks and other special uses; and the remaining 25 percent was used for a variety of nonagricultural purposes (Table 10.1). Usage of the nation's land resource has been remarkably stable over the past one-third century. One frequently hears the lament that prime farmland is fast disappearing as cities grow and new roads and airports are built. In point of fact, land used for nonagricultural and nonforest purposes grew at a compound annual rate of 1.01 percent between 1949 and 1987—hardly an alarming rate.

In some states and regions the Bureau of Census records a rather large decline in cropland since 1950. This does not necessarily mean, though, that the nonagricultural sector has bid land away from agriculture. In Pennsylvania, for example, the available evidence suggests that significant amounts of farmland have simply been *abandoned* since 1950 and that this accounts for much, if not most, of the decline in cropland recorded by the Bureau of Census (see, e.g., Hallberg and Partenheimer 1991).

The distribution of cropland among the various regions has also remained quite stable since 1949 (Table 10.2). Cropland in the Northeast, Southeast, and southern Plains has declined slightly while that in the northern Plains, Corn Belt, Delta States, and Mountain States has increased slightly. Some of these changes were due to differences in land idled under government programs in the two years. In general, there has been little change in regional cropland use over some thirty years.

The percentage of total cropland harvested for each of the principal crops in 1949, 1982, and 1987 in the forty-eight coterminus states is

Table 10.1. Land usage in the United States, 1949, 1982, and 1987

Usage	1949	1982	1987
Total land area (million acres)	2,273	2,265	2,265
		(percentage)	
Agricultural use	63.0	54.0	53.4
Cropland	21.1	20.7	20.5
Harvested	15.5	15.3	12.9
Failed	0.4	0.2	0.3
Summer Fallow	1.1	1.4	1.4
Idle	1.0	0.9	3.0
Used only for pasture	3.1	2.9	2.9
Grassland pasture & range	27.8	26.3	26.1
Grazed forest land	14.1	7.0	6.8
Nonagricultural use	37.0	46.0	46.6
Forest land not grazed[a]	19.3	21.9	21.8
Transportation uses[b]	1.0	1.2	1.2
Farmsteads & rural roads	0.7	0.4	0.3
Recreation & wildlife	1.2	9.3	0.9
National defense	0.9	1.1	0.9
Urban & miscellaneous	13.9	12.1	12.5

Source: Frey and Hexem 1985; and Daugherty 1991.
[a]Excludes forestland in parks and other special uses.
[b]Highways, railroad rights-of-way, and airports.

Table 10.2. Distribution of cropland used for crops, 1949, 1981, and 1987

Region	1949	1981	1987
		(percentage)	
Northeast	4.4	3.5	3.6
Lake States	9.9	10.4	9.7
Corn Belt	20.2	22.6	22.3
Northern Plains	24.3	24.2	26.4
Appalachian	5.8	5.0	5.0
Southeast	5.2	3.8	3.1
Delta States	4.3	5.1	4.7
Southern Plains	11.6	9.8	8.8
Mountain States	9.0	9.8	10.7
Pacific States	5.4	5.7	5.6

Source: Frey and Hexem 1985; and Daugherty 1991.

presented in Table 10.3. With a few exceptions, one is again struck by the relative stability in cropland usage. The increase in soybeans is quite understandable in view of the sustained strong demand for soybeans worldwide over this thirty-three-year period. The rather substantial reduction in oat acreage is also understandable in view of the rapid substitution of machine power for horse and mule power on farms between 1949 and 1982.

The data just reviewed suggest that the supply of land for crops is very inelastic in the short run and only slightly less inelastic in the long run. In fact, using a rather simple aggregate model and data for the period 1970–90, I estimate the elasticity of cropland supply to be 0.05 in the short run (over a 1-year period) and 0.08 in the long run (over many years).[1] Apparently the opportunity cost of keeping cropland in crop production is very low, and the cost of cropping land not previously used for crops is relatively high. Thus, supply is very inelastic for both falling and rising crop prices. The supply of grassland for grazing can also be expected to be quite inelastic. Much of the land for grazing is controlled by the Bureau of Land Management, which limits the number of animals that can be grazed on publicly owned, but privately operated, lands in federal grazing districts.

Land Preservation

In recent years, various groups have become increasingly concerned about preserving prime agricultural land. This has been particularly true in the more densely populated areas of the country—areas surrounding large cities and in the northeastern states. Pennsylvania is the most recent state to enact a prime agricultural land preservation statute. In this case a "purchase of development rights" option was chosen.

Preservation of prime agricultural land is a goal some pursue on the basis of the belief that the nonagricultural sector is bidding away significant

Table 10.3. Cropland harvested for principal crops, 1949, 1982, and 1987

Crop	1949	1982	1987
		(percentage)	
Wheat	21.8	22.3	19.3
Soybeans	3.0	19.9	19.8
Rice	0.5	0.9	0.8
Peanuts	0.7	0.4	0.5
Corn	24.6	23.3	22.6
Sorghum	3.2	4.5	3.8
Oats	11.3	2.9	2.4
Barley	2.8	2.6	3.5
Hay	20.5	17.1	20.8
Cotton	7.9	2.8	3.5

Source: Frey and Hexem 1985; and Daugherty 1991.

quantities of prime agricultural land from farmers and that this process is jeopardizing our capacity to produce food. Although nonagricultural interests are clearly bidding land away from farmers in some areas (near large metropolitan areas), as we saw in the previous section of this chapter, this phenomenon is by no means significantly curtailing our ability to produce food.

More commonly, agricultural land preservation is motivated by a combination of several considerations: (1) there are significant local economic benefits associated with sustaining a viable agricultural industry, (2) there are social as well as economic benefits to be derived from maintaining an efficient, orderly, and fiscally sound urban development program that includes agriculture, (3) there are benefits to both rural and urban residents from maintaining open spaces and other environmental amenities that go along with preserving agricultural land, and (4) the present generation has an obligation to protect the land resource base for future generations. Those in favor of land preservation schemes argue that the private market, which only accounts for private interests, fails to allocate land in a manner that will ensure achieving all these objectives.

It is clear that open space and environmental amenities are collective goods (see Glossary) of which the private market will not ensure the optimal quantity. Furthermore, the private market does not always ensure that acceptable choices are made regarding land to be developed. Consider, for example, a parcel of land (parcel A) that will yield annual returns of $50 per acre in its best agricultural use and $100 per acre in its best nonagricultural use. If these returns can be expected into perpetuity and the relevant discount rate is 10 percent, this parcel is worth $500 per acre (i.e., 50/0.10) to farmers and $1,000 per acre (i.e., 100/0.10) to developers.

Now consider two other parcels of land, B and C. Parcel B will yield annual returns of $90 per acre in its best agricultural use and $150 per acre in its best nonagricultural use. Parcel C will yield annual returns of $60 per acre in its best agricultural use and $101 per acre in its best nonagricultural use. The variation in agricultural returns on these three parcels is due to productivity differences. The variation in nonagricultural returns on these three parcels may be due to a variety of factors including location of the parcel relative to an urban center, size of the tract relative to development needs, drainage characteristics of the land, water availability, etc.

In this hypothetical case and assuming no other complications exist, the competitive market would be expected to allocate parcel B for development purposes if there is sufficient demand for only one parcel, because the parcel B returns to nonagricultural usage are the greatest of the three parcels. Land preservation advocates would probably not want this to happen. They would argue, "Why let the best agricultural land go for development purposes when poorer agricultural land may do quite well (parcel *A*) or about as well (compare parcels *A* and *C*)?" An agricultural land planning council would likely seek to encourage the development of parcel *A* even though this is not consistent with the allocation of parcels that the competitive market would make.

Many schemes have been used by local authorities to influence the choice of land going into nonagricultural uses and thus to attempt to preserve prime agricultural land. Five of the more popular schemes are listed below (see Conklin 1980 for a more complete listing and assessment of these schemes).

Under the *preferential assessment* scheme, land in agricultural use is assessed for its *use* value rather than for its *market* value. The assumption is that use value of farmland is less than fair market value. The expectation, then, is that a lower property tax would encourage farmers to keep their land in agricultural uses even in the face of other incentives to convert the land to nonagricultural uses. The idea is to prevent property taxes from pushing land out of agricultural uses.

The *capital gains taxation* option provides for the taxing of capital gains on farmland so heavily that speculation and rapid turnover of land lose their attractiveness.

Zoning and agricultural districting identify specific land areas as unique and irreplaceable agricultural areas. Within these areas landowners' options are strictly limited to agricultural production.

Local governments may *purchase development rights.* Under this option, the farmer retains the farmland for agricultural purposes but gives up the right to develop it for a specified number of years or in perpetuity, in return for a one-time local government payment equal to the difference between the value of the land for development purposes and its value in agricultural uses. This option is currently in use in New York, New Jersey, Massachusetts, New Hampshire, Connecticut, Maryland, and Pennsylvania.

Under the *fee simple purchase and lease-back* program, land is acquired by the local government by eminent domain and then leased back to farmers for agricultural uses. This option is used in Canada but not in the United States.

All of these options have achieved some success in slowing down the development of agricultural land near developed areas. Thus, land preservation advocates are heartened. It would be hard to argue, however, that any of these policy options have been successful in *preventing* the conversion of agricultural land. It would be equally hard to argue from a food production or food security perspective that further conversion should be prevented. Certainly also there is little social justice in preventing farmers from selling out to developers who are willing to pay a price for land consistent with that land's current best use.

Finally, there is a distributive issue here. Assume a developer really prefers parcel A but the local development authorities prevent him or her from buying parcel A via a purchase of development rights scheme. The developer will likely go slightly further out of town (or, worse yet, to another region) and purchase parcel B. Thus, the land preservation scheme not only failed to prevent land conversion, it increased the demand for parcel B and thus redistributed the benefits of development from the owner of parcel A to the owner of parcel B.

Chemical Use

Costs and Benefits. Farmers have from the earliest of times attempted to boost crop yields by controlling weeds and plants that compete with the cultivated crop for moisture and nutrients, by controlling insects and plant diseases that retard yields, and by using fertilizers to supplement the plant nutrients available in the soil. Early pest-control techniques included burning of fields and rotation of crops from year to year, biological control such as the introduction of natural enemies of pests, and the use of the chemical arsenic. The introduction of DDT and other chlorinated hydrocarbons in the 1940s and 1950s and more recently organophosphates has provided the basis for the now widely used insecticides, herbicides, and fungicides. In the last twenty-five or so years, herbicide use has increased sixfold. Insecticide and fungicide use increased at a much lower rate until the mid-1970s and then declined as more toxic materials requiring lower application rates were developed and adopted (Antle and Capalbo 1986). Between 1960 and 1986 and on a per harvested acre basis, nitrogen use has quadrupled, phosphate use has nearly doubled, and potash use has nearly tripled (U.S. Department of Agriculture 1988).

It is generally agreed that use of pesticides[2] in most instances reduces crop losses significantly and causes gross farm income to increase more than do farm costs increase. Lichtenberg and Zilberman (1986) cite the widely quoted (average) returns of from $3 to $5 for every $1 spent by farmers on pesticides. Returns to the use of commercial fertilizers are no doubt at least as high. With average returns as large as these, it is clear why pesticides and commercial fertilizers have become key inputs in crop production.

The payoff from the use of pesticides and commercial fertilizers would appear at first blush to be positive to both farmers and consumers. Figure 10.1 helps us understand why this is true for consumers but not necessarily true for farmers. Use of these chemicals by all corn producers means that any given quantity of corn can now be produced at less cost than could the

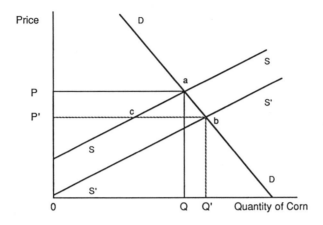

Figure 10.1. Benefits from pesticide and commercial fertilizer use

same quantity of corn without chemicals. Hence the supply curve which represents the sum of corn producers' marginal (private) costs shifts to the right, or from SS to S'S'. The result is that, in equilibrium, price falls from P to P' while output increases from Q to Q'.

The early-adopting farmers are quite eager to use the new chemicals. So long as they do so before corn price falls to P', they stand to gain handsomely because their costs per bushel of corn produced decline while their revenue per bushel of corn changes only slightly if at all. The early adopters' individual supply curves (marginal private cost curves) will shift to the right, but there are so few of them that when all farmers' individual supply curves are aggregated into the market supply curve, its position will have changed imperceptibly. As more and more farmers get the same idea, however, the aggregate supply curve will shift further and further to the right until it eventually reaches S'S'. When supply curve S'S' is realized, corn price will fall to P' so that excess profits from corn production are again eliminated. The process at work here is the "technological treadmill" I described in Chapter 3. Farmers cannot afford to get off the treadmill even though in the long run they may accomplish little by adopting the new technology. Thus, in the long run, farmers may be no better off than they were before the use of chemicals.

It is instructive to examine producer and consumer welfare after full adoption of the technology by estimating producer and consumer "surplus." Producers' gross revenue after all have adopted pesticides and commercial fertilizers is given by the area 0Q'bP'. Their variable costs of production are measured by the area under the supply curve and above the horizontal axis, or area $0Q'b$. Hence producers' aggregate returns above variable costs are given by the area $0bP'$. This area is termed producers' surplus since it represents net rent or returns above all *variable* costs of production. Thus, it can be seen that by using chemicals, corn farmers' "surplus" will be greater by the area $(SS'bc - PacP')$ than if chemicals are not used. This area may or may not be positive! With a highly inelastic demand curve, as is the case for most agricultural commodities, this area is likely to be negative. Thus, in terms of now standard welfare measures, farmers are likely worse off in the long run from adoption of pesticides and chemical fertilizers even though there are strong short-run incentives for them to use these chemicals to the fullest.

To show that consumers benefit from adoption of this cost-reducing technology, note first that the area under the demand curve (which may also be termed the marginal social benefit curve) and out to the equilibrium quantity Q' represents the aggregate benefit or utility (measured in money terms) consumers derive from consumption of quantity Q' of corn. Subtracting from this aggregate benefit or utility the total amount of money consumers spend on corn (i.e., $0Q'bP'$), we have as a measure of "consumer surplus" the area above the price line P' and to the left of the demand curve. A comparison of this surplus with and without chemical usage reveals that the gain to consumers from farmer adoption of chemicals is positive and equal to the area $PabP'$.

Clearly consumers and early-adopting farmers gain from farmer adoption and use of pesticides and commercial fertilizers. Upon aggregating consumers' and producers' surplus, we find that collectively society is better off at the new equilibrium defined by P' and Q' even though farmers as a group may be worse off. It is in this sense that taking advantage of a technology such as pesticides and chemical fertilizers is a means to progress in an economy—that is, adopting existing technology so as to move the supply curve out to the right, which causes consumer prices to fall.

Unfortunately, chemical usage involves costs (social costs) in addition to the private costs reflected in the producers' supply function. Continued use of pesticides, for example, leads to genetic resistance of plant pests to pesticides and to destruction of natural pest enemies and thus to partial destruction of the natural ecosystem. Furthermore, pesticides are a major contributor to non-point-source pollution (see Glossary) of both underground and surface water. Finally, pesticides are poisons and as such pose health risks to humans and animals. Fertilizer usage also imposes costs on society. Leaching and surface runoff of nitrogen and phosphate fertilizers increase the nitrate and phosphate content in surface and ground water. Thus, fertilizer use causes (1) a hazard to human health, (2) a hazard to fish life in lakes and streams, and (3) destruction of bacterial action that cleans out organic wastes in streams and lakes naturally.

The problem is these social costs are not reflected in the supply function as usually constructed (or, what is the same thing, the marginal *private* cost function), so the private market does not account for them. What is necessary here is the marginal *social* cost function that incorporates *all* costs associated with chemical use—private costs as well as the social costs described above. Figure 10.2 compares the equilibrium solution when the social costs are incorporated into the equilibrium solution and when they are not. Here MSC represents the marginal *social* cost function whereas MPC represents the marginal *private* cost function. The latter is the usual supply function and is identical to S'S' in Figure 10.1. The vertical distance between these two functions represents the social cost of chemical use at each level of corn output. The socially optimal equilibrium solution is given by the intersection of the market demand curve, DD, and the marginal *social* cost function, MSC. Clearly the socially optimal solution results in a higher equilibrium corn price as both private and social costs are taken into account. The socially optimal equilibrium level of corn produced and marketed will fall to Q'' and this reduced level of output will (it is hoped) cause a shift in the demand curve for chemicals so that chemical use will also decline. The equilibrium price will rise to P'' to reflect the social costs of pollution.

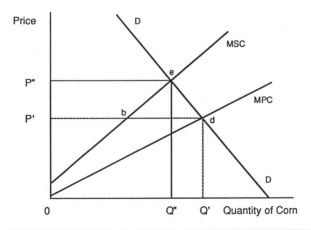

Figure 10.2. Optimal corn production with pesticide use

Policy Controls. The inquisitive student will ask, "How do we estimate these social costs?" A thorough discussion of this subject would take us far afield. Suffice it to say that there are ways of getting rough estimates of at least *some* of these costs. But to be quite truthful, we must admit that our analytical and/or measurement tools are not adequate to generate precise estimates. We do know, however, that these costs are significantly greater than zero. Thus, many argue, we must begin now to develop allocation methods that recognize their existence. Exclusive reliance on the private market with its sole concern for *private* interests will clearly lead to a nonsocially optimal level of pesticide and fertilizer usage and hence of agricultural commodities like corn.

The inquisitive student will also ask, "Even if we had precise estimates of these social costs so that we could determine socially optimal price and output levels, how do we encourage corn farmers to produce at the socially optimal level using the socially optimal quantity of pesticides and commercial fertilizers?"

In the main, the United States has opted for voluntary programs in efforts to achieve environmental objectives. To be successful, voluntary programs require farmers to be aware of the problem so that they will be induced to use practices consistent with maintaining long-run soil productivity and limiting off-farm damages. The Soil Conservation Service and the Cooperative Extension Service are principal agencies with the responsibility for so educating farmers. These agencies provide information to farmers about the consequences of various levels of pesticide and chemical applications under different climate and soil conditions on the assumption that such information can help alleviate some of the social costs indicated above.

Another solution to the problem is to impose a tax on producers equal to the social costs. This solution "internalizes" the social costs and forces producers to operate as if their supply function were MSC rather than MPC (Figure 10.2). This solution is consistent with the general view that polluters are responsible for the environmental damages that occur because they consider only private costs, so polluters should be expected to bear the costs. Alternatively, tax policies are consistent with the notion that polluters are responsible for achieving the socially optimal position and should pay whatever penalty is necessary to accomplish that end. Although this is the solution generally preferred by economists, it is not without its limitations.

A third solution is to limit to the socially optimal level the amount of pesticide and/or commercial fertilizer that can be applied. Such a limitation would be imposed by direct government regulation. This is the solution that producers prefer because less of their "surplus" is eroded away under regulatory control than under taxation. This is also the policy solution that, at least in a restricted way, has been adopted in the case of pesticides. Pesticide use is regulated by the Environmental Protection Agency (EPA). The EPA is authorized to register all pesticides to ensure labeling of contents and usage instructions, to certify applicators of hazardous pesticides, to cancel or restrict usage when the pesticide presents an "unreasonable" risk, to establish tolerance limits for pesticide residues on food, and to set maximum contaminant levels and recommended maximum contaminant levels for drinking-water pollutants. Under the Water Pollution Control Act of 1972 as amended in 1977, states are required to develop programs to reduce non-point-source pollution. In most cases, states have opted for regulatory programs coupled with technical assistance and cost-sharing efforts designed to reduce the incidence of pollution.

A fourth solution is to restrict the amount of corn that can be produced in the hopes that this will lead to the socially optimal use of chemicals. Voluntary acreage reduction programs and the conservation reserve program are two ways of implementing this solution. As we have already learned, however, when acreage is restricted, farmers are prone to substitute nonland inputs for the restricted land input. Both pesticides and fertilizers have in the past been used to increase yields on restricted crop acreages. Further, if the past is a good guide to the future, some fear that when food supplies again become short, emphasis will shift from voluntary acreage reduction programs and/or the conservation reserve program to planting fencerow to fencerow irrespective of the environmental consequences.

Curiously enough, this fourth solution implies that the responsibility for pollution control lies not with the polluter but rather with the pollutee. Acreage reduction programs as implemented currently are voluntary programs. In order to encourage producers to participate in these programs, an incentive in the form of a land rent must be offered. Thus, under programs of this type, taxpayers pay farmers (actually, we may say, "bribe" farmers) to reduce acres planted and thereby, it is hoped, reduce the amount of chemicals to the socially optimal level.[3]

Soil Erosion

Cost of Soil Erosion. Soil erosion is the movement of soil by water or wind from one place to another. It has been of major concern to federal, state, and local governments since the dust bowl years of the 1930s. Soil erosion is of concern because it is costly to farmers to the extent that it leads to crop productivity losses and/or increased production costs, and it imposes social costs that are not accounted for by farmers and ranchers. Crosson (1986, 40) describes these costs as follows:

The soil moved by erosion carries nutrients and organic matter from the eroded site and makes the crop rooting zone more shallow. This may lower the productivity of the eroded site by reducing its nutrient supply, by restricting its ability to infiltrate and hold water, and by making the soil more difficult to work by lessening tilth. Gullies, if allowed to grow, can interfere with the efficient use of farm machinery. Productivity in places of deposition may also be reduced if the deposited soil is less fertile than that on which it is deposited and deposited soil may smother seedlings just emerging after planting. Soil carried by high-velocity winds can cut through and destroy standing crops. Soil that reaches streams and is carried to lakes, reservoirs, harbors, and estuaries lessens recreation values, damages fish spawning grounds, increases costs of cleaning and transporting municipal water supplies, reduces capacity of reservoirs used for irrigation, electric power production, flood control, and water supply, and imposes costs of dredging to keep rivers and harbors clear for transportation.

Crosson's rough estimates of the "on-farm" costs of soil erosion are in the neighborhood of $500–$600 million—less than 1 percent of cash receipts from farm marketings. Clark, Haverkamp, and Chapman (1985) estimate the annual "off-farm" costs of erosion to be in the range of $4.2–$16.9 billion, with a "best-guess" of about $8.0 billion! A more recent study by Economic Research Service (U.S. Department of Agriculture 1987) estimates the off-farm costs to be in the range of $4.1–$15.0 billion. How precise these estimates are is no doubt a subject for a great deal of debate. Even if available estimates of both on-farm and off-farm costs are in error by 50 percent, though, it is clear that off-farm, or social, costs are currently much more significant than on-farm costs. To ignore these off-farm costs is to put a significant burden on society.

Extent of Soil Erosion in the United States. Soil erosion is considered "agronomically" tolerable when annual rates are at or below the level needed to maintain long-term soil productivity. This "tolerance" level is established by the Soil Conservation Service on the basis of physical characteristics of the soil, such as topsoil depth and structure and subsoil structure. For most soils this tolerance level, or T-value, is set at 5 tons per acre per year; but it can be much lower than 5 on thinner and less well drained soils. Adoption of this criterion as a guide to soil conservation policy assumes that the principal aim of such policy is to maintain soil productivity so that a given generation does not pass on to succeeding generations higher costs of food production that result from productivity losses.

Actual soil erosion rates vary greatly among and within regions as a result of differences in physical characteristics of the soil, climate of the area in which the land is located, crops grown on the land, crop management practices followed, and the extent of long-term investments made in soil conservation structures and practices. Bills and Heimlich (1984) provide the following taxonomy for classifying lands by their erosive characteristics:

> *Nonerosive land* is land for which physical characteristics are such that regardless of the crop or practices applied, it will not erode at rates greater than the tolerance level.
>
> *Moderately erosive land managed below tolerance* is land that has the potential for erosion rates greater than tolerance, but is managed via crop choice and practices in a manner that keeps erosion below the tolerance level.
>
> *Moderately erosive land managed above tolerance* is land on which erosion rates could be restricted to tolerance levels, but management practices cause erosion rates greater than the tolerance level.
>
> *Highly erosive land* is so inherently erosive that almost no management except permanent sod cover can keep the erosion rate at or below tolerance.

Based on this taxonomy, Bills and Heimlich found that land highly susceptible to sheet and rill erosion (see Glossary) is relatively more prevalent in the Northeast, Appalachian, and Corn Belt regions. Relatively large amounts of cropland moderately susceptible to sheet and rill erosion are managed above tolerance in the Appalachian, Corn Belt, Delta, and Southeast regions. Rates of wind erosion on cropland are highest in the northern and southern Plains and Pacific regions. More than two thirds of the cropland most susceptible to sheet and rill erosion was committed to row crop production during the 1977 crop year. More than three fourths of all cropland moderately susceptible to sheet and rill erosion eroding at more than the tolerance level was planted to corn, soybeans, and other row crops. Approximately half of the cropland moderately susceptible to sheet and rill erosion was committed to wheat, hay, or other close-grown crops and grasses.

The extent of soil erosion and the distribution of erosive lands by regions is shown in Table 10.4. In 1982 nearly 47 percent of the cropland in the United States was eroding at or above tolerance levels. Table 10.4 also shows the location of the "highly erodible" lands, which are the target of the conservation programs of the Food Security Act of 1985 as will be described in more detail below.

Since soil erosion can lead to a reduction in farm profitability, one might expect farmers to adopt recommended soil-conserving practices rather than use more erosive practices, particularly on those farms with moderately

Table 10.4. Soil erosiveness and soil erosion by regions in the Untied States, 1982

Region	Cropland	Cropland eroding at rates above tolerance	Soil loss per acre per year	Highly erodible cropland currently eroding at rates of:			
				<1T	1T–2T	2T–3T	>3T
	(*mil. acre*)	(*mil. acre*)	(*tons*)	(*mil. acre*)			
Northeast	17.3	6.0	3.9	2.1	1.3	0.9	3.2
Appalachia	22.6	9.9	5.8	2.2	1.2	1.0	5.3
Southeast	18.0	7.9	8.4	1.0	0.5	0.5	1.9
Delta	21.9	9.0	5.5	0.9	0.4	0.5	1.0
Corn Belt	92.4	50.4	8.0	5.0	2.5	3.0	12.6
Lake States	44.0	21.6	5.4	2.3	1.4	0.9	1.4
Northern Plains	93.6	38.9	5.7	11.2	7.2	4.0	2.6
Southern Plains	44.8	24.3	13.1	3.2	1.9	1.3	8.4
Mountain	43.3	22.4	8.8	4.1	2.4	1.7	11.5
Pacific	22.7	6.8	4.4	1.7	0.6	0.9	2.1
U.S. total	420.6	197.2	69.0	34.0	19.3	14.7	50.0

Source: U.S. Department of Agriculture 1986b.

[a]Highly erodible cropland is land with an inherent potential to erode at rates of 8 tons per acre per year based on physical characteristics of the soil and rainfall intensity and duration. *T* here means tolerance level of erosion, or 5 tons per acre per year.

erosive land. But, of course, this will only be true if the profitability of using soil-conserving practices in the current year equals or exceeds the profitability of using the more "conventional" practices and delaying the adoption of soil-conserving practices another year. As Walker (1982) points out, a farmer's decision will tend to lean toward delaying adoption of the soil conserving practices as the price of agricultural output increases and as the rate of discount applied to future income streams increases. Hence the decision to adopt is not only dependent on farm profitability, it is also dependent upon policies that affect prices of farm products and/or discount rates.

Farmer Use of Soil-Conserving Practices. Several studies have been conducted in an effort to determine whether or not farm profitability could be enhanced through adoption of soil-conserving practices at existing prices. Most of these studies have used what is known as the "universal soil loss equation" to estimate soil loss on a given type of soil.[4] Conservation practices are then designed so as to reduce this loss to the tolerance level or to a level recommended by the Soil Conservation Service. These practices would include some combination of crop rotation, cropping method such as minimum tillage and contouring, and use of terraces. Finally, profits are simulated under the various alternatives possible and compared with profits simulated under "conventional" practices.

As might be expected, results differ for different farms within the same region and across regions. White and Partenheimer (1980) studied twelve different types of dairy farms in Pennsylvania. They found that by using the Soil Conservation Service recommended soil-conserving practices, returns to fixed factors on two farms would have increased, returns to fixed factors on four farms would have decreased by 5 percent or less, and returns to fixed factors on six farms would have decreased from 7 to 30 percent compared with the "conventional" plan. Work done by Lee, Narayanan, and Swanson (1975) in Illinois also suggests that adoption of soil-conserving practices frequently "does not pay." In fact, this is the conclusion of most such studies. One exception is the study by Crowder et al. (1984), who used a more recent simulation model which accounted for soil losses as well as loss of pesticide, nitrogen, and phosphorus in surface waters and nitrates in soil percolate. The Crowder et al. study suggests that when these additional losses are considered, there may be more farm situations in which non-point-source pollution could be reduced without reducing farm income.

Farm Policy and Soil Conservation. As Osteen (1985) points out, soil erosion on a given farm will depend on (1) the physical characteristics of the land (soil type, slope, etc.), (2) climatic conditions in the area, (3) soil-conserving practices employed on the farm, and (4) crops grown and cultivation practices used for these crops. Farm programs do not affect the physical and climatic characteristics of the farm. Farm programs can, however, have an impact on decisions relating to long-term investment in soil-conserving practices, on crops grown, and on crop cultivation practices used.

The Agricultural Conservation Program was authorized under the Soil Conservation and Domestic Allotment Act of 1936. This program is administered by the Soil Conservation Service and the Agricultural Stabilization and Conservation Service (ASCS). It is designed to encourage farmers and ranchers to carry out resource-conserving practices. Specific objectives include (1) restoring and improving soil fertility, (2) minimizing erosion caused by wind and water, and (3) conserving resources and wildlife. Cost-sharing is offered for farm conservation measures that are considered necessary to meet the most urgently needed conservation problems and that would not otherwise be carried out to the extent needed to meet the public interest. To be eligible, the farmer must request cost-sharing before beginning the practice. In lieu of cash reimbursement, cost-sharing assistance may be in the form of partial payment by ASCS for the purchase price of materials and services needed by the farmer for carrying out approved practices. The farmer bears the balance of the cost (about 50 percent) and supplies the labor and management necessary to carry out the practices.

The types of soil-conserving practices supported under this program in 1986 are shown in Table 10.5 along with the acreage supported and number of agreements consummated with farmers. The Agricultural Conservation Program increases farmers' ability and incentive to invest in long-term land improvements of a soil-conserving nature. This program, then, goes at least partway toward encouraging farmers to manage their land resource in a way that is consistent with societal welfare.

Total ASCS assistance to farmers under this program amounted to $138.9 million in 1986. This is a sizable amount in absolute terms but amounts to only about 0.3 percent of the total cost of farm and related programs (see Table 5.5). Indeed, as a percentage of the total cost of farm and related programs, conservation expenditures have been much lower during the 1980s than at almost any time since 1950.

Table 10.5. *Agricultural conservation program agreements for soil conservation and pollution abatement, 1986*

	Annual and long-term agreements[a]
Diversions (1,000 acres)[b]	71
Conservation tillage (1,000 acres)	631
Permanent wildlife habitat (1,000 acres)[b]	20
Sediment retention, erosion, or water control structures (number)	12,655
Sod waterways (1,000 acres)[b]	176
Permanent cover establishment (1,000 acres)	542
Improvements to permanent cover (1,000 acres)	1,011
Tree planting (1,000 acres)	98
Timber stand improvement (1,000 acres)	27
Water impoundment reservoirs (number)	4,783
Stripcropping (1,000 acres)	92
Terrace systems (1,000 acres)[b]	342
Cropland protective cover (1,000 acres)	637
Grazing land protection (number)	6,664
Contour farming (1,000 acres)	21

Source: U.S. Department of Agriculture 1988a.

[a]Long-term agreements are for periods of from six to ten years.

[b]Total acres served.

Farmers Home Administration emergency loans, other subsidized credit programs, subsidized crop insurance, and disaster payments also increase farmers' ability and incentive to invest in long-term land improvements of a soil-conserving nature. Price support and related programs, on the other hand, tend to work at cross-purposes. Deficiency payments, acreage reduction and diversion payment programs, nonrecourse loans, and export enhancement programs are all designed to increase the economic attractiveness of the commodities to which they apply. As a result, farmers' economic incentives to adopt soil-conserving practices are reduced. Further, farmers' incentives to produce the program commodities relative to nonprogram commodities are enhanced. As Table 10.6 shows, the program commodities are, in general, those for which soil erosion is more severe!

Commodity Program Participation and Soil Erosion. Studies of farm program participation reveal the following (Reichelderfer 1985):

1. Program participants typically have larger farms than do nonparticipants.
2. Program participants typically have a larger portion of their land in crops and obtain a larger share of sales from crops than do nonparticipants.
3. Program participants typically crop their farms more intensively and include a higher proportion of commodity crops than do nonparticipants.
4. More producers in the Plains States, North Central States, and the South participate in the farm programs than in the Northwest, Southwest, and Northeast.

These findings suggest that larger farmers contribute more to erosion than smaller farmers and that those who participate in the farm programs contribute more to erosion than do those who do not participate. On the other hand, these findings also suggest that farm programs do not induce farmers in all of the more erosive areas of the country to participate and thereby plant the more erosive crops.

Reichelderfer (1985) also studied farm program participation rates in sixty-eight counties in eleven land resource areas of the United States that the Soil Conservation Service identifies as critical "cropland erosion control" or critical "water conservation and salinity control" areas. Reichelderfer's aim was to determine more definitively whether or not farm programs discourage the use of soil conservation practices on highly

Table 10.6. Erosion potential of selected agricultural commodities and eligibility for farm program benefits by commodities

Commodity	Erosive condition	Eligible to receive:		
		Deficiency, diversion, or disaster payments	Nonrecourse loans	Crop insurance
Cotton	Most	yes	yes	yes
Peanuts	Most	yes	yes	yes
Tobacco	Most	yes	yes	yes
Sugar beets	Most	yes	yes	yes
Soybeans	Most	yes	yes	yes
Potatoes	Most	no	no	no
Corn	Moderately	yes	yes	yes
Grain sorghum	Moderately	yes	yes	yes
Wheat	Less	yes	yes	yes
Barley	Less	yes	yes	yes
Oats	Less	yes	yes	yes
Rice	Less	yes	yes	yes
Grassland	Least	no	no	no
Hayland	Least	no	no	no
Range and pasture	Least	no	no	no
Timber	Least	no	no	no
Tree crops	Least	no	no	no
Vineyards	Least	no	no	no

Source: Reichelderfer 1985.

erodible land. Generalizing from her sample, she reaches four major conclusions. First, between 28 and 45 percent of the cropland in the United States eroding at rates of 5 tons per acre per year or more is operated by participants in the commodity programs and/or agricultural conservation programs. Second, roughly 24–38 percent of the cropland in the United States eroding at rates below 5 tons per acre per year is operated by participants of commodity programs and/or agricultural conservation programs. Third, from one quarter to one half of the soil erosion above tolerance levels in the United States could be reduced by changing commodity program options and/or conservation program options so as to be more responsive to soil erosion. Fourth, operators of one half to three quarters of the cropland eroding at rates above tolerance do not participate in these programs. Thus, we must look to other reasons to fix blame for soil erosion, and we must look to farm policies other than those now in place to encourage more soil-conserving practices.

Conservation Provisions of the Food Security Act of 1985. The conservation provisions authorized under the Food Security Act of 1985 and continued under the Food, Agriculture, Conservation, and Trade Act of 1990 are designed to reduce erosion from cropland and ensure that current commodity price support and production adjustment programs are more consistent with conservation programs than they were in the past. The "conservation compliance" provision of this act makes all persons already cultivating highly erodible cropland *ineligible* for price support loans, for Farmers Home Administration loans, for federal crop insurance, and for disaster payments unless they implement a conservation plan on their highly erodible land. The "sodbuster" provision of the act makes all producers who convert highly erodible land to cropland without implementing an approved conservation plan *ineligible* for the same price support benefits of farm programs.

Highly erodible land under this act includes all land with an erodibility index (EI) of 8 tons per acre per year or more. This index is a measure of the inherent potential of a soil to erode. It considers only length and steepness of slope, rainfall intensity and duration, and soil characteristics. Management variables are ignored since these can change as ownership or use of the land changes. Of the total cropland acres in 1982, 25.2 percent (118 million acres) met the EI criterion (Table 10.4). Of this total, 84 million was eroding at a rate greater than 5 tons per acre per year, about 65 million was eroding at a rate greater than 10 tons per acre per year, and 50 million was eroding at more than 15 tons per acre per year.

Conservation compliance will, of course, be costly to farmers. This has been estimated to range from $7 per acre in the southeastern and Delta States to $17 per acre in the Pacific region (U.S. Department of Agriculture 1986b). However, failure to comply will be in the vicinity of three times more costly! That is, a farmer of highly erodible land who fails to comply with the conservation provisions will be ineligible for farm program benefits estimated to range from $37 per acre in the Southeast to $62 per acre in the southern Plains (U.S. Department of Agriculture 1987). Thus, although compliance will be costly, under the 1985 act there are strong incentives to comply.

For those farmers farming the 49 million acres eroding at rates greater than 15 tons per acre per year, another option is available. This is the conservation reserve program. Under this program, farmers may submit bids to receive an annual rental fee for their highly erodible land in return for agreeing to remove the land from crop production and put it into soil-conserving uses as approved by the local conservation district. The contractual arrangement is to be for a period of ten to fifteen years.

Water Use

Throughout the seventeen states west of the ninety-fifth meridian, water is generally the most limiting factor in agricultural production. Natural rainfall will not support the levels of crop production potentially possible from the soils of the region. Areas in this region with the warmest climate and longest growing seasons are generally the most deficient in natural water supply.

Because development of agriculture in the West was a precondition for the economic development of the region, water for irrigating crops was a first necessity. Some early private efforts to provide irrigation water in the region were successful on a limited scale, but it was federal legislation around the turn of the twentieth century that provided irrigation water on a grand and sustainable scale. The Homestead Act of 1862 and the Desert Land Act of 1877 provided the foundation for agricultural settlement in much of the West and stimulated efforts at securing subsequent irrigation development. The Carey Act of 1894 was the first attempt at reclaiming lands by means of irrigation. This act granted to each state a tract of land not to exceed 1 million acres and directed each state to encourage the investment necessary to stimulate irrigation projects. The Land Reclamation Act of 1902, however, gave the most significant impetus to irrigation development. This act provided for direct federal funding of irrigation projects. Settlers were to receive land free, but they were to repay within ten years (subsequently changed to twenty years) their appropriate portion of the cost of the irrigation system put in place by the federal irrigation project and from which they drew water for irrigating the land. Their appropriate portion of the cost was generally quite minimal.

To ensure the priority of a family-farm policy and to limit the amount of subsidy that an individual landowner might lawfully receive from federally funded irrigation, the Land Reclamation Act provided "that no right to the use of water for land in private ownership shall be sold for a tract exceeding 160 acres to any one landowner, and no such sale shall be made to any landowner unless he is a bona fide resident upon such land, or occupant thereof residing in the neighborhood." This is the provision that touched off the so-called 160-acre limitation debate of a few years ago (see Paarlberg 1980) and resulted in changing the limit to 960 acres.

Unfortunately, the West's seemingly insatiable appetite for water is causing drawdown of the region's water supply faster than it is being replenished (otherwise known as "water mining"). This is probably nowhere more apparent than in some portions of the Ogallala aquifer, which lies under parts of South Dakota, Wyoming, Colorado, Kansas, Oklahoma, New Mexico, Texas, and most of Nebraska. Groundwater "mining" is also occurring from the Arizona, California, and Washington aquifers. Obviously if water mining continues, the West will eventually run out of existing supplies of water. Of more immediate concern, however, is the fact that water mining causes subsidence, saltwater intrusion, and higher pumping costs to mine what water remains.

Irrigated agriculture is the largest consumer of water in the West, consuming nearly 90 percent of all water transferred from surface or ground sources (Whittlesey 1986). The federal government subsidizes 80–90 percent of the irrigation projects, with the states picking up the rest. Water recipients are obligated for less than 20 percent of the investment in structures and conveyance systems (Paarlberg 1980). A widely quoted estimate is that farmers pay, at best, one third of what the water they get from irrigation costs to produce or would bring on the open market. Clearly incentives are tilted in the direction of overuse and by that sector—farming—that is the biggest user. One solution to this problem is to let the private market price water and in this way remove the favoritism toward agriculture.

Water Rights. In addition to the development of water to sustain the economic development of the West, a fundamental component of agricultural development in the West has been the establishment of a complex system of water rights favoring agriculture. Two basic doctrines govern the use of water in the West. The "riparian doctrine" accords to the owner of land a right to the use of water contiguous to that land. This doctrine was adopted as part of the common law of England and is the dominant water rights law adopted for all of the United States. The holder of a riparian right can make any "reasonable use" of water as long as he or she uses it on riparian land. Under a riparian system, there is no priority in right among users. That is, in case of shortage, all users must share in reducing consumption.

The riparian doctrine, however, proved inadequate for development of lands not adjacent to a watercourse. To cover this case, the doctrine of "prior appropriation" grew out of the need for flexibility in the legal framework to encourage investment in irrigation systems. Settlers had little incentive to develop irrigation systems if the legal framework allowed future upstream development priority in the use of the water supply created. Consequently, the western states developed the principle of "first in time, first in right," which accords a person who first develops and uses a water supply the right to continue this use in preference to those who come later.

The prior appropriation doctrine is unique in that rights are not necessarily tied to ownership of land or use of water on land bordering the water source. The right so acquired is a real property right. It can be sold, transferred, mortgaged, or bequeathed, but frequently only if attached to the land for which it was obtained. The right pertains to a specific use and is fixed to a specific location of use and amount of time. Prior appropriation facilitated western expansion by providing a framework for the parceling of water to a large number of irrigators.

Who Should Pay? The issues of who should have the right to water, for what purposes, and when are not the only ones. One other is whether consumers living east of the ninety-fifth meridian should be required to pay a lion's share of the huge federal subsidy for irrigation projects designed to help sustain the western economy in general and western agriculture in particular? Taxpayers in the West would probably say yes! Indeed a case can be made for answering in the affirmative in general. Consider Figure 10.3. Here SS represents the supply curve for western agricultural output without irrigation water and S'S' represents the supply curve for western agricultural output with irrigation water. Curve S'S' is drawn on the assumption that any given output of agricultural production is less costly under irrigation. With irrigation, equilibrium agricultural output from the West increases and equilibrium agricultural prices in the West fall. Using the concepts of consumers' surplus and producers' surplus, with which the student should now be familiar, we see that consumers gain the area P'bdeP'' while producers gain the area S'dbS but lose the area P'beP''. The gains to producers need not be positive. If the demand curve is perfectly elastic, gains to producers will clearly be positive. If the demand

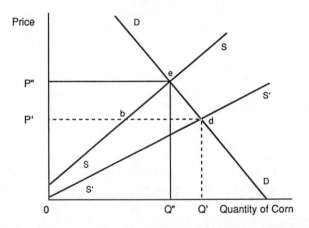

Figure 10.3. Distribution of benefits from irrigation

curve is highly inelastic, as is commonly the case for agricultural products, gains to producers may be negative. In either case, however, the irrigation water will be used since it is available on the farm, so output will increase and consumers will benefit.

Clearly consumers might well be expected to pay some of the costs of the irrigation projects, if not a lion's share of these costs. Indeed we might expect eastern consumers to pay some of these costs because not only has irrigation in the West provided products at lower cost, but it has provided some products (fruits, nuts, vegetables, melons) that may not have been available otherwise, and certainly not at such low cost.

There are, though, other tradeoffs that need to be considered here. In particular, we must ask whether producers in other regions of the country have been disadvantaged because of the high subsidies provided western agriculture. Clearly these irrigation projects have meant that some eastern agricultural products are now producible in the West. Land currently irrigated by these projects constitutes only about 2.5–3.0 percent of total U.S. cropland but includes a significant percentage of vegetable, fruit, nut, cotton, sugar beet, potato, barley, rice, and alfalfa acreage. Thus, some farmers in the East have undoubtedly been forced out of production of one or more of these crops into a second-best alternative. From society's standpoint, this is not necessarily objectionable. From the standpoint of the southern cotton producer or the Wisconsin milk producer, however, the issue may not be so clear.

Another way to examine the topic of irrigation is to ask whether irrigation has contributed to the oversupply of agricultural output. Martin (1979) examined this issue in considerable detail by estimating the incremental output of agricultural commodities from federal investment as a percentage of total U.S. production. His conclusion is that irrigation has not been the major culprit in oversupply of cereal crops and that the problem of oversupply would exist even without irrigation. Other issues are more important here—mechanization, hybrid seeds, substitution of nonland inputs for the land input, etc.

Perhaps a better way to look at this issue is to ask, What would be the impact on the rest of the country if western farmers were required to pay about three times as much for irrigation water as they now pay, as would be the case under a free-market policy with respect to water use? Clearly this would raise the cost of agricultural production and therefore the price of food. It would also limit the variety of food products available or, alternatively, make the cost of this variety extremely high to consumers because we would need to rely on foreign countries exclusively for such products as nuts, kiwi fruit, and so on. Production of some of the field crops would shift to eastern regions that rely on natural water—for example, corn, cotton, and soybeans. Land values in areas relying on natural water would increase while land values in the West would fall. Some nonagricultural industries would probably be encouraged in the irrigated areas of the country. In view of the incremental estimates of Martin (1979), however, we should not anticipate very large changes in any of the variables mentioned here as a result of this free-market water policy.

Summary

Agriculture has historically been viewed as a major instrument in the development of the United States. Accordingly, farmers have traditionally been given priority rights to the ownership or use of natural resources. The market was viewed as the chief instrument for effecting adjustments in land use and land ownership, and therefore, private interests were permitted to guide allocation decisions, not "public" interests as reflected through the regulatory process. Water, necessary for development of the West, was to be provided by the public sector since it was too expensive to develop on a private basis. Farmers were given access to this resource and at highly subsidized rates. Adoption of new technology by farmers was encouraged as a means of increasing social welfare and well-being.

In recent years several issues have been raised concerning natural resource use and exploitation that call for changing priorities, policies, and/or institutions. Not all of these issues can be easily resolved. Sound economic analyses, though, can provide useful insight and policy guidance.

One area of recent concern relates to the alleged disappearance of prime farmland and to the preservation of agricultural land in densely populated areas. We have seen that prime agricultural land has *not* been disappearing at significant rates in recent years as some would have us believe. In fact, cropland has *increased* slightly since 1949. Land used for nonagricultural and nonforest purposes has increased since 1949, but not at alarming rates. Various schemes for preserving agricultural land in the more populated areas of the country have been formulated and implemented. One might question, however, the wisdom of attempts to prevent land from moving into its best (nonagricultural) alternative use, particularly in view of the fact that there is already a surplus of agricultural production of the major crops and milk.

One of the significant technological developments since World War II has been the production of pesticides and commercial fertilizers, application of which has brought forth significant increases in crop yields. Use of these chemical-based inputs has clearly been spurred on by farmers' continuing search for profit-increasing technologies. Use of these inputs has also increased as farmers have sought to increase yields on fewer acres in response to the acreage restriction programs of farm policy designed to reduce farm surpluses. Unfortunately there are social costs associated with the use of these chemical-based inputs. Since farmers cannot be expected to consider these social costs when collectively deciding how much to apply, usage extends beyond the socially optimal level. Extramarket considerations must then be employed to encourage socially optimal application rates. Pesticide and/or fertilizer use taxes may be necessary to restrict usage. Quantity restrictions on farm production and/or input use are a viable alternative, and the option preferred by policymakers to date.

Soil erosion is another issue that deserves the attention of today's policymakers. Soil erosion is of concern because (1) it reduces on-farm productivity and hence farm profits, (2) it imposes off-farm costs on society by relocating topsoil where it is not wanted and by carrying nitrates and phosphates into underground and surface water, thereby imposing costs on society, and (3) it imposes costs on future generations in the form of reduced soil productivity. Farmers, driven by the desire to maximize profits, most frequently are not motivated to use soil-conserving practices because incremental costs associated with using such practices exceed incremental returns. Thus, as in the case of pesticides and chemicals, extramarket considerations must be brought to bear on allocation decisions and/or farming practices. Here again penalties may be used to encourage less pollution. This option is authorized under the Food Security Act of 1985 in the form of making farmers ineligible for program benefits who plant program commodities on highly erodible land. Another option is the conservation reserve program of the same act, which provides long-term rental payments to farmers who voluntarily take highly erodible land out of production. A third option, provided by the Agricultural Conservation Program administered by the Agricultural Stabilization and Conservation Service, is to provide technical assistance and cost-sharing to encourage the use of soil conservation practices. Policies designed to enhance farmers' incomes need to be reexamined in light of their propensity to encourage farmers to use soil-depleting practices.

A final issue examined in this chapter was water use in the West where natural rainfall is insufficient to sustain crop and some livestock production. Here the issues are even more complex. Subsidized irrigation projects encourage overuse of water, causing water mining of the underground aquifers, soil subsidence, and salinity as well as increased costs to mine additional water. Further, subsidized irrigation contributes at least in part to the generation of surplus agricultural commodities and brings about adjustments in other regions of the country where irrigated crops can also be grown without irrigation.

Notes

1. These estimates were derived by multiple regression using a distributed lag model in which acres planted in the United States was the dependent variable and acres idled and the index of prices received by farmers for crops were the independent variables. Preliminary analysis suggests that both short-run and long-run elasticities have become slightly larger over time, but not greater than those shown here. Although this analysis does not fully capture the returns from farming or reflect the alternative uses of cropland, the results appear to be reasonable and consistent with those of other studies. Further, given the different crops grown and the different intensities of nonagricultural demand for farmland in different regions of the country, one should expect the elasticity of cropland to differ in different regions. A more thorough analysis would thus examine regional differences in supply elasticities.

2. The term *pesticide* as used in the remainder of this section will include insecticides for direct control or eradication of insects; herbicides to selectively kill various weeds and unwanted plants; fungicides for the control of rusts, mildews, and molds; rodenticides for the control of animal pests (and sometimes birds); and a variety of other chemicals used to control ticks, mites, nematodes, snails, slugs, and certain unwanted bacteria.

3. Coase (1960) argues that it makes no difference from an efficiency point of view whether the polluter compensates the pollutee for damages imposed or the pollutee "bribes" the polluter to reduce pollution levels so long as the socially optimal position is reached. For a criticism of Coase's argument see Baumol and Oates 1975.

4. The universal soil loss equation provides an estimate of the soil loss from an acre of land based on the following factors: (1) rainfall for the type of farming area or crop region, (2) soil erodibility, (3) slope and length of slope, (4) crop management, including crop rotations and tillage method used, and (5) erosion control methods adopted.

Suggested Readings and References

Anderson, Terry L. 1983. *Water Crisis: Ending the Policy Drought.* Baltimore: John Hopkins University Press.

Antle, John M., and Susan M. Capalbo. 1986. "Pesticides and Public Policy: A Program for Research and Policy Analysis." in *Agriculture and the Environment*, edited by Tim T. Phipps, Pierre R. Crosson, and Kent A. Price. Washington, D.C.: Resources for the Future.

Bills, Nelson L., and Ralph Heimlich. 1984. "Assessing Erosion on U.S. Cropland: Land Management and Physical Features." U.S. Department of Agriculture. Economic Research Service. AER-513. July.

Clark II, E. C., J. A. Haverkamp, and W. Chapman. 1985. *Eroding Soils: The Off-Farm Impacts.* Washington, D.C.: Resources for the Future.

Conklin, Howard E., ed. 1980. "Preserving Agriculture in an Urban Area." Cornell

University Agricultural Experiment Station Bulletin no. 86.

Crosson, Pierre R. 1986. "Soil Erosion and Policy Issues." In *Agriculture and the Environment*. edited by Tim T. Phipps, Pierre R. Crosson, and Kent A. Price. Washington, D.C.: Resources for the Future.

Daugherty, Arthur B. 1991. "Major Uses of Land in the United States: 1987." U.S. Department of Agriculture. Economic Research Service. Agricultural Economic Report no. 643. January.

Eleveld, Bartelt, and Harold G. Halcros. 1982. "How Much Soil Conservation is Optimum for Society?" In *Soil Conservation Policies, Institutions and Incentives*, edited by Harold G. Halcros, Earl O. Heady, and Melvin L. Cotner. Ankeny, Iowa: Soil Conservation Society of America.

Frey, H. Thomas, and Roger W. Hexem. 1985. "Major Uses of Land in the United States: 1982." U.S. Department of Agriculture. Economic Research Service. Agricultural Economic Report no. 535. June.

Halcrow, Harold G. 1984. *Agricultural Policy Analysis*. New York: McGraw-Hill Book Co.

Hallberg, M. C., and E. J. Partenheimer. 1991. "Character and Recent Trends of the Pennsylvania Agricultural Economy." Pennsylvania State University, Agricultural Experiment Station Bulletin 869.

Lee, M. T., A. S. Narayanan, and E. R. Swanson. 1975. *Economic Analysis of Erosion and Sedimentation: Upper Embarras River Basin*. IIEQ Document no. 74-41. Chicago: Institute for Environmental Quality.

Paarlberg, Don. 1980. *Farm and Food Policy*. Lincoln: University of Nebraska Press.

Reichelderfer, Katherine H. 1985. "Do USDA Farm Program Participants Contribute to Soil Erosion?" U.S. Department of Agriculture. Economic Research Service. Agricultural Economic Report no. 532. April.

U.S. Department of Agriculture. 1986a. "An Economic Analysis of USDA Erosion Control Programs: A New Perspective." Economic Research Service. Agricultural Economic Report no. 560.

_____. 1986b. "What Will it Cost Farmers to Comply with Conservation Provisions?" *Agricultural Outlook*. Economic Research Service. October, pp. 27–30.

_____. 1987. "Soil Erosion: Dramatic in Places, But Not a Serious Threat to Productivity." *Agricultural Outlook*. Economic Research Service. April, pp. 28–30.

_____. 1988a. *Agricultural Statistics*. Washington, D.C.

_____. 1988b. "Economic Indicators of the Farm Sector: Production and Efficiency Statistics, 1986". Economic Research Service. EPIFS-6-5.

Advanced Readings

American Agricultural Economics Association. 1986. "Soil Erosion and Soil Conservation Policy in the United States." Soil Conservation Policy Task Force Occasional Paper no. 2. January.

Baumol, William E., and William E. Oates. 1975. *The Theory of Environmental Policy: Externalities, Public Outlays, and the Quality of Life*. Englewood Cliffs,

N.J.: Prentice-Hall.

Coase, R. H. 1960. "The Problem of Social Cost." *Journal of Law and Economics* 3 (4): 1–44.

Crowder, Bradley M., Donald J. Epp, Harry B. Pionke, C. Edwin Young, James G. Beierlein, and Earl J. Partenheimer. 1984. "The Effects on Farm Income of Constraining Soil and Plant Nutrient Losses." Pennsylvania State University, Agricultural Experiment Station Bulletin 850. May.

Lichtenberg, Erik, and David Zilberman. 1986. "Problems of Pesticide Regulation: Health and Environment versus Food and Fiber." In *Agriculture and the Environment*, edited by Tim T. Phipps, Pierre R. Crosson, and Kent A. Price. Washington, D.C.: Resources for the Future.

Martin, William E. "Returns to Public Irrigation Development and the Concomitant Costs of Commodity Programs." 1979. *American Journal of Agricultural Economics* 61 (5): 1107–14.

Osteen, Craig. 1985. "The Impacts of Farm Policies on Soil Erosion: A Problem Definition Paper." U.S. Department of Agriculture. Economic Research Service. Staff paper AGES841109. January.

Walker, David J. 1982. "A Damage Function to Evaluate Erosion Control Economics." *American Journal of Agricultural Economics* 64 (4): 690–98.

White, Gerald B., and Earl J. Partenheimer. 1980. "Economic Impact of Sedimentation-and-Erosion-Control Plans on Selected Commercial Dairy Farms in Pennsylvania." Pennsylvania State University. Agriculture Experiment Station Bulletin 830. January.

Whittlesey, Norman K., ed. 1986. *Energy and Water Management in Western Irrigated Agriculture*. Boulder: Westview Press.

11 Stabilization Policy

In Chapter 3, we saw that American farmers are subjected to quite high levels of crop price variability. Careful study of agriculture in colonial America and even in Europe since at least the Middle Ages will reveal that this has generally been the case in agriculture. Given the state of the art in the field of economics, we are unable to argue unambiguously that the levels of price variability indicated in Chapter 3 are excessive. Further, it will be recalled that several agricultural and nonagricultural sectors *not* protected by government price and income support programs have levels of price variability equally as high or even higher than in, say, wheat and the feed grains, and producers of these commodities still manage to cope.

Nevertheless, high levels of price variability in agriculture are unattractive to both producers and consumers. Large fluctuations in farm commodity prices that lead to large fluctuations in farm incomes can materially complicate and frustrate farmer and farm family decision making. Similarly, large fluctuations in farm prices that lead to large fluctuations in retail prices can make consumer decision making difficult. To the extent that high farm price variability is the result of fluctuations in available supplies, consumers are additionally concerned about unstable food supplies. Thus, there is broad support from both sides of the market for government action aimed at moderating price variability and/or compensating producers for associated losses and protecting consumers against unreliable supplies.

The purpose of this chapter is to examine this subject in some detail. First, we ask why price variability is a problem and whether we can have too little variability as well as too much. Next, we investigate alternative sources of price variability. Finally, we outline the policy options available for dealing with this problem and look at the advantages and disadvantages of each of these policy options.

We will confine our attention to year-to-year, or longer, price fluctuations. Seasonal fluctuations, while very substantial for some commodities, are more regular and thus more predictable. Since seasonal price fluctuations are more regular and more predictable, farmers and the marketing institutions available to farmers are reasonably able to cope with them. Most marketing orders, for example, have some direct or indirect method of dealing with seasonal fluctuations, as we saw in Chapter 8. It is thus the more unpredictable or unanticipated sources of variability (primarily variations in weather patterns) on which our attention will be focused.

How Much Price Instability Is Acceptable?

Prices play an important role in rationing scarce supplies and in guiding production decisions. For example, in competitive equilibrium, we normally expect the profit-maximizing farmer to use inputs in amounts that will result in the equality of the price of any single input with the value of marginal changes in use of that input. All else equal, the farmer is expected to use less of an input as the price of that input increases and to produce greater amounts of a commodity as the price of that commodity increases.

If prices are stable when demand is rising or falling, then adjustments in production needed to meet this change in demand may be prevented or at least delayed. If input prices are stable as new technological developments change the relative productivities of different inputs, then farmers would not be encouraged to alter their mix of inputs optimally. If prices change too slowly, there will be little incentive for private firms to store agricultural products and government will be forced to carry this burden. If prices do not rise in years of weather-reduced output or fall in years when aggregate supplies are high, then producer incomes cannot be stabilized without direct income support.

Some price flexibility, then, would appear to be in order. This quickly became clear during the wage and price freeze of the Nixon administration in 1973. It also become clear to Chinese economic planners who instituted economic reforms in 1979 that permitted prices to vary (somewhat, at least) according to supply and demand conditions. More recently, planners in the countries of Eastern Europe have discovered that years upon years of stable prices have led to a variety of distortions and to a general lack of economic progress.

On the other side of the coin, too much price variability can have equally or even more deleterious consequences. If prices fluctuate by more than enough to compensate for production shortfalls or surplus supplies, producer incomes may also become highly unstable—more so for some farms than others depending on their size, enterprise mix, degree of specialization, etc. Table 11.1 provides some perspective on the variability in net income per farm from farm sources since 1964. The two smallest sales categories are not shown here since net farm incomes on farms in these groups has been negative over most of this period.

Here we see that net income from farm sources was less volatile on the commercial-sized farms during the 1970s than were farm prices (compare the indexes of variability for prices shown in Table 3.10). Interestingly, net farm income from farm sources for all sizes of farms was even more volatile during the 1980s, when farm prices were generally declining. During this period, the farm income of the midsized and small farms was decidedly more volatile than that of the larger farms. During the mid- to late 1960s, on the other hand, net farm income on the smaller farms was *more* stable than on the larger farms.[1] The reader must again be cautioned, however, that because of inflation and technological changes, a farm selling X dollars' worth of product during the 1960s was of a quite different size than was one selling X dollars' worth of product in the 1980s.

Unfortunately, the data used in this analysis are much to aggregative for us to draw definitive conclusions about these apparently contradictory observations. One possible explanation, though, is that today there is much more enterprise diversity on the smaller farms than on the larger, commercial farms than was true in the 1960s, and this greater diversity results in greater income variability. This, however, is a subject for further research.

It has long been argued that excessive fluctuations in farm incomes can lead to capital rationing—that is, to below optimal or below desirable levels of capital investment (Johnson 1947).[2] Robinson (1975) contends, on the contrary, that during periods of high prices substantial investment (perhaps

Table 11.1. Indexes of variability of net income per farm from farm sources by sales class, 1964–88

Sales class	Index of variability		
	1964–72	1973–79	1980–88
$500,000 and over	na	2.6	6.2
$200,000–$499,999	na	7.3	8.5
$100,000–$199,999	13.4	8.5	19.1
$ 40,000–$99,999	9.9	8.2	73.4
$ 20,000–$39,999	6.6	20.6	109.4
$ 10,000–$19,999	5.5	60.8	505.1

Note: See Chapter 3 for a definition of the index of variability and for the method of calculation of this index. Net farm income per farm here excludes government payments to farmers.

na—not available.

even "overinvestment") occurs, whereas during periods of low prices investment may be delayed because of a shortage of funds. On the positive side, however, Robinson suggests that during periods of low prices, inefficiencies may be squeezed out. Robinson's argument is consistent with the general conclusion that in the long run, rather than capital being rationed, the sector may be even better off with price variability. Unfortunately both of these contentions are difficult to verify and few satisfactory efforts at doing so have been attempted. The weight of the evidence currently available, though, would seem to refute the idea that price instability in agriculture has led to capital rationing.

Excessive price and income fluctuations can, though, lead to complications in farmers' shorter-run production decisions. When prices are highly variable, farmers tend to incur extra costs in discovering or estimating what prices will be during the next production period. They may also incur extra costs in protecting themselves against such wide price fluctuations. This, of course, leads to higher than necessary production costs. Further, when prices fluctuate too widely, they can transmit erroneous signals, leading to overadjustments, which in turn lead to unstable production-consumption balances.

Unstable prices have parallel consequences for consumers. When food prices are higher than necessary, the consumer has more difficulty in deciding upon the appropriate allocations of her food budget and must expend more time, energy, and money in discovering what food prices are or will be when she would ordinarily want to buy. Furthermore, to the extent that farm price fluctuations are due to fluctuations in production or other elements of available supplies, consumers are concerned about shortages of food.

Sources of Price Instability

In Chapter 3 we saw that both the demand schedule for agricultural products and the supply schedule for agricultural output are highly inelastic (i.e., very steep). When a random shift in either of these two schedules occurs, the equilibrium price will likely change by a substantially greater percentage than does the equilibrium quantity.

Variations in weather which affect domestic crop yields (and even harvested acreages in some cases) make random shifts in the supply curve more likely than random shifts in the demand curve. Such shifts can be substantial. For example, average U.S. wheat yields for a year preceding and for the year during adverse weather in the Plains States and Southwest for selected years were as follows:

27.5 bu/acre in 1958 to 21.6 bu/acre in 1959
32.7 bu/acre in 1972 to 27.3 bu/acre in 1974
38.8 bu/acre in 1984 to 34.5 bu/acre in 1986
37.7 bu/acre in 1987 to 34.1 bu/acre in 1988

We should note here that average U.S. cotton yields were also significantly affected by these same adverse weather conditions except in 1987–88.

Similarly, average U.S. corn yields for a year preceding and for the year during adverse weather in the Corn Belt and South for selected years were as follows:

67.9 bu/acre in 1963 to 62.9 bu/acre in 1964
85.9 bu/acre in 1969 to 72.4 bu/acre in 1970
97.0 bu/acre in 1972 to 71.9 bu/acre in 1974
109.7 bu/acre in 1979 to 91.0 bu/acre in 1980
113.2 bu/acre in 1982 to 81.1 bu/acre in 1983
119.8 bu/acre in 1987 to 84.6 bu/acre in 1988

Here again average U.S. soybean yields were significantly reduced in the last four of these years.

But variations in domestic yields resulting from variations in weather are not the only reason for variations in prices of agricultural commodities. To see this it is useful to consider the basic identity between aggregate supply and aggregate demand for a given commodity. If all domestic *production* (P_t) is *consumed* domestically (C_t) in a given year (t), then we have as our basic identity

$$P_t \equiv C_t \tag{11.1}$$

and domestic production variations must of necessity be translated into domestic consumption variations. This, of course, is not a very realistic relation for any of the agricultural commodities. Production is not simultaneously consumed, nor is all production in a given year consumed in that same year. In general, there is storage activity as well. That is, there is *"carry-in"* (S_{t-1}) from the previous year that must be added to current production to find total availability for year t, and there is *"carry-out"* (S_t) for the current year that must be added to current consumption to find total use for year t. Thus, our basic identity now becomes

$$P_t + S_{t-1} \equiv C_t + S_t \tag{11.2}$$

which emphasizes that domestic production variations could be at least partially moderated by variations in carry-in and/or carry-out.

Reserve stocks of commodities are held for different purposes. First, some minimum level of stocks is necessary to keep the "pipelines" full. That is, even if production throughout the year were continuous and smooth, the product would take time moving through the marketing system from farms to processors and grain merchants. Thus, processors and grain merchants need to have ready access to some reserve supply in order to meet day-to-day, week-to-week, or month-to-month variations in demand. But production of agricultural commodities is not continuous and smooth throughout the year. Hence, producers and others have the opportunity to speculate by storing grain at harvest time with the hope that when the grain is finally sold they receive a price higher than the price at harvest plus storage costs. Thus, there is a "speculative" demand for commodity storage in addition to the demand associated with keeping the pipelines full.

We do not ordinarily expect private firms to develop a storage policy for the purpose of smoothing out or preventing price fluctuations to producers. This is generally assumed to be the responsibility of governments. As we know from previous chapters, for the major field crops and for dairy products there is not only private storage activity but also government storage activity. Thus, we need to disaggregate total carry-in and total carry-out into that portion attributable to the private sector (S_{t-1} and S_t) and that portion attributable to the government sector (G_{t-1} and G_t):

$$P_t + S_{t-1} + G_{t-1} \equiv C_t + S_t + G_t \tag{11.3}$$

Government sector storage may arise because (1) there is an excess of production over commercial use at government supported prices, (2) the government needs a stockpile of reserves to meet its donation program commitments, or (3) buffer stocks are needed for the operation of a market or price stabilization program.

Finally, in an open economy such as the United States, available supplies can be expanded by imports (Im_t) and current use includes exports (Ex_t). Thus, the final form of our basic identity is

$$P_t + S_{t-1} + G_{t-1} + Im_t \equiv C_t + S_t + G_t + Ex_t \tag{11.4}$$

The point to be observed here is that random shocks to any of the variables on the right side of the identity sign must be countered with adjustments in one or more variables on the left side. Clearly, once the new year (t) comes around, carry-in can no longer be adjusted. Further, when changes during year t become evident, it may be too late to change production. Reactions on the left of the identity sign to changes in any of the variables on the right will thus most likely be limited to imports.

Similarly, random shocks to any of the variables on the left side of the identity sign must be countered with adjustments in one or more variables on the right side. Abnormal weather causing yield reductions such as noted earlier will cause very large reductions in P_t on the right. The slack must be taken up through decreased consumption, decreased private sector or government sector carry-out, or decreased exports.

If P_t is abnormally small due to unfavorable weather and S_{t-1}, G_{t-1}, and Im_t are at normal levels, prices will likely be higher than normal so as to ration the smaller amount available among the various product users (domestic consumers, stockholders, and foreign buyers). If P_t is abnormally large due to favorable weather or if carry-in is abnormally large due, say, to favorable weather in a previous year, prices will likely be lower than normal so as to encourage greater use by domestic consumers, stockholders, and/or foreign buyers. High or low prices can similarly be expected to accompany abnormal increases or decreases in final product usage. Thus price fluctuations can be seen to result from abnormal levels of product availability or product demand.

Yield variations are certainly a common source of such price fluctuations, but they are not the only source. Certainly we have had changes in per capita consumption. Changes in per capita consumption, however, are generally not as disruptive as yield variabilities, because they tend to be gradual changes over a long period of time rather than random, unpredictable, year-to-year fluctuations. A more likely secondary source of price fluctuations is weather-induced variations in yields in *other countries* making imports to the United States either less available or more costly or both, or making the demand for exports more or less intensive, which would put upward or downward pressure on prices.

Consider a commodity for which the United States is a major exporter, such as wheat or soybeans. Assume also that there is a short crop or a large crop of this commodity in the country of a large trading partner of the United States such as the USSR. If this trading partner has its own storage policy with which to absorb the shock of its own production shortfall or surplus of wheat or oil crops, the impact of the shock will be absorbed internally and there will be little change in world market prices for wheat or soybeans and hence on wheat or soybean prices in the United States. If, on the other hand, there is no storage policy in the USSR, the USSR will tend to bid up the world price for wheat or soybeans during periods of production shortfall in the USSR or cause world prices for wheat or soybeans to fall during periods of production surpluses in the USSR as USSR demand for wheat or soybeans on the world market falls. In other words, if the USSR fails to put in place a buffer stock policy of its own, it will "export" price variability due to events in the USSR to the United

States. As the United States has become more interdependent with other nations through trade policies and capital flows and floating exchange rates in the last couple of decades, and since very few other countries have a buffer stock program of their own for grains, events almost anywhere in the world quickly get translated into price effects in the United States.

Policies for Dealing with Price Instability

Equation (11.4) now makes clear the limitations (and opportunities) for government reaction to price variability. Since private sector and government sector carry-in are determined prior to year t, S_{t-1} and G_{t-1} cannot be adjusted in time period t. In a closed economy (i.e., in an economy with $Im_t = Ex_t = 0$) any production variability will require adjustments in domestic consumption or carry-out—it cannot be willed away or reacted to in any other way such as by variable export restrictions, variable import duties, commodity trade agreements, or foreign food aid. In an open economy, on the other hand, an individual country can make use of trade and trade policies to stabilize consumption as domestic production fluctuates even though this option is not available to the world at large.

Trade Policy. A variety of trade policies are available and have been used in an effort to moderate the effects of fluctuations in supplies that would otherwise cause unstable prices. Most of these policies are only marginally successful. Most do not constitute a suitable long-range program for dealing with the problem. Further, as we saw in Chapter 8, for many their social costs probably outweigh their price stabilizing benefits. In this section I identify the more important of these policies.

IMPORT CONTROLS. By tightening or relaxing import controls (import duties, import quotas, or voluntary trade agreements covering imports) domestic supplies can be augmented or curtailed so as to reduce the amplitude of domestic price fluctuations. The cassava-soybean episode discussed in Chapter 9, for example, resulted in a voluntary agreement between the European Economic Community (EEC) and Thailand for restraint of Thai cassava coming into the EEC when supplies of feed grains in the EEC were high. In the mid-1960s, U.S. beef producers were successful in their bid to have quotas placed on beef imports to stem the decline in U.S. beef prices as domestic production remained high (see Chapter 9). In the mid-1970s, the United States relaxed its import quotas on dairy products and sugar in an effort to check the rise in domestic prices resulting from domestic production shortfalls. Clearly these options work best when residual world supplies are sufficiently large that the increased U.S. demand does not cause a significant rise in import (world) prices.

EXPORT POLICIES. On the export side, export taxes and subsidies, voluntary export restraint agreements, and export embargoes have all been used to discourage exports when domestic supplies are low and prices high. Many countries have used export taxes for this purpose, although the United States has generally avoided this option. The United States, however, has used trade embargoes for this purpose. The most celebrated trade embargo of recent vintage (the wheat embargo to the USSR) was not imposed to protect local supplies or curb U.S. wheat price inflation but rather for political reasons (see Chapter 2). The more general soybean trade embargo used in the early 1970s, however, was instituted in an effort to stem the rise in soybean prices as increased export demand and protein shortfalls around the world occurred. Unfortunately, trade embargoes tend not to be very effective because of the adverse international press attending their application, because other producing countries will supply the commodity that the nation imposing the embargo is attempting to withhold from the international market, and because the nation imposing the embargo tends to lose markets that are difficult to get back once the embargo is lifted.

Finally, export subsidies (or the removal of export subsidies) can be used when domestic supplies are abundant (or tight). As we saw in Chapter 8, the United States has frequently used export subsidies for this purpose even though this policy is officially applied to "make the United States more competitive" or to retaliate against "unfair trade practices" of other nations. The Food for Peace Program (Public Law 480) was initiated for the expressed purpose of disposing of farm surpluses, although as I argued before, it is no longer used primarily for this purpose.

TRADE AGREEMENTS. Bilateral and multilateral trade agreements are also used from time to time in an effort to buffer price fluctuations at home.[3] Bilateral agreements have generally not proved successful in reducing price variability. They are difficult to negotiate and impossible to enforce. Both parties to the agreement will be guided by local conditions and demands which always supersede the conditions of a two-country agreement. Multilateral or international commodity agreements, on the other hand, do have the force of multicountry sanctions against a country that defaults on the conditions of the agreement. Further, multilateral agreements are often negotiated under the auspices of the General Agreement on Tariffs and Trade (GATT), so they also have the backing of a worldwide organization to which all parties to the agreement belong.

Multilateral price stabilization agreements attempt to ensure importers access to specified minimum quantities in periods of world shortages and high prices and to protect exporters by guaranteeing a market for specified minimum quantities when supplies are large and prices low. Signatory exporting countries are obligated to deliver specified minimum quantities to importers at the maximum price only when the market price rises to that level or above. Signatory importing countries are obligated to purchase specified quantities when prices fall to or below the lower price boundary.

Multilateral agreements then serve to stabilize exports and imports. They do not prevent market prices from rising above or falling below the respective price band limits at which guaranteed purchases or sales take place. Importing countries are free to purchase more than the minimum quantities, but they will have to pay the prevailing market price to get these quantities. Similarly, exporters can sell more than the signatory nations have agreed to purchase, but they will have to accept the prevailing lower market price to dispose of these quantities. The impact of production or demand variations will thus be transmitted to prices in that portion of the market not covered by the agreement. If the latter market is small, there will be little consequence on aggregate market prices or quantities. If, however, the latter market is large, substantial price instability may occur despite the efforts of the signatory nations.

International commodity agreements have proven to be extremely difficult to negotiate even when there is consensus among exporting nations that something needs to be done to stabilize prices. One of the initial problems is to persuade all the major exporters and importers to participate. If one or more of the major suppliers or importers are excluded or opt out for political or other reasons, there is little prospect that the agreement will be successful in holding the line on prices. Obviously, control over a major proportion of total supplies and total use is essential to the success of any arrangement of this sort.

A number of international agreements have been established on an informal or formal basis over the years (see World Bank 1986). Many have failed. Five agricultural commodity agreements that have survived for relatively long periods of time at least in some form are those for wheat, sugar, coffee, cocoa, and olive oil (see Glossary for a description of these international agreements). In addition, an international agreement on dairy products was negotiated under the auspices of GATT in 1979. The coffee agreement has been most successful. The cocoa agreement has been almost totally unsuccessful, in part because the largest importer (the United States) and the largest exporter (Côte d'Ivoire) generally fail to agree on price limits. The sugar agreement also has been unsuccessful, largely because the major importer (the United States) does not participate in pricing decisions, and the agreement has had to deal with another large actor (the EEC) shifting from being a large importer of sugar to that of being a large exporter of sugar. Consequently the sugar arrangement now exists basically to collect and share data and to foster discussions.

A second major problem with these agreements is that once an agreement has been reached, all parties to the agreement do not abide by the agreed-upon rules, and there are few means to force them to do so. The dairy agreement is a case in point even though it was negotiated under GATT. In late 1984, the EEC began violating the major purpose of the agreement by selling butter and other dairy products at prices below the negotiated minimum export price in order to reduce the EEC's enormous stockpile of dairy products. In protest to this action the United States withdrew from the agreement in 1985, and thus, the agreement broke down.

The wheat agreement too has had only limited success. The initial agreement broke down when prices fell below the agreed-upon limit during the depression of the 1930s and export quotas were ignored when the major exporters failed to cut back on production. The 1949 agreement worked fairly well because economic conditions were fairly stable and generally compatible with the price range and quota arrangements established in the agreement. Subsequently, however, the agreements have been of little significance because the agreed-upon conditions have generally been incompatible with existing market conditions.

Most all such agreements have the expressed purpose of stabilizing prices. But if this aim is also confounded by a secondary (or even a primary) aim of raising the general level of producer prices, the arrangement is likely to fail for three reasons: (1) there will be strong incentives for supplier members to cheat by selling more than their quota or agreed-upon market share since there are few effective sanctions against cheating, (2) it will then be profitable for new suppliers to enter the industry and thus disrupt any attempts at keeping market supplies in check, and (3) there will be strong incentives for consumers to find cheaper substitute products. These sources of failure are perhaps most dramatically illustrated by the frustrations the OPEC oil cartel has faced since its inception.

In negotiating a stabilization agreement, the dominant issue is usually the price band that will trigger changes in quotas or the release and acquisition of stocks. Importing countries want to maintain a lower price band than do exporting countries. The lower price boundary is more critical to exporting nations while the upper boundary is more critical to importers. The wider the price band, the greater the incentive for private storage and the less need for intervention by a buffer-stock agency. The price band should be centered around the long-run equilibrium price so as to avoid the accumulation of excess stocks. In practice, the equilibrium price is extremely difficult to forecast. Funding for a buffer-stock type of agreement can become a major issue. One way to resolve this issue is to tax exports. But whether such a fund is financed by an export tax or an import fee will depend on who should shoulder more of the burden—exporters or importers. If demand and supply are equally elastic (or inelastic), the burden of the assessment will be shared equally by consumers and producers. But if supply and demand elasticities are unequal, the group with the more inelastic schedule will bear the largest share of the burden.

Storage or Reserve Policy. As was pointed out in the previous section, the private sector does have an interest in storing agricultural commodities. The private sector is *not*, however, motivated to hold stocks to protect consumers against future food shortages or to moderate farm price fluctuations due to variations in available supplies. Satisfying these objectives is the purview of government.

The data in Tables 11.2 and 11.3 show the relative importance of both private and government stocks for various agricultural commodities since 1950. It appears that pipeline levels of stocks for the grains and cotton were reached in the mid-1970s when surpluses were low and Commodity Credit Corporation (CCC) stocks were practically nonexistent. From the mid-1950s to the mid-1960s, on the other hand, grain and cotton surpluses were very troublesome and CCC stocks were quite large. When one nets out CCC stocks, it becomes clear that private stocks are much less variable from year to year than are CCC stocks. Further, Tables 11.2 and 11.3 make clear that for those commodities for which exports are a high percentage of total use (wheat, rice, corn, and cotton) stocks constitute a much higher percentage of use than commodities which the United States does not produce in sufficient quantities to be a major exporter (e.g., dairy products and beef). Even so CCC stocks of dairy products became quite large in the early 1980s as the dairy price support program encouraged more milk production than would clear the market at supported prices.

The concept of insuring farmers against losses due to fluctuating prices via government policy began to emerge during 1935 and 1936. Henry Wallace, President Franklin Roosevelt's secretary of agriculture, advocated use of the "ever-normal granary" by which surpluses would accumulate in government storage facilities during "the fat years" so as to be available for release on the open market during "the lean years." This plan was thus to serve to moderate price fluctuations brought about by variations in production and in this way provide farmers a measure of protection against widely fluctuating prices. As Wallace envisaged the ever-normal granary, it

Table 11.2. Ending total stocks of selected agricultural products in the United States as a percentage of total U.S. use, 1950–90

Year	Wheat	Rice	Corn	Cotton	Milk	Beef
1950	**46.6**	**11.7**	**25.8**	**15.2**	**4.1**	**1.5**
1951	28.4	4.2	16.9	18.7	3.2	2.5
1952	68.7	3.1	28.5	44.4	4.4	2.7
1953	116.9	16.0	33.7	78.2	9.0	2.0
1954	125.2	59.2	56.0	90.6	11.2	1.5
1955	**118.9**	**71.9**	**42.5**	**129.0**	**7.4**	**1.5**
1956	88.3	30.9	50.3	69.8	4.5	1.7
1957	96.7	40.4	49.1	62.2	5.2	0.9
1958	130.2	33.0	46.2	76.0	3.9	1.3
1959	125.0	21.0	50.2	45.5	3.4	1.5
1960	**120.6**	**17.7**	**54.8**	**47.2**	**4.4**	**1.2**
1961	107.7	9.0	41.7	55.1	7.9	1.3
1962	101.8	12.1	35.0	88.4	9.6	1.2
1963	69.7	10.7	39.9	84.4	7.7	1.7
1964	67.8	10.4	29.6	105.3	4.2	1.7
1965	**41.9**	**10.8**	**19.1**	**134.1**	**3.6**	**1.4**
1966	35.3	10.0	19.7	84.7	4.1	1.6
1967	45.3	7.4	25.9	48.1	6.9	1.4
1968	70.4	17.1	24.8	57.9	5.7	1.4
1969	71.9	18.1	20.9	52.7	4.5	1.7
1970	**54.4**	**22.4**	**14.8**	**34.5**	**5.0**	**1.6**
1971	67.4	12.2	21.7	27.6	4.3	1.7
1972	30.9	5.6	11.8	32.0	4.6	1.6
1973	17.3	8.7	8.2	27.8	4.5	2.1
1974	25.7	6.2	7.5	58.2	5.1	1.7
1975	**35.1**	**37.4**	**11.0**	**34.6**	**3.3**	**1.5**
1976	65.3	36.1	19.6	25.3	4.7	1.8
1977	59.4	24.4	23.1	44.5	7.0	1.3
1978	45.5	24.5	24.4	24.9	7.2	1.7
1979	41.8	18.6	26.7	19.0	7.0	1.6
1980	**43.1**	**10.6**	**19.1**	**22.3**	**10.2**	**1.4**
1981	44.3	32.5	36.4	55.8	14.1	1.1
1982	62.7	54.0	48.6	73.6	15.2	1.2
1983	55.1	37.5	15.0	22.4	16.3	1.3
1984	55.3	52.8	23.4	34.7	12.5	1.4
1985	**105.5**	**62.1**	**62.2**	**111.9**	**9.7**	**1.2**
1986	82.9	31.7	65.9	35.5	9.0	1.6
1987	47.0	20.6	54.9	40.8	5.2	1.5
1988	29.3	15.8	26.6	51.1	5.6	1.7
1989	24.1	16.5	16.6	18.3	5.8	1.4
1990	**34.9**	**15.0**	**17.8**	**13.4**	**6.1**	**1.7**

Source: U.S. Department of Agriculture, *Agricultural Statistics,* various annual issues.
Note: Total use here includes exports.

Table 11.3. Ending CCC stocks for selected agricultural products in the United States as a percentage of total U.S. use, 1950–90

Year	Wheat	Rice	Corn	Cotton	Milk
1950	**15.9**	**1.0**	**13.9**	**0.5**	**1.4**
1951	6.9	0.5	13.5	0.0	0.1
1952	29.6	0.0	10.4	1.9	0.2
1953	83.3	2.5	13.3	1.1	6.2
1954	111.4	40.9	23.4	13.6	8.6
1955	**99.6**	**56.8**	**27.6**	**52.8**	**4.5**
1956	70.9	19.5	34.9	29.6	1.6
1957	80.7	26.7	34.9	6.7	2.2
1958	102.3	19.9	35.4	8.5	0.8
1959	108.0	11.8	36.1	30.0	0.4
1960	**102.6**	**7.3**	**35.7**	**11.2**	**1.0**
1961	92.8	0.5	20.4	10.5	3.9
1962	86.1	2.9	14.6	31.9	6.2
1963	77.2	2.0	21.2	30.0	4.4
1964	58.9	1.4	13.4	49.4	0.8
1965	**40.3**	**0.8**	**5.6**	**77.8**	**0.4**
1966	20.6	0.3	3.3	46.8	0.0
1967	8.8	0.1	4.0	4.2	3.4
1968	7.8	6.7	6.6	0.2	2.6
1969	10.2	7.1	5.3	17.4	1.2
1970	**18.3**	**11.4**	**2.3**	**2.2**	**1.8**
1971	24.2	2.9	3.1	0.0	1.3
1972	18.4	0.2	1.3	0.0	1.7
1973	0.3	0.0	0.1	0.0	0.4
1974	0.1	0.0	0.0	0.0	0.3
1975	**0.0**	**19.5**	**0.0**	**0.0**	**0.1**
1976	0.0	16.7	0.0	0.0	0.3
1977	0.0	9.6	0.1	0.0	3.0
1978	2.4	6.4	1.4	0.0	3.5
1979	2.3	1.2	3.4	0.0	2.6
1980	**8.2**	**0.0**	**3.3**	**0.0**	**5.7**
1981	7.6	11.6	4.0	0.0	10.1
1982	7.9	16.8	15.8	3.7	11.7
1983	7.6	20.0	3.0	1.3	12.8
1984	7.3	36.1	3.2	1.0	8.8
1985	**19.3**	**35.0**	**8.4**	**9.1**	**6.6**
1986	27.4	5.4	19.5	0.5	6.3
1987	30.9	0.1	9.7	0.0	2.5
1988	11.8	0.0	5.5	0.4	3.0
1989	8.5	0.0	4.5	0.2	3.9
1990	**4.7**	**0.0**	**3.0**	**0.0**	**3.7**

Source: U.S. Department of Agriculture, *Agricultural Statistics,* various annual issues.
Note: Total use here includes exports.

would serve first and foremost as a device for stabilizing supplies and prices rather than as a device to protect or elevate farm incomes. Unfortunately, the plan was only loosely conceptualized, and no hard and fast rules for accumulating and releasing stocks were specified. This together with the general depressed conditions of the time made it perhaps inevitable that the ever-normal granary plan should be turned into a mechanism for supporting farm incomes in times of stress rather than maintained solely as a price stabilization mechanism. In any event, the ever-normal granary program almost immediately became a price support program implemented with the nonrecourse loan program.

Another attempt at instituting a price stabilization program came about with the Food and Agriculture Act of 1977. The world food crisis of the early 1970s precipitated much discussion focused on the creation of international grain reserves. Unfortunately no agreement could be reached on a worldwide plan. The 1977 act authorized a U.S. plan in the form of the farmer-owned reserve (FOR) program (see the Glossary) The FOR is an extension of the nonrecourse loan program in the sense that it covers a commodity loan for a period of up to three years. Farmers placing commodities in the FOR receive a loan rate higher than the regular nonrecourse loan rate. In return, participating farmers agree not to market the grain until the market price reaches a specified *release* price level. At the release price, farmers may, but are not required to, sell their FOR grain. Incentives in the form of interest penalties and storage charges are, however, provided for encouraging them to do so. If farmers do not remove their grain from the FOR, the secretary of agriculture originally had the authority to recall the loan or require payment of the loan at a specified *call* price. The 1981 act rescinded the idea of a call price and gave the secretary of agriculture authority to increase the interest rate when market price rose above the release price.

When supplies are abundant, commodities are accepted into the reserve at the entry price, privately held stocks are reduced, and a fall in market price is staved off. Once the reserve is in place, and if market supplies and use are at normal levels, market price will fluctuate between the regular nonrecourse loan rate and the FOR release price. When supplies are low relative to use, market price will rise to the release price, at which time at least some FOR stocks will be released.

As noted in Chapter 2, Salathe, Price, and Banker (1984) argue that the program has had a positive impact on commodity prices and farm incomes but has not significantly reduced price variability. Other studies also confirm that the FOR has enhanced wheat and corn prices but attribute a slight price-stabilizing impact to the program (see Meyers and Smyth 1988). In those cases where price variability was estimated to be reduced by the FOR, however, the reduction was generally very slight and perhaps not beyond the bounds of modeling and estimation error. It appears that as with the ever-normal granary plan, FOR has for the most part been transformed into a price-enhancing mechanism. Indeed, following the 1980 embargo of grain to the USSR, the Carter administration opened the FOR to as much grain as farmers wished to store and offered a higher FOR loan rate and storage payment as incentives for them to do so. The aim was to improve prices by taking large amounts of grain which could no longer be sold to the USSR off the market. Clearly this changed the FOR from a buffer-stock mechanism to a supply management/price-enhancing program.

WHO GAINS FROM STABILIZATION? Assume there is a bumper corn crop in a certain year due to exceptionally good weather. This phenomena can be depicted on our usual supply-demand diagram as a rightward shift of the aggregate supply curve. The market consequences of this situation can then be seen as shown in Figure 11.1. The supply curve shifts from SS (its "normal weather" position) to S'S', and equilibrium price falls from P_e to P' while equilibrium quantity produced and marketed increases from Q_e to Q'. Unfortunately, this is to farmers' disadvantage since as we saw in Chapter 3, with an inelastic demand curve, gross farm income will fall.

Assume, then, that the government removes from the market and stores temporarily in warehouses the excess above "normal" corn quantity (i.e., Q' − Q_e) so as to drive market price up to P_e instead of the lower level, P'.

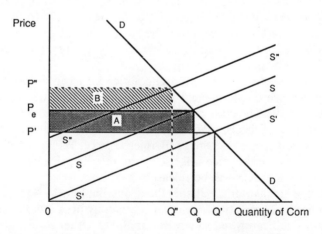

Figure 11.1 Economic consequences of price stabilization in the corn market

Thus farmers will receive P_e rather than P' for all the corn they produce (i.e., for Q' corn).

Now assume some years later a severe drought occurs, causing such a poor corn harvest that the reduction from normal equilibrium output is equal to the increase in production over normal equilibrium output during the bumper crop year. In this case, the aggregate supply curve for corn shifts to the left to position $S''S''$, and equilibrium price rises to P'' and equilibrium quantity falls to Q''. Corn farmers would complain about the weather but would nevertheless be better off because with an inelastic demand curve their aggregate gross income would be higher with the smaller-than-normal corn crop and higher-than-normal price.

Regardless of the drought's impact on farmers' incomes, though, the government agreed some years ago to implement a buffer-stock scheme to smooth out corn prices, and indeed it already has corn stored in public warehouses to be used to this end. Thus the government will release (Q_e − Q'') corn stocks onto the open market, driving market price down to P_e so that farmers receive this lower price for their corn instead of price P''.

It has long been known that under conditions quite typical in agriculture, producers gain and consumers lose from this type of price stabilization scheme (see Massell 1969).[4] This can be seen from Figure 11.1. During the bumper crop year, when the government takes part of the crop off the market thus driving market price up to P_e, producers' surplus increases by the shaded area labeled A. During the year of the drought, when the government releases stocks to be sold on the open market along with the corn actually produced in that year to drive market price down to P_e, producers' surplus declines by the shaded area labeled B. It is clear from this diagram that area A is greater than area B. Thus assuming droughts and bumper crops are equally likely over the long term, farmers will clearly benefit from a buffer-stock scheme such as that outlined here.

The student should, with the aid of Figure 11.1, also be able to show that by tallying up the gains and losses in consumers' surplus in the bumper crop and drought years, consumers will lose by this type of buffer-stock scheme. As Massell points out, however, the gains to producers exceed the losses to consumers, so that society as a whole is better off with a buffer-stock scheme.

The size of the gains and losses here will depend on the elasticity of the demand curve both when stocks are released and when stocks are accumulated. For example, consumers gain and producers lose from a buffer-stock scheme that operates to withhold supplies in low-price years and to release stocks in high-price years if demand is inelastic at high prices and elastic at low prices. This is probably not characteristic of the demand curve for most agricultural products, but it illustrates the importance of knowing something about the actual shape of the demand curve. Another consideration is, of course, the cost of storage. This will depend, among other things, on the length of time necessary to store the product. Most likely a perpetual storage operation will be required. Finally, a full assessment of the issue of who gains and who loses from a buffer-stock scheme must consider participants in industries that are affected by corn production and/or corn consumption. In general, we can conclude that farmers are likely to gain from a price stabilization scheme. It may help some farmers survive periods of low prices even if it does not lead to an increase in total farm revenue in the longer term.

MANAGEMENT RULES FOR BUFFER STOCKS. The first decision to be made in designing a buffer-stock policy is whether a price rule or a quantity rule for acquisition and release of stocks is to be adopted. To the extent that the United States has had or currently has a price stabilization program, pricing rules have been adopted for governing stock acquisition and release decisions and these rules have clearly not served to stabilize farm prices (see Table 3.6). As I argued before, however, price stabilization has not been a primary objective of U.S. farm policy despite some rhetoric to the contrary. Halcrow (1984) argues that during the 1980s, stocks would have been lower had we had a quantity rule governing stock acquisition and release. He further suggests that a stabilization program based on a pricing rule prevents prices from moving soon enough to allow expansion of exports in good crop years. Clearly a quantity rule would permit market prices to guide production within the stocking rules in addition to satisfying a reasonable social welfare criterion. Furthermore, a quantity rule would tend to divert attention away from concentrating on what constitutes a politically acceptable price band. For these reasons it probably makes more sense to adopt a quantity rule. Unfortunately quantity rules cannot guarantee that prices will be kept within politically acceptable limits. This is no doubt why an effective stabilization policy has generally not been sustained for long periods of time in the United States.

Nicholls (1945) argued for adoption of a program under which Congress would designate an operating range for a buffer-stock program in terms of specific minimum and maximum carryover, adopt a five-year moving average of production as the criterion for storage within this range, and give specific percentages of the excess of actual production above or below this average by which ending total stocks would be increased or decreased. Several researchers have since constructed rather sophisticated models from which more precise decision rules can be derived for the operation of an optimal storage policy (see, e.g., Gustafson 1958). Obviously the rules so derived are dependent upon the objectives to be optimized, and these must be clearly and unambiguously stated and continuously adhered to by policymakers before further progress can be made.

Crop Insurance. All-Risk Crop Insurance offered through the Federal Crop Insurance Corporation (FCIC) was initiated during the New Deal era and has generally been available to U.S. farmers ever since in most counties in which a significant amount of agricultural production takes place. Coverage under the program is available for over thirty crops and has paid benefits on a wide variety of losses but most frequently drought, floods or excessive moisture, frost, hail, wind, and insects.

Farmers who want to purchase All-Risk Crop Insurance must sign up for the program early in the annual cropping cycle. The contract specifies dates by which the farmer must complete various agricultural practices and offers three price and three yield elections from which to choose. The yield choices available are based upon the established yield for the class of land the farmer wishes to insure. The farmer can insure at 50 percent, 65 percent, or 75 percent of this established yield. Price elections based on expected market conditions determine the rate at which the farmer will be compensated if a loss is suffered.

To see how the program works, consider a corn producer whose established yield is 150 bushels per acre on 500 acres of corn land insured at the 75 percent level with a $2.50 per bushel price election. Assume a drought occurs and that only 90 bushels per acre are realized. This farmer will receive $28,125 compensation from the FCIC: [(0.75 × 150 bu) − 90 bu] × $2.50 × 500 acres. The farmer will not recoup the entire loss of $75,000 (the difference in expected gross income at $2.50 per bushel from a 150 bushel/acre crop versus a 90 bushel/acre crop), but the compensation of $28,125 will certainly help in weathering a serious cash flow problem.

In return for this coverage, the farmer pays a cash premium based on the expected indemnity for the class of land being insured. The cost of the premium is subsidized by the Federal Deposit Insurance Corporation, which pays all administrative and operating costs of the program. The subsidy amounts to about 25 percent of the total cost of operating the program (U.S. Department of Agriculture 1989). On the above policy, for example, a Corn Belt farmer may be expected to pay a premium of $3–$5 per acre—a relatively small price to pay for a potentially large payout.

Revenue Insurance. Although the United States has never had a farm revenue insurance program, it has been offered as an alternative at various times, and Congress mandated in the Agriculture and Food Act of 1981 that such an alternative be studied. Since such a program has never operated in this country, we must guess how it might operate. A good guess is that it would operate much like the current All-Risk Crop Insurance program, with participants being eligible for payments whenever gross revenue per acre fell below a guaranteed (insured) level of return. The revenue guarantee would most likely be based on some percentage of the most recent (say, five-year) average of gross returns. Premiums would be based on estimated losses, although, of course, these would be very difficult to forecast on an actuarial basis.

If the program were financed entirely or primarily by premiums paid by participating farmers, the average level of returns would not be increased but farmers would be saved from bankruptcy during a period of low prices and/or crop disasters. Premiums paid in years of good yields and/or high prices would be equivalent to forced savings—savings which would be returned to farmers in years of low returns.

Hedging and the Options Markets. The futures market can be used to speculate on price (indeed, a futures market would not exist if there were no speculators), but it can also be used by a seller to fix a price for a commodity that is to be purchased or sold later. Thus, a futures contract can be used to transfer risk of price fluctuations from the risk-averse farmer to the risk-loving speculator. A corn farmer can sell corn futures when he plants his corn. Later, when the corn is harvested, he can sell the corn and simultaneously buy back the futures. The whole process, which is called hedging, is equivalent to a forward sale in that the price that the farmer will receive is determined at the time the crop is planted.

Use of the futures market is an option available to most farmers for insuring against price changes, but few farmers take advantage of this option. There is a lack of understanding of how these markets operate. Not all farm commodities are covered by futures or options markets.[5] Hedging provides protection against adverse price movements for only a limited period of time, usually twelve months or less. Hedging reduces, but does not eliminate, risk. If a farmer sells forward 100,000 bushels of corn but the crop fails completely or partially, the farmer may have to buy on the cash market at high prices to meet the commitments of the futures contract.[6]

Many farmers do not wish to lock in a fixed price, since that precludes potential gains *as well as* potential losses. Instead, farmers may like to insure against extremely low prices. This insurance can be accomplished by trading in *options* on futures contracts. Options are available on sugar, cotton, soybeans, corn, hogs, and cattle. A farmer can insure against low prices by purchasing a "put" option to sell at a specified "strike" price. If the market price falls below the "strike" price, he exercises the option. If the market price rises above the "strike" price, he loses what he paid for the option but sells his crop for a higher cash price.

Options markets are somewhat more complex than futures markets. A seller (buyer) of a commodity purchases a put (call) option conveying the right, but not the obligation, to sell (buy) a futures contract at a given value on a specified date. In return for this right, a premium is paid to the party writing the contract. The size of the premium depends upon the length of time the protection is offered, the past volatility of the market, the level of protection sought, and expectations about future price movements. The longer the period covered, the more variable the market has been or is expected to be, or the higher the level of protection, the greater the premium demanded. While options offer somewhat longer protection than do futures contracts, they still only offer protection for up to eighteen months. Further, the size of the contracts is quite often out of reach of many farmers. Finally, futures contracts and options can ensure that a price that is low relative to recent historical standards can be locked in as a price floor, but they cannot provide relief from the low price itself.

Summary

Although it is unclear how much price instability is too much, high levels of price instability are a concern to both producers and consumers. High levels of annual volatility in producer prices and thus retail prices can lead to misallocation of resources and excessive costs in both production and consumption. To the extent that wide variations in supplies also at times accompany price volatility, consumers are additionally concerned about food shortages. Thus, there is wide support for policy aimed at price stabilization in agriculture.

Both the supply schedule for farm output and the domestic demand schedule for farm products are highly inelastic, at least in the short run. Consequently random shifts in either of these two schedules can be expected to produce relatively greater price changes than quantity changes. Variations in demand from year to year do occur, but they are likely more smooth and predictable than are variations in domestic supply. Supply of most agricultural commodities, on the other hand, is heavily influenced by such natural events as droughts, floods, early frosts, and disease. Thus, the most obvious source of price instability in agriculture is variations in domestic supply.

But changing weather patterns are not unique to the United States. Thus, when crops are poor in the countries of our trading partners, demand for U.S. (as well as other countries') exports increases, which drives world market prices and thus U.S. prices upward. When crops are good in the countries of our trading partners, the reverse takes place. Since few of our trading partners have a buffer-stock program of their own and since the United States is so dependent upon exports for disposal of such a high portion of its food grains, feed grains, and soybeans, another major source of price instability in the United States is from production variations in the rest of the world.

Two sets of policies are available and have been used in efforts to moderate price instability—domestic buffer-stock programs and trade policies. The latter consist of various import and export controls, including export embargoes, as well as bilateral trade agreements and international commodity agreements. The former are more effective because control is maintained locally. The latter have generally been of limited effectiveness because multiple objectives must often be addressed, it is difficult to achieve agreement on specific details of a stabilization strategy, individual member objectives change over time as domestic conditions and priorities change, and there is little to prevent cheating.

At the domestic level implementation of a buffer-stock program can be an effective means of moderating price instability. Here the issues are (1) how much protection is desired, (2) should a pricing rule or a quantity rule be used with which to trigger stock accumulation and stock release, and (3) who should shoulder the greatest burden of the costs—consumers or producers. The United States has attempted two such programs during the last half-century—the ever-normal granary program and the farmer-owned reserve program. The former was implemented by the nonrecourse loan program and almost immediately became a program to support farm incomes with stabilization being a secondary objective. The farmer-owned reserve program similarly appears to have been transformed into a price enhancement program with price stabilization being a secondary objective.

In addition to these types of policies, there are pure insurance or insurance-type programs available to the individual producer. These include (1) the federal All-Risk Crop Insurance program initiated during the New Deal and (2) hedging on the futures market or trading in the options market. A third, which could be made available by government action, is a revenue insurance program. This would most likely operate much like the current crop insurance program but could cover income from the whole farm rather than from a single crop.

Notes

1. The variability of *total* farm family income (i.e., net income from the farm operation plus government payments and income from nonfarm sources) exhibits a quite different pattern. In this case, the variability is highest for farms in the $20,000–$99,999 sales classes, and only slightly greater for farms in the smaller sales classes than for those in the larger sales classes. This pattern is the same for each of the time periods shown in Table 11.1. Further, the variability of total farm family income was still highest in all sales classes during the 1980s, but the difference from previous years was only slight.

2. Johnson's explanation of capital rationing rests on the proposition that in the face of the uncertainty in farm income generated by highly fluctuating prices, the farmer will be unable to obtain all the capital funds desired at the going rate of interest. There is no presumption here that farmers' desire for capital funds is decreased.

3. This section borrows heavily from Robinson 1989.

4. Massell (1969) also shows that when the source of price instability is from the demand side of the market, consumers gain and producers lose from a buffer stock scheme designed to smooth out prices. However, price variability due to random variations in aggregate demand from year to year is typically much less of a problem than is price variability due to random variations in aggregate supply from year to year.

5. Currently cotton, sugar, orange juice, corn, oats, soybeans, soybean oil, soybean meal, wheat, live cattle, live hogs, and pork bellies are traded on the futures market.

6. See Tomek and Robinson 1981 for a thorough discussion of futures markets and hedging.

Suggested Readings and References

Halcrow, Harold G. 1984. *Agricultural Policy Analysis*. New York: McGraw-Hill.

Johnson, D. Gale. 1947. *Forward Prices for Agriculture*. Chicago: University of Chicago Press.

Nicholls, William H. 1945. "A Price Policy for Agriculture, Consistent with Economic Progress, That Will Promote Adequate and More Stable Income from Farming." *Journal of Farm Economics* 27 (4): 737–56.

Robinson, K. L. 1975. "Unstable Farm Prices: Economic Consequences and Policy Options." *American Journal of Agricultural Economics* 57 (4): 769–77.

_____. 1989. *Farm and Food Policies and Their Consequences*. Englewood Cliffs, N.J.: Prentice-Hall.

Tomek, William G., and Kenneth L. Robinson. 1981. *Agricultural Product Prices*. 2d ed. Ithaca: Cornell University Press.

Trechter, David. 1985. "The Potential Role of Insurance in U.S. Agricultural Policy." *The Farm and Food System in Transition: Emerging Policy Issues*. FS-49. East Lansing: Michigan Cooperative Extension Service, Michigan State University.

U. S. Department of Agriculture. *Agricultural Outlook*. Economic Research Service. AO-156. September, pp. 2–4.

World Bank. 1986. *World Development Report 1986*. New York: Oxford University Press.

Advanced Readings.

Burnstein, Harlan. 1977. "Welfare Implications of Instability in Agricultural Commodity Markets." Department of Agricultural Economics and Rural Sociology, Pennsylvania State University. A.E. & R.S. no 132.

Eaton, David J. 1980. "A Systems Analysis of Grain Reserves." U.S. Department of Agriculture. Economic Research Service. Technical Bulletin no. 1611. January.

Gustafson, R. L. 1958. "Carryover Levels for Grains." U.S. Department of Agriculture. Economic Research Service. Technical Bulletin no. 1178. October.

Johnson, D. Gale. 1947. *Forward Prices for Agriculture*. Chicago: University of Chicago Press.

Knapp, Keith C. 1982. "Optimal Grain Carryovers in Open Economies: A Graphical Analysis." *American Journal of Agricultural Economics* 64 (2): 197–204.

Massell, Benton F. 1969. "Price Stabilization and Welfare." *Quarterly Journal of Economics*. 83 (2): 285–97.

Meyers, William H., and D. Craig Smyth. 1988. "Evaluations of the Reserve Program." In *The Farmer-Owned Reserve After Eight Years: A Summary of Research Findings and Implications,* edited by William H. Meyers. Iowa State University Agricultural and Home Economics Experiment Station Research Bulletin 598. July.

Salathe, Larry, J. Michael Price, and David E. Banker. 1984. "An Analysis of the Farmer-Owned Reserve Program, 1977–82." *American Journal of Agricultural Economics*. 66 (1): 1–11.

Subotnik, A., and J. P. Houck. 1976. "Welfare Implications of Stabilizing Consumption and Production." *American Journal of Agricultural Economics*. 58 (1): 13–20.

Waugh, Frederick V. 1944. "Does the Consumer Benefit from Price Instability?" *Quarterly Journal of Economics* 58 (4): 602–14.

IV

Prologue to the Future

12 Policy Analysis and Farm Policy Choices for the Future

Choosing policy objectives is no simple matter in a plurality like the United States. It is a time-consuming and evolving process, quite alien to a smaller, relatively more monolithic society like Sweden, for example. In certain areas, broad social goals are readily apparent and widely affirmed. The maintenance of an agricultural sector consisting of efficient, family-sized farm units provides one such example in spite of the fact that a "family-sized" unit is difficult to define to everyone's satisfaction. In other areas, lengthy dialogue may be necessary to sort out and distill society's aspirations and to develop a consensus. A current issue that appears to fit this latter category is that variously referred to as "alternative," "low-input," "sustainable," "regenerative," or "organic" agriculture (see, e.g., Schaller 1988). Important elements of the dialogue will involve a sharing of information on the need for policy action and of the consequences of proposed policy solutions: how will the policy be implemented, who will benefit and by how much, who will lose and by how much, how much will it cost, and what other social aims will need to be sacrificed partially or totally in order to achieve this one? There will likely be much debate, many divergent views expressed, and many participants to the debate. Applied economists can be expected to play an important role in the process.

This final chapter focuses on the choice of policy objectives for U.S. agriculture in the years to come. It does so on the basis of the evidence provided by economic analysis rather than on the basis of formal surveys of the voting public. It is not intended to provide an exhaustive listing of policy goals. It merely aims to focus attention on perceived social needs and potentials as well as on possible new approaches, rather than on the merits and demerits of a continuation of past approaches.

Since the choices discussed here are based on evidence provided by economic analysis, an initial task of this chapter is to review the strengths and limitations of the conceptual model usually employed by economic analysts. In particular, I focus on the applicability of some of the assumptions of this model and on issues of relevance that this model is incapable of addressing. As we will see, there is no substitute for insights accumulated from extensive study of and familiarity with the sector being examined and the actors in it. Unenlightened, purely theoretical arguments alone will not be sufficient. One of the areas in which this is most apparent and most crucial has to do with the extent to which resources will or can be expected to adjust to various economic and/or noneconomic stimuli. This is an area of sufficient importance to merit particular attention before proceeding to our discussion of policy goals.

The Character of Policy Analysis

The Perfectly Competitive Model. Policy analysis assesses the social consequences of specified policy choices. In economics, the analytical approach most commonly used for such an assessment is rooted in the assumptions of the perfectly competitive model and in comparisons of the estimated outcomes of policy choices with the presumed outcomes of a competitive equilibrium.

Indeed throughout this book, the perfectly competitive model or the concept of national and international "free trade"[1] has served as our norm for evaluating the consequences of policy choices for agriculture. The basis for this choice is that the perfectly competitive mode of economic behavior (which assumes [1] no one firm large enough to influence output prices, [2] perfect factor mobility, [3] perfect information, [4] no externalities, [5] free entry and exit, and [6] no market distortions caused by government intervention) leads to the most *efficient* use of scarce resources. This means that under perfect competition, consumers can expect to have access to food made from agricultural products that were produced at the lowest cost possible with existing technology. Any intervention by government that causes the configuration of market prices, mix of inputs used, and mix of outputs produced to deviate from the configuration that would prevail under perfect competition or free markets is, according to this norm, undesirable from society's standpoint. A summary measure of this deviation is the "social cost," to which I have repeatedly had occasion to refer. This summary measure involves a comparison of consumers' and producers' surplus under perfect competition with that under the intervention strategy being evaluated net of any government (taxpayer) costs incurred.

All this is standard fare for the applied welfare economist and nearly universally accepted as an appropriate analytical tool. It is not absolutely free of flaws, but the consensus is that the flaws are in the main not serious enough to invalidate this approach to assessment of the *efficiency* aspects of a market, industry, or economy (Just, Hueth,and Schmitz, 1982). It provides a solid conceptual base for analyzing policy choices, and this is the sense in which the approach has been used in studying policy choices in the preceding chapters of this book.

When one attempts to estimate empirically the relationships needed to quantify the conceptual model, however, more serious concerns arise. On one level, these concerns revolve around (1) whether the data revealed to us and available for empirical estimation are accurate enough to yield reliable estimates of needed parameters and relationships, and (2) whether the assumptions implicit in deriving these parameters and relationships from existing data are consistent with the conceptual ideas. For example, a major issue is whether the data produced by a system under the influence of substantial government involvement, as has been true in U.S. agriculture for the past forty years, can reveal insights on how the system and the actors in that system would behave in the *absence* of government intervention. On another level, there are concerns that any particular application may be much too oversimplified to capture all of the interrelationships of relevance so as to yield an accurate assessment of the economic system being analyzed. Specific questions here are (1) does the application include all relevant activity (i.e., farming, marketing, transportation, consumption, etc.), and (2) does it capture all relevant spatial and temporal dimensions of the phenomenon under study (i.e., interregional trade, international trade, cargo preferences, storage operations, etc.)?

Although both of these sets of concerns are of vital importance, the major issues and questions can generally be resolved by careful analysis and thought. Furthermore, criticisms of any particular application (or, indeed, of all such applications) do not invalidate the basic conceptual approach, only its implementation.

Limitations of the Model. But policy analysis is tricky business and must go far beyond tinkering with and generating solutions to standard conceptual models. What is needed is a thorough, painstaking, and seldom very neat verification of the assumptions underlying these models in light of observed facts, plus an assessment of issues standard models are incapable of addressing.

EQUILIBRIUM AND DISEQUILIBRIUM. Is *equilibrium* in the sense implied by this conceptual model applicable to agriculture? Clearly, it is not necessary that a perfectly competitive equilibrium without government intervention be attainable in order for the conceptual approach outlined here to be valid. That is, if a competitive equilibrium is the most desirable market outcome which can be conceived, then perfect competition should serve well as our base of comparison regardless of what outcome we may or may not expect to see in the real world in the absence of government intervention.

Unfortunately, economic analysts and critics of government intervention often rather blithely *assume* that if the policies and intervention currently in place were removed, a competitive equilibrium would attain or that there are natural tendencies that would move the system being analyzed toward a competitive equilibrium once shocked away from a previous equilibrium. If this is the basis on which government intervention is criticized, it is little wonder that so many are disenchanted with government intervention. If free-market conditions would prevail in the real world, then by definition government intervention that prevents attainment of free-market conditions is counter-productive. We have already established that perfect competition leads to the most *efficient* market outcome. Thus, if the system would gravitate toward equilibrium when left to its own devices, the appropriate admonition would seem to be "Don't 'fix' it with unnecessary government intervention"!

On the other hand, if a competitive equilibrium that involves an equation of marginal *social* benefits with marginal *social* costs (recall our discussion in Chapters 1 and 10) would not occur in the real world without government intervention, then it is also counterproductive to criticize policy because, for example, it is so costly in terms of budget outlays or because it causes farmers to behave differently than they otherwise would. Furthermore, if such an equilibrium would not naturally occur without government intervention, then whatever budgetary costs are needed to force this equilibrium to occur are necessary costs for achieving *efficiency*.

An important question, then, is, Would a competitive equilibrium in agriculture be achieved without government intervention? The answer is a qualified no. Agriculture probably comes closest of any of our industries to the textbook conditions necessary for the existence of a competitive equilibrium. There has been a great exodus of farms and farm people from agriculture in the last nearly forty years (see Tables 3.2, 3.3, and 3.9). Nevertheless, the farms still remaining are so numerous that no one of them has a large enough share of the market to significantly influence price or marketing conditions. In some cases, agricultural cooperatives may be large enough to influence the market (e.g., in cranberries, some citrus fruits, and hops), but these are exceptions rather than the rule. Information is not necessarily perfect but is adequate for the most part to guide allocation decisions. Relatively speaking, entry and exit are free. For some, however, the large amount of capital needed to get started in farming (see Chapter 3) may serve as an effective entry barrier.

On the other hand, whether at any one time a competitive equilibrium is attainable in agriculture and whether the sector can reach a new equilibrium once it receives a disequilibrating shock are much more complex issues. As we saw in Chapter 10, there are significant externalities in agriculture that are not likely to become internalized without some kind of outside influence. Farmers have little incentive to reduce the costs they impose on citizens outside farming as a result of water pollution that accompanies the spreading of animal manure and use of chemicals on cropland. Similarly, farmers in general have little incentive to prevent or retard erosion of the soil, which also contributes to environmental pollution and additionally contributes to a reduction in soil productivity likely to have its greatest impact on future generations. In this sense, then, equilibrium is unattainable without outside interference.

In Chapter 3 we saw that because of the competitive structure of agriculture and given the tremendous technological advances we have had over the course of the last nearly forty years, farmers have been in a constant struggle to adopt new technology so as to produce more output at less cost and thereby increase their net incomes. The end result of this process is, unfortunately, that in the long run (when all farmers adopt the new technology), prices fall as the aggregate supply curve shifts to the right, so that they are no better off than before despite their efforts. In Chapter 3 this process was referred to as the "technological treadmill."

The question here is, Will resources adjust to a new equilibrium in the face of new-technology adoption? In a subsequent section, I present evidence suggesting that equilibrating adjustments of agricultural inputs and outputs *can* be expected to occur in agriculture in the long run (although, of course, externalities may still exist). Unfortunately, we do not always know how long the long run is, nor do we know very much about the dynamics of the adjustment process itself. This is an area about which the static, perfectly competitive model can provide little insight since it cannot deal with dynamic issues. In point of fact, the agricultural economy is always evolving and is never stationary. If it ever attains a competitive equilibrium, it is likely only there momentarily because it is always adjusting to new stimuli. Some questions that should be of interest to policymakers in this area include: Who gets hurt in the process of adjusting from one equilibrium to another and by how much? Is there something that government can and should do to ease the pain? Will appropriate adjustments leading to the next competitive equilibrium be made by the sector *on its own?*

The development of the mechanical tomato harvester is a case in point. In the 1960s, plant scientists at the University of California developed a uniformly ripening processing tomato. Subsequently, engineers developed a machine that would mechanically harvest this tomato without serious harvesting losses. This machine ultimately saved tomato producers much expense on the previously labor-intensive, handpicking harvesting methods and, consequently, resulted in processed tomatoes available to consumers at considerably lower retail prices. In a very careful analysis using standard, applied welfare methods Schmitz and Seckler (1970) showed that the gross social rate of return to the research on this technology was in the range of 929 percent to 1,282 percent!

This analysis was right and proper. Still, had it stopped here it would have been grossly inadequate. The reason is that the mechanical tomato harvester also displaced several workers who were previously employed in handpicking operations. True, these were back-breaking jobs providing little glory for the worker. But they were jobs that are no longer! The point that must be made is that here is a group of people whose lives were seriously affected. In the final analysis, social rate of return calculations based on the perfectly competitive model must include consideration for the displaced workers. Schmitz and Seckler go on to suggest that just compensation to such displaced workers not only should be deducted from the calculations, *it must be made*. Compensation might include job retraining as well as relocation of the family. Obviously this raises many questions: What kind of retraining? Where to relocate? Who must pay for the expenses incurred? These are issues that cannot be ignored by a just and caring society.

One must also be more than a little curious about whether a competitive equilibrium in U.S. agriculture or one of its subsectors is even desirable when related sectors are not competitively organized. If, for example, there are significant barriers to trade in milk or beef, does it make sense from an efficiency point of view for the United States to seek a perfectly competitive solution for its dairy or beef sectors? Similarly, if there is strong intervention in the U.S. feed grain and/or soybean sectors, does it make sense from an efficiency point of view for the United States to seek a perfectly competitive solution for its dairy sector, which uses as inputs the output of the feed grain and soybean sectors? The theory of the second-best (see Henderson and Quandt 1980, pp. 315–16) suggests that when it is impossible to attain one of the conditions for a perfectly competitive solution, it is not necessarily a good idea to try to satisfy all of the remaining conditions. If, for example, there are restrictions preventing farmers from equating the value of the marginal product of one input with the competitive equilibrium price of that input, then a second-best solution (i.e., one that sacrifices some efficiency) may be optimal.

A fairly strong case can be made for global free trade. I have made such a case in Chapter 9 of this book. A host of researchers have concluded that global free trade in agriculture would benefit all of the world not just the United States (see Baker, Hallberg, and Blandford 1989 for a brief review of several such studies). The signatory countries to the General Agreement on Tariffs and Trade (GATT) have been working toward this end for forty years. The U.S. trade representative, secretary of agriculture, and president have been pushing for global free trade in agriculture for the past several years. The United States and Canada signed a bilateral free-trade agreement in January 1989 after a couple years of negotiations to give the impression, at least, that they were serious about free trade.

Yet, is global free trade good for everyone? It unquestionably is if adjustments everywhere necessary to accommodate to free trade can or are likely to occur. Consider, for example, most of the countries of Africa. They clearly stand to lose from agricultural free trade in the short run as prices of cereals and of livestock products rise above current subsidized levels. They would, presumably, gain in the long run by making adjustments in resource use such that they (1) reduced their dependence on the rest of the world for food products and (more important) (2) produced more of the products for which they have a comparative advantage and that other countries want to buy with their increased incomes. But will these adjustments be made? If not, the residents of these countries will be even worse off under free trade than they are now! This is an issue of crucial importance. It is also one that has received little attention from the global free-trade modelers and limited attention from anyone else. It is conceivable, for example, that global free trade in agriculture may well serve only to place added burden on international development agencies such as the World Bank, the International Monetary Fund, and the Food and Agricultural Organization of the United Nations.

It is certainly true that global free trade for agriculture *is* a desirable goal for the world. It may be a mistake, however, to argue for global free trade in agriculture without addressing the question of whether all countries can make the adjustments necessary to accommodate to free trade so that the lot of their residents is ultimately increased.

ECONOMIES OF SIZE. An important concept emanating from the theory of the firm under perfect competition is that of economies of size. Economies of size refer to the cost or return advantages or disadvantages enjoyed by different-sized farms.[2] If farms are, on average, too small to capture available size economies, production efficiency will be sacrificed. If farms get larger than necessary to capture available size economies, this too is important because it is a signal that farms are getting beyond the size of the family farms that society has up to now deemed the most desirable sized unit with which to populate the agricultural sector.

If size advantages exist, then a larger farm is able to purchase the greater quantity of inputs it needs at lower per unit costs, is able to exploit existing technology more fully, is able to combine lumpy inputs more effectively to lower per unit production costs, is able to more fully utilize its fixed resources (including management), etc. This is depicted in Figure 12.1 by the long-run average cost curve ($LRAC_0$) declining continuously up to the point where farm size equals Q and constant thereafter. Technical economies of size refer to the cost savings that result from increased size and/or specialization. Pecuniary economies of size result when larger farms are able to buy greater quantities of inputs at a lower per unit cost or sell greater quantities of output at a higher per unit price. Diseconomies are said to have set in when the cost curve begins to turn up, as does $LRAC_1$ beyond the farm size indicated by Q on the horizontal axis.

It is interesting to note in this connection the fate of the so-called bonanza farms of the late 1800s (see Drache 1964). Between 1840 and 1860, the federal government granted huge tracks of public lands to companies in an effort to stimulate railroad construction across this country. These companies had little need for all this land, but they did have need for the cash this land would bring on the open market, so much of it was sold to private developers. Some investors purchased tracts as large as 50,000 acres or more in North Dakota, Illinois, Iowa, and Nebraska and attempted to operate these acreages as single-unit farms under one manager with hired labor and substantial investments in power and equipment. By the turn of the century, however, these tracts had all been broken up into small, family-sized units (generally 160-acre units) and rented or sold outright to individual farm families because the original tracts were simply too unwieldy to operate as originally planned. The implication here is that these bonanza

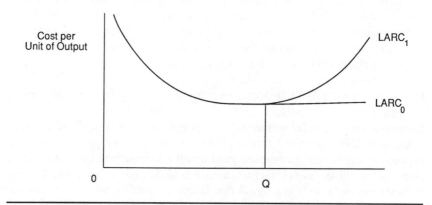

Figure 12.1. Economies of size in agriculture

farms were operating on the upward-sloping portion of the long-run average cost curve where diseconomies of size had set in. Although the minimum point on the long-run average cost curve is likely much further to the right today than it was in 1900, it is still nowhere near 50,000 acres for any of the field crops.

Estimating economies of size relationships is not as straightforward as it may appear, and not everyone agrees on the most appropriate methodology. Thus, there is room for disagreement on exactly where the curve should be positioned. There is little disagreement on its general shape, however. A very rough idea of the nature of size economies in agriculture can be obtained from an examination of the 1988 ratio of inputs to outputs on farms in different sales classes:

Sales class	Average gross farm income	Input/output ratio
$500,000 and over	$1,042,879	$0.97
$250,000–$499,999	$314,299	$1.05
$100,000–$249,999	$155,973	$1.06
$ 40,000–$99,999	$72,192	$1.20
$ 20,000–$39,999	$34,413	$1.59
$ 10,000–$19,999	$22,589	$1.66
Less than $10,000	$11,580	$1.98

The measure of inputs here consists of the 1988 dollar cost of all farm inputs including a charge for depreciation and an estimate of the opportunity cost of equity capital (assessed at the rate of 9 percent of total farm equity). The measure of output (average gross farm income) includes the 1988 dollar value of all farm sales, estimated nonmoney income, and government payments. The data used in this analysis are from the U.S. Department of Agriculture's Farm Cost and Returns Survey (U.S. Department of Agriculture 1989).

This tabulation suggests that per unit costs decline rapidly as farm size increases and that the cost curve becomes quite flat for the larger sized farms. Those farms with annual gross sales of less than $40,000 would have an input-output ratio of near 1 only if their return on equity capital was in the vicinity of 2.5 percent! Clearly the operators of these farms are willing to accept such a low return on capital. But this is nowhere near a competitive return (the rate of interest on AAA-rated corporate bonds in 1988 was 9.71 percent). It is in this sense that many would argue farms selling less than $40,000 of farm produce in 1988 were "inefficient."

The distribution of per unit costs by farm size just shown, however, masks much detail about costs on U.S. farms. It does not consider farm location, type of farming, degree of specialization, etc. Several studies aimed at a more accurate assessment using more refined techniques of analysis have documented the existence of decreasing average costs for farm firms (see, e.g., Madden 1967; Miller, Rodewald, and McElroy 1981; and Buxton et al. 1985). The consensus is that average costs decline very rapidly with increasing size in the smaller size categories and then flatten out for the larger size categories of farms—quite like the curve labeled $LARC_0$ in Figure 12.1 and the distribution of input-output ratios shown above. There is no evidence of costs increasing with farm size that would produce a curve like the one labeled $LARC_1$ in Figure 12.1.

Buxton et al.'s (1985) work, for example, points to considerable economies of size in dairy. It appears that the average dairy farm in the Northeast and the Lake States is probably somewhat below optimal size. Those in the West and Southwest, on the other hand, which typically have larger dairy herds, have probably captured most of the technical economies achievable. Tobacco farms have undoubtedly been maintained at a size well below the most efficient sized unit because of the allotment policies that have been adopted for this subsector over the years. Miller, Rodewald, and McElroy (1981) studied size economies on farms producing wheat, feed grains, and cotton in seven regions of the United States. They conclude that economies of size do exist on these crop farms but their importance is not great. The estimated long-run average cost curves declined significantly at first and then became quite flat over a wide range of farm sizes. Medium-sized farms in most regions, though, were found to be nearly as efficient as large farms.

As Table 12.1 shows, farms in the United States have on average gotten significantly larger whether one measures size in terms of acres per farm or deflated cash receipts per farm per year. Deflated cash receipts per farm is a somewhat less desirable measure of farm size because it varies so widely with variations in output prices. On the other hand, when growth of farms is measured in acres, the influence of those farms producing land-intensive products like broilers or milk is not adequately reflected. By either measure of farm size we must conclude that farms in the United States have gotten substantially larger since 1950. One might argue this growth in farm size is due to the "technological treadmill" thesis of Cochrane (see Chapter 3), or one might simply argue this has been a normal, expected response to declining deflated prices of farm output in the face of relatively constant deflated prices of farm inputs (see Hallberg 1988). It is interesting to note that the average annual increase in deflated sales per farm is slightly greater post-1970 than pre-1970, while the average annual increase in acres per farm is slightly greater pre-1970 than post-1970. Apparently nonland inputs have become somewhat more important in the growth of farm firms in

Table 12.1. Number of farms and average farm size in the United States, 1950–90

Year	Number of farms	Acres per farm	Deflated cash receipts per farm[a]
1950	5,648	213	$8,998
1951	5,428	222	9,172
1952	5,198	232	9,933
1953	4,984	242	11,107
1954	4,798	251	11,514
1955	4,654	258	12,424
1956	4,514	265	13,470
1957	4,372	272	13,326
1958	4,233	280	14,370
1959	4,097	289	15,793
1960	3,963	297	16,619
1961	3,825	305	17,679
1962	3,692	314	18,292
1963	3,572	322	19,796
1964	3,457	332	20,764
1965	3,356	340	21,722
1966	3,257	348	22,993
1967	3,162	355	24,620
1968	3,071	363	25,691
1969	3,000	369	27,689
1970	2,949	374	28,546
1971	2,902	378	29,317
1972	2,860	382	31,420
1973	2,823	385	31,406
1974	2,795	388	31,482
1975	2,521	420	34,915
1976	2,497	422	37,439
1977	2,456	427	39,184
1978	2,436	429	40,109
1979	2,437	428	46,527
1980	2,440	426	47,724
1981	2,440	424	41,755
1982	2,407	427	44,532
1983	2,379	430	42,586
1984	2,334	436	43,081
1985	2,293	441	49,102
1986	2,250	447	48,852
1987	2,213	451	50,401
1988	2,197	453	49,538
1989	2,171	457	49,876
1990	2,140	461	51,930

Source: U.S. Department of Agriculture, *Agricultural Statistics,* various annual issues.
[a]Deflated by index of prices received by farmers, 1977 = 100.

recent years than has the land input or inputs that would encourage greater land use per farm, such as large tractors and combines.

Although it is hazardous to generalize from the size distribution of farms by sales classes, the data in Table 3.9 together with the input-output ratios presented above suggest that the vast majority of farms in the United States are below the most efficient size. Of all farms in 1988, 34.2 percent sold less than $5,000 of farm products, 46.9 percent sold less than $10,000, 59.4 percent sold less than $20,000, and 70.8 percent sold less than $40,000. Certainly, these farms did not have a sufficient volume to earn a decent living solely from the farm for an average-sized farm family (see Table 3.9). It is often argued that farm programs have kept the small farmer in business longer than would have been the case without government programs. The implication here is that farm programs have prevented many farms from becoming large enough to achieve the major economies of size currently available.

The operators of the majority of these farms likely have a full-time or part-time nonfarm job which provides most of the family income, and most operate the farm for reasons other than making a good family living. They care little whether government intervention exists. They will remain in farming so long as they can find a nonfarm job nearby.[3] Since they collectively do not account for a high percentage of total agricultural output, they probably have little impact on total agricultural efficiency. In contrast, commercial farmers have collectively taken advantage of new technology to increase size and in this way have moved closer and closer, if not all the way, to that size that results in minimum average total costs.

TECHNOLOGY ADOPTION. An assessment of the performance of agriculture under either free-market conditions or government interference would not be complete without asking: How progressive is the sector in terms of its tendency to adopt new methods and practices and in this way seek to maintain its competitive edge as well as provide consumers the benefits of cost-reducing technology? As we have already seen, deflated market prices of the major farm commodities have declined significantly since 1950—by as much as one half or nearly so for oats, sorghum, soybeans, peanuts, steers, lamb, and milk, and by one fourth for wheat, corn, rice, and broilers (Table 3.4). Deflated prices paid, on the other hand, have decreased slightly, if at all, and in fact were higher during the 1970s and early 1980s than in the 1950s and 1960s. It is clear that farmers have been facing a seemingly never-ending price-cost squeeze since 1950.

Farmers have been able to survive this price-cost squeeze by rapidly adopting those technologies that enable them to increase productivity (Table 3.5). Cochrane (1958) uses the treadmill thesis to explain not only the persistence of chronic disequilibrium in agriculture but also to explain the fact that farmers have been forced to adopt the technology bringing about this chronic disequilibrium in order to survive. Whatever the motive, great increases in productivity have occurred in U.S. agriculture since 1950.

Clearly the farm sector has been progressive in the sense that its individual members have adopted the new technology available. It is difficult to imagine that this would have happened any more rapidly without the price and income support programs that have been in place since the 1930s. It is equally difficult to argue that this rapid rate of technology adoption was due primary to the existence of these programs. The Agricultural Experiment Stations and the Cooperative Extension Service (both federally funded programs), however, have been a strong force in both agricultural technology discovery and in the delivery of this new technology to farmers (see, e.g., Ruttan 1982).

EQUITY. Much of the time of policymakers is taken up with considerations of equity. Is the distribution of wealth on which a competitive equilibrium is based equitable? Are income levels of farm families commensurate with income levels of nonfarm families? Should they be? Is the distribution of farm program benefits across farm sizes, commodities, regions, age-groups, and resources equitable? Who should share in the cost of farm programs and in what proportions? To what extent should today's farm policy place a burden on future generations? How much *efficiency* should be sacrificed in order to achieve a more *equitable* distribution of benefits or burdens? Should farmworkers who are displaced by technological advances be compensated? If so, in what ways and who should bear the cost?

All of these are questions that must be resolved outside the framework of the perfectly competitive model. Indeed, they can only be resolved with reference to a social preference function that is very elusive to both the economic analyst and the policymaker. It is the responsibility of the policymaker to attempt to determine this social preference function through the political process and thus to resolve these equity questions. The analyst can and should contribute, though, by objectively weighing the evidence to highlight the existence or absence of any such inequities. A good deal of the discussion in Chapter 4 was devoted to providing just such evidence.

The Capacity of Agriculture to Adjust

In the previous section I inquired as to whether a competitive equilibrium was attainable in agriculture without government policy or controls. I answered the question partially. A complete answer must anticipate whether resources would adjust consistent with a competitive equilibrium. This is the objective of the present section. In particular, I examine the historical record of adjustments in both inputs and outputs since 1950.

History of Adjustments

INPUT USE. Table 12.2 shows the trends in quantities of key farm inputs since 1950. Among other trends, this table highlights the steep decline in labor use and the simultaneous steep increase in tractor horsepower and fertilizer and other chemical use. The total *number* of tractors and crawlers per 100 acres planted has not changed greatly since 1950. Total tractor *horsepower* used per 100 acres planted, however, has nearly quadrupled since 1950.

Per acre use of nitrogen has increased tenfold and per acre use of potash has increased fivefold since 1950. Per acre use of phosphate has also increased, but on a much less dramatic scale—about two and one-half times since 1950. Increased fertilizer use has brought about much of the

Table 12.2. Farm inputs used in U.S. agriculture, 1950–90

	1950	1955	1960	1965	1970	1975	1980	1985	1990
Index of quantity of farm inputs (1977 = 100)									
Farm labor	265	220	177	144	112	106	96	85	75
Fertilizers and pesticides	19	26	32	49	75	83	123	123	113
Feed, seed, and livestock	58	66	77	86	96	93	114	106	118
Power and machinery	72	83	83	80	85	96	101	80	72
Taxes and interest	83	89	95	101	102	100	100	91	94
Farm real estate	109	108	103	103	105	97	103	95	93
All farm inputs	101	102	98	96	97	97	103	92	89
Hours of farm work/acre planted	55	45	39	34	27	22	19	17	17
Commercial fertilizer and lime use acre/planted (lbs.)									
Nitrogen	6	11	17	31	51	52	64	67	69
Phosphate	11	13	16	24	31	27	31	27	27
Potash	6	11	13	19	28	27	35	32	33
Liming materials	169	117	139	189	177	187	193	152	156
Wheel and crawler tractors/100 acres planted									
Number of tractors	1.0	1.2	1.4	1.6	1.6	1.3	1.3	1.4	1.4
Horsepower	26	36	47	59	69	67	86	91	97

Source: U.S. Department of Agriculture, *Agricultural Statistics,* various annual issues.

crop yield increases observed in Table 3.5 (as well as many of the externalities discussed in Chapter 10).

Table 12.3 points to the input substitutions farmers have made since 1950. Land input in agriculture has remained relatively constant since 1950 while labor use has declined by a factor of 3.5 (Table 12.2). Thus, the steady increase in the ratio of land to labor use shown in Table 12.3 since 1950 is hardly surprising. It is also well-known that machine and fertilizer use in agriculture has increased quite rapidly at the expense of labor through the 1950s, 1960s, and early 1970s.

The substitutions in inputs indicated here were in large part influenced by corresponding changes in their relative prices, as the competitive model predicts. For example, as the ratio of machine prices to wage rates falls, farmers are expected to substitute the relatively less expensive machines for the relatively more expensive labor. Through at least the 1970s, this is precisely what happened in U.S. agriculture.

Kislev and Peterson (1982) put this in clearer perspective. Increasing urban incomes relative to income from farming for many farm families or to wage incomes for many farmworkers, increased education of rural

Table 12.3. Ratios of quantities of farm inputs and ratios of prices of farm inputs in the United States, 1950–90

	1950	1955	1960	1965	1970	1975	1980	1985	1990
Ratios of input quantities (1977 = 100)[a]									
Land/labor	41	49	58	72	94	92	107	112	124
Machine/labor	27	38	47	56	76	91	105	94	96
Nitrogen/labor	4	8	15	30	63	76	112	127	139
Nitrogen/land	9	17	25	42	67	83	104	111	112
Nitrogen/machine	13	22	31	54	82	84	106	135	144
Machine/land	66	77	81	78	81	99	98	84	77
Ratios of input prices (1977 = 100)[b]									
Land/labor	53	54	61	73	65	87	124	82	59
Tractors/labor	114	107	103	103	86	96	108	116	106
Fertilizer/labor	236	207	167	145	86	141	106	88	69
Fertilizer/land	448	384	271	197	133	162	86	107	117
Fertilizer/tractors	208	193	162	141	100	146	99	76	65
Tractors/land	215	199	168	140	133	111	87	141	181

Source: U.S. Department of Agriculture, *Agricultural Statistics,* various annual issues.
[a]Nitrogen input measured by total pounds of nitrogen applied. All other inputs here are measured by reported quantity indexes.
[b]Price of land measured by per acre land value. All other prices are measured by reported price indexes.

people, and removal of other barriers to the mobility of rural people have led to the outmigration of farm labor. At the same time a reduction in the price of mechanical inputs relative to the opportunity cost of farm labor encouraged the substitution of machinery for labor. The mechanical inputs, in turn, permitted farm families to cultivate larger acreages purchased or rented from those who left farming. The end result has been not only input substitutions but also larger farm sizes.

The data in Table 12.3 provide additional insights into adjustments farmers make. Notice that from the mid-1950s to the beginning of the 1970s the fertilizer/labor, fertilizer/machinery, and machinery/labor price ratios declined fairly steadily, encouraging farmers to substitute fertilizer and machinery for labor and to use increasing amounts of fertilizer relative to machine inputs. During the first half of the 1970s, however, these price ratios began moving in the opposite direction, the first two for the duration of the oil crisis and the last until 1983–84. Based on the changes in the ratios of input quantities shown in Table 12.3 it seems clear that farmers responded to the changes in relative input prices with a reduction in the use of machinery relative to labor and land in the latter half of the 1970s and the early 1980s. The reduction in fertilizer use relative to labor and machinery during this same period was most likely also related to changes in the relative prices of inputs. Following 1975, the fertilizer/machinery and fertilizer/labor price ratios resumed their downward path, and the substitution of fertilizer for machinery and labor once again trended upward.

Of additional significance here is the fact that the land/labor and machinery/labor price ratios declined throughout most of the 1980s. An examination of the ratios of quantities of inputs reveals that farmers' rapid substitution of machinery for labor in the previous two decades was halted and even declined significantly through the early 1980s. If this trend continues, we can expect the demand for large farm machines to remain somewhat weak and the growth in farm size to slow down. If farmers refuse to buy more and larger machines, they are not likely to be able to farm larger acreages, and thus farm sizes will stabilize. Indeed, judging from the trend in acres per farm shown in Table 12.1, one most conclude that growth in farm size has slowed markedly since 1975.

FARM ENTERPRISE MIX. A growing total population in the United States has bid away some farmland acreage from the farming sector, but the decline in farmland since 1949 has been minuscule (see Chapter 10). Cropland acreage has actually increased slightly over this period. The proportion of farm acreage planted to different crops has also remained fairly stable over the period (see Chapter 10). The proportion of acreage devoted to wheat and the major feed grains has changed very little since

1949. Oat production has declined since less of this commodity has been needed for work animal feed. Cotton production has declined slightly over the period as the demand for cotton has decreased. Soybean production increased significantly in the early years of the period as the demand for protein feed increased and soybeans became a more popular crop. Sugar acreage (beet plus cane) has increased only slightly; tobacco acreage has decreased by about one half; and peanut acreage has remained quite stable since 1949.

The proportion of total cash receipts from farming derived from the various farm enterprises has also been remarkably stable since 1950 (Hallberg 1989). In some cases (hogs, eggs, tobacco, and cotton), noticeable but small declines are evident, while in other cases (poultry, feed grains, and oil crops), small increases are noted. For most enterprises the changes in relative importance are quite small or nonexistent over the period. The explanation for this stability is to a large degree to be found in the stability of relative output prices (see Table 3.10) due in large part to farm programs.

PART-TIME FARMING. Another form of "enterprise adjustment" that can be and has been made by farm families relates to their decisions regarding allocation of available family labor among farm and nonfarm occupations. Part-time farming has been a part of agriculture for a long time. Posnan (1966) cites evidence of part-time farming in Italy (pp. 350 and 420), Germany (p. 482) and England (p. 568) in the early Middle Ages! Hood (1936) studied part-time farming in New York in 1936 and concluded that it was a response to the "rural-urban" movement which had begun at least a half-century earlier. In addition, it has long been recognized that part-time farming is, under certain situations, consistent with utility-maximizing behavior of farmers. Lee (1965), for example, demonstrates theoretically that as long as the nonfarm wage rate is *below* the marginal rate of return from farming, the farmer will allocate his labor to farming. Because of diminishing returns to farm work, however, there will be a point at which the marginal rate of return to farm work will fall below the nonfarm wage rate. From that point on the rational farmer will allocate any remaining work time to a nonfarm job. Before taking up nonfarm work, the farm will be operated in a labor-intensive way, the marginal utility of work will be low, and there will be much leisure time. With a combination of farm and nonfarm work, the marginal utility of farm work will rise, and leisure time will fall. Most likely this dual occupation will necessitate adjustments of farm enterprises, technology use, etc.

Research reported elsewhere (Hallberg 1989) suggests that farm families do indeed adjust to the relative earning power of the farm by reallocating family labor between farm and nonfarm work when income from the farm is low. For the midsized farms the relationship between nonfarm income and the ratio of farm earnings to household income in the United States was fairly strong. For the three largest size categories the relationship was not significant, suggesting that on these farms little surplus labor exists and the farm operation is dominant. For the two smallest sized farms the relationship was significant but not very strong. Here the farm makes such a small contribution to family income (it has actually detracted from family income in recent years) that relative farm earnings are of little consequence.

Resource Mobility

LABOR. Historically there have been severe impediments to the adjustment of farmworkers out of agriculture and into nonfarm jobs in order to prevent the persistence of low returns (Chapter 3). These impediments have included lack of education and skills, lack of knowledge about nonfarm job opportunities, employment barriers created by organized labor, and lack of availability of nonfarm jobs in the area where these surplus labor resources could feasibly relocate.

For the most part these types of impediments have by now been removed or at least eased so that much of the surplus labor in agriculture has migrated to the nonagricultural sector. Nevertheless, there are still instances in which nonfarm jobs requiring the skills possessed by surplus agricultural workers are not available. In these instances, the excess labor resources cannot move out of agriculture. Also, changing to a different type of agricultural production involves major shifts in capital, which is never easy and sometimes simply not possible. It is relatively easy for the midwestern grain farmer to shift production from corn to soybeans or even from corn to hay. It is quite another matter for the dairy farmer to shift from milk production to vegetable production or even to shift from milk production to beef production. The intensity of resource use in dairy and beef is vastly different so that with the same resources, a Wisconsin dairy farmer simply could not produce the same level of net (or gross) receipts with beef as he or she can with dairy.

CAPITAL. As I noted in Chapter 3, some argue that many resources in agriculture are subject to "asset fixity" and that there is a wide range over which output price will vary before changes are made in input use and in product output. A variety of agricultural assets have been so characterized and, so the argument goes, will be used to produce specialized products until they are completely depreciated, almost without regard to product prices.

As we saw in Chapter 10, land has a low salvage value in nonfarm uses in most areas and it responds very little to product prices. Apparently the opportunity cost of keeping land in crop production is very low, and the cost of cropping land not previously used for crops is relatively high. Thus, supply is very inelastic for both falling and rising crop prices.

As a result of these considerations, it is frequently argued that over a wide range of output prices and in the short to intermediate run, aggregate supply for farm products tends to be quite inelastic. Even so, durable inputs fixed to the industry tend to be less fixed on individual farms and much less so for individual commodities. The land input may be fixed in the aggregate, but there is in most areas of the country little to prevent changes in the crops or livestock grown on that fixed input. This explains why supply tends to be more elastic for individual farm products than for aggregate farm output. Furthermore, farm labor does not appear to be as fixed in agriculture as it once was. Finally, operating inputs (chemicals, fuels, seed, repairs, etc.) which have a low (even zero) salvage value once committed to farming are consumed quickly and their usage level can be adjusted in the subsequent production period. Thus, asset fixity is likely only of major concern in the very short run. Given a slightly longer period of time, significant adjustments are made. Farmers certainly did respond to favorable output prices between 1970 and 1979 with increased input use. Similarly they responded with reduced input use during 1981–86 when output prices fell! Further, as the data in Table 12.3 suggest, farmers have responded to changing relative input prices with changes in input substitutions.

Policy Objectives

In Chapter 1 I discussed several different types of policy that have been pursued for agriculture: development, regulatory, credit, conservation and environment, stabilization, and compensation policy. Many instruments have been used to implement these policies. Some have been quite successful, such as our land distribution schemes and transportation subsidies; others have been less so, as evidenced by failure of the Federal Farm Board created in 1929 to solve the then-existing farm problem. In some cases we have stayed the course over many years, as, for example, in the development of an agricultural sector consisting of small, family-sized farm units. In other instances, we have strayed from original policy objectives, as exemplified by turning the "ever-normal granary" idea from a farm price and food supply stabilization program into a price enhancement mechanism. In still other instances, the "solution" has not necessarily been superior to the "problem"—for example, there has been an uneven distribution of the benefits of farm programs among farm people.

The type of policy pursued, however, depends on the needs and conditions of the times, and needs and conditions change over time. A first requirement to charting a course for future policy for agriculture is to identify and prioritize the various policy objectives. This can only be done, of course, in light of available evidence on broader social aims and objectives, results of past policy approaches, and needs of farm families. Unfortunately, well defined conceptual models are not available for this purpose. Rather, we must rely on the less elegant, but no less rigorous, inductive methods anticipated in the last section of Chapter 1.

It is not the intent of this section to suggest the ultimate configuration and ranking of policy goals for agriculture. Others would no doubt offer a different set of objectives and a different ranking. Rather, I offer a modest beginning toward articulation of policy objectives that focus on current and likely future needs and/or are consistent with perceived, broad social aims and objectives. I hope this discussion will stimulate a more general dialogue leading to viable future policy for agriculture based on current and future needs and prospects rather than on retention of old approaches. I believe it will be more useful to students and policymakers than would a simple shopping list of problems or issues. From a practical point of view, priority problems exist and they deserve attention. Government resources are limited, so some priority ranking must be invoked to determine how scarce government resources are to be allocated.

Farm Income Enhancement. As is abundantly clear, farm income enhancement has in the past received very high priority by the framers of farm policy. Farm income enhancement has been implemented with both direct income payments and farm product price supports as well as with production controls. The need for farm income enhancement today, however, must be seriously questioned. As we saw in Chapter 3 (see especially Tables 3.8 and 3.9), much progress has been made in an overall sense toward farm-nonfarm income equality. The data in Table 12.4 emphasize the point that the sum of farm and nonfarm income of farm families in the $40,000 sales class and above is now on a par with or greater than the money income of all U.S. households. Most of the "commercial-sized" farms (those with annual sales of $100,000 or more) appear to do very much better than the average U.S. household. They have little need for further income support. For these farms, direct government payments (not

Table 12.4. Net income per farm from farm and nonfarm sources as a percentage of money income of all U.S. households, 1960–88

	1960	1965	1970	1975	1980	1985	1988
Under $5,000	66	79	71	84	75	84	82
$5,000–$9,999	88	91	77	78	74	51	71
$10,000–$19,999	117	111	95	76	67	50	63
$20,000–$39,999	199	157	138	90	60	63	70
$40,000–$99,999	288	267	238	161	82	82	87
$100,000–$249,999[a]	556	573	477	354	178	171	183
$250,000–$499,999	na	na	917	779	374	403	395
$500,000 and over	na	na	6,317	4,471	2,244	2,154	2,000

Source: U.S. Department of Agriculture 1989; and U.S. Bureau of the Census, *Current Population Reports,* various issues.

Note: Net farm income plus nonfarm income (estimated for 1950–59) less the value of commodities produced and consumed on the farm, rental value of farm dwellings, net change in farm inventories, and direct government payments.

[a]This sales class is $100,000 and over in 1960 and 1965.

na—not available.

included in the data tabulated in Table 12.4), as we saw in Chapter 4, constitute a very small percentage of total family income. The farm is the chief source of family income.

Several of the smaller farms do have money incomes lower than that of the average U.S. household. Interestingly enough, the very smallest farms seem to fare better than do the middle-sized farms (those with cash sales of $10,000–$40,000). Nevertheless, if we account for the nonmoney and psychic benefits of rural living, most of the smaller farms appear to do reasonably well in comparison with nonfarm families. In large part they are able to do so because of their nonfarm jobs. Again direct government payments are of little consequence to these farm families. *The state of the nonfarm economy providing nonfarm work is of much greater concern to them.*

There is also the question of equitability of income support for farmers. As we saw in Chapter 4, direct income payments as well as benefits from price enhancement are for the most part distributed as equitably as can be expected across different-sized farms. However, these farm program benefits are *not* distributed equitably across commodities or regions. Some producers receive no benefits whatever—for example, potato and apple producers. Some producers who receive no benefits from price supports, import quotas, or export subsidies actually suffer from a feed grains policy that causes them to pay higher prices for their feed—for example, broiler and egg producers. Finally, to the extent that farm program benefits get capitalized into land values, the income enhancement aspects of current farm programs result in inequitable treatment of landowners and tenants.

Competitiveness and Trade Policy. Maintaining competitiveness in the international arena should receive very high priority. *Subsidizing exports, however, is not the way to accomplish this objective.* Subsidies simply result in a transfer of welfare from U.S. citizens to foreign citizens (Chapter 9) while doing little to encourage U.S. producers to adopt up-to-date technology or management practices or otherwise become more efficient and progressive producers. We have seen that in order to improve their relative income position, farmers do readily adopt new technology and, when prices of inputs change, do readily change the mix of inputs accordingly. Although relative output prices have not changed significantly in the last fifty years, there is little reason to expect that farmers would *not* adjust their output mix in response to the appropriate signals that would be sent if output prices were not fixed by rigid government policies.

In general, there is also little reason to expect that in the absence of government price supports farmers would not make the adjustments necessary for a progressive agricultural sector that is able to compete in the international sphere. If U.S. agriculture is not competitive in one or more products, such as wool or sugar, we can be reasonably sure the resources released from production of these commodities will be productively employed in other areas, albeit after some delay. In case this were for some reason prevented from happening (e.g., restrictions on capital or labor mobility), *then it should be the function of policymakers to devise appropriate mechanisms to encourage such adjustments rather than to protect a noncompetitive industry that results in inefficient use of resources.* It is not enough to say that resource adjustments will automatically occur. That, I have already argued, is probably true. But the time that elapses before this happens may in some cases be intolerable! Thus, a policy approach to ensure that appropriate adjustment incentives exist may be the most appropriate. We should not be satisfied with policy that encourages resources to be continuously employed in the production of some commodity that is unneeded or can be obtained elsewhere when these resources would be more productive in other pursuits.

The United States, along with most developed countries having a strong agricultural base, has long had a love affair with export subsidies designed to encourage increased exports and with import controls designed to protect local industries from foreign competition. As we have seen, such policies accomplish little in terms of encouraging a greater degree of competitiveness on the part of the industries aided or protected. U.S. export subsidies involve a substantial transfer of welfare from the United States to foreign consumers. Import restrictions serve to encourage production inefficiencies, raise consumer prices of affected products, and retard alternative best uses of resources.

Yet second-best approaches are sometimes called for. When one of our trading partners engages in export dumping of a particular commodity, one can no longer argue (e.g., on efficiency grounds) that the United States should not also subsidize exports of that commodity or, in retaliation, impose high import barriers on a commodity imported from that trading partner. Thus, while our primary focus for the long term should be on encouraging competitiveness and removing export subsidies and import controls, authority to retaliate against selective and serious unfair trade practices should probably be retained. This authority should be broad enough to cause severe economic hardship over sustained periods of time if necessary, rather than limited to merely creating a nagging irritation. Further, this authority should probably be sanctioned through the framework of the GATT. *However, export subsidies or import controls should not be a permanent feature of U.S. agriculture policy.*

Resource Protection and Conservation. Resource protection and soil conservation are objectives that should be placed very high on the policy agenda. Pollution of groundwater and surface water in the United States (as well as elsewhere) is of utmost concern and will have impacts on the lives of people of current as well as future generations. Obviously agriculture is not the only source of this pollution. It is, though, a large contributor in many areas of the country. It may well be that policies to deal with pollution control will increase the cost of agricultural production and therefore the price of food. It is not likely that this will seriously affect our competitive edge, however, since other countries have the same problem (albeit to different degrees) and will have to face up to the same costly solutions.

Preservation of the productivity of the soil and prevention of drawdown of underground water supplies are of more concern to future generations. We must begin now to adopt appropriate policies to deal with these issues. Economic analysis can provide insights into whether controls, subsidies, taxes, or some other approach is optimal here. The conservation reserve feature of the Food Security Act of 1985 is one positive approach. Means of making it even more effective need to be continuously sought.

Price Stabilization. Price stabilization is another policy objective that deserves fairly high priority. Although the incidences of price instability are difficult to pinpoint, when excessive it most certainly complicates the lives of agricultural producers and detracts from production efficiency. In the past, we have initiated programs designed to deal with instability but quickly turned them into income enhancement programs as political pressures became too great and/or as the economic status of farmers changed. What is needed is a solid commitment to farm price and/or income stability with rules of implementation clearly spelled out and adhered to. Perhaps a stabilization policy built around quantity rules rather than price rules would make it less easy to turn such a program into a price enhancement program. Ultimately, though, it is the sustained commitment that will have to evolve. Although implementing a price stabilization policy will be costly both in terms of budgetary outlays and in terms of gross farm revenue lost, it will likely produce efficiency gains as well as equity benefits. In contrast, *farm income enhancement is costly, is of questionable merit in terms of social needs, causes inequities of its own, and does little to stabilize prices.*

Rural Development. Many farm families do not have the resources necessary to generate a level of income from the sale of farm produce alone sufficient to sustain the farm family. Some farm families with somewhat more resources still suffer periods of low family income due to fluctuations in prices and/or yields. Other farm families, often for the same reasons, find that they are at times temporarily short of the cash needed to meet current debt obligations. In most of these instances, the farm operator or his or her spouse or both seek nonfarm jobs with which to supplement their farm income on either a permanent or a temporary basis. In fact, for all but the very largest sized farms, nonfarm sources of income appear to be *very* important to the nation's farmers. Clearly if nonfarm jobs are not readily available to them, they will suffer severe hardships. Policy designed to assist the rural community to survive and prosper, then, will be of considerable significance to many of our farm families. *Indeed, policy aimed at maintaining the economic viability of the rural community and of the jobs for farm people that the rural community can provide will be much more important to many of these farm families than will be the more traditional price and income support policy of the past forty years.*

Targeted Support. Since the 1930s, agricultural policy in the United States has been oriented to commodities rather than specifically to people or resources in need. Further, it has been directed to those commodities that were readily storable, which also just happened to be the commodities grown in the more politically powerful regions of the country: the dairy regions, the Corn Belt, the Great Plains, and the Old South. The result has been a legacy of support for agriculture that not only left many people out but provided support for many who had little need.

In the 1930s, 1940s, and perhaps even 1950s, this was not such a problem, because almost all farmers everywhere were so highly diversified that support for a feed grain or cotton farmer also meant support for a hog, cattle, or poultry producer. Today, farmers are rather more highly specialized, so we can no longer make this claim. Many of these highly specialized farmers are very progressive businesspeople who need little outside income support. Providing income support to all corn, cotton, wheat, milk, or rice producers just to assist the few who are really in need of assistance is a questionable approach. A more appropriate policy approach may well be to develop means tests just as we do in other welfare programs and provide assistance on this basis. In this way we could help the potato and apple and egg producers in need just as well as the feed grain producers in need or, for that matter, just as well as the single-parent family whose income level is below the poverty line.

Organizing such a program would be difficult and costly, just as it has been in the case of other welfare programs. It is, though, probably the most cost-effective way of singling out those in need and providing assistance accordingly. The need here might be income support or it might be adjustment assistance, including job retraining. Perhaps it is the closest we can come to *decoupling,* a term recently popularized to denote the separation of farm program benefits from farm production decisions (see, e.g., Agricultural Policy Working Group 1988).

Once the concept of *targeting* is in place, then it is easy to generalize from targeting individuals in need to targeting resources, regions, countries, etc. Indeed the concept is by no means new. The Food Security Act of 1985 has targeted highly erodible cropland and swampland for inclusion in the conservation reserve program. This act also provided for targeting countries for which retaliation for unfair trade practices was deemed to be in order.

Resource Adjustments. When input prices change, U.S. farmers are expected to change their mix of inputs accordingly. Similarly, when new technologies are developed, U.S. farmers are expected to adopt these technologies in order to produce more efficiently. As we have seen, farmers do change their input mix and they do adopt new technologies. It is curious, then, that when output prices *should* change in response to shifting demand conditions or shifting comparative advantages, the usual policy response in the past has been to discourage adjustments by maintaining output prices at arbitrarily high levels or by setting up other disincentives. *The more appropriate policy response would appear to be to encourage adjustments so that resources are employed in their best use—both physical and human resources.* I have suggested that the idea of asset fixity in agriculture is probably overstressed. Nevertheless, there are instances in which adjustments are difficult because assets are sticky, farm people do not have the skills needed for an alternative occupation, or alternative occupations within or outside of agriculture are not readily available. If this is the case, a more rational approach to the problem would be to develop policies aimed specifically at easing the adjustment process. This may take the form of job retraining, development efforts designed to create nonfarm jobs, farm asset buyout programs, developing new crops for agriculture, developing and maintaining viable marketing channels so farmers will have access to needed markets for traditional or new products, etc.

Summary

The conceptual framework commonly used by economists for analyzing policy choices leads to valuable insights about the workings of an economic system as well as about intervention strategies to which an economic system is or might be exposed. Allocative efficiency is the cornerstone of this framework. It yields results relevant to a static equilibrium. Additional approaches, though, must be invoked for a full assessment of policy choices and impacts. Equity considerations cannot be assessed with the above conceptual framework, yet they are part and parcel of policy prescriptions. Whether the real system can actually achieve equilibrium (tacitly assumed by the above framework) or can tend in this direction and the character of the adjustment process itself are also of vital concern to the policymaker. Whether the system is progressive in the sense that it readily adopts new technology and achieves the size economies currently available is also an important consideration.

Policy that has evolved for the agricultural sector appears to have been founded on basic tenets that no longer hold or are no longer relevant: (1) resources in agriculture are relatively fixed even in the long run, so equilibrating adjustments would not occur naturally; (2) farm income enhancement was needed to make farm family incomes comparable to those of nonfarm families; and (3) supporting farm commodities was sufficient to aid those farm people in need. Competitiveness, resource protection and conservation, price stabilization, and rural development appear to be the more important objectives that should be pursued today. Targeting policy actions to people and resources rather than to commodities seems to be more consistent with the needs of today and of the future. In this way specific low-income problems among farm people, specific resource adjustment problems, and specific environmental problems can be more effectively resolved.

Notes

1. By *free trade* I mean the absence of government intervention of any kind that would prevent attainment of a competitive equilibrium in a spatial context—that is, interregionally or internationally. Thus, free trade and perfect competition might be used interchangeably even though the latter is often abstracted from spatial considerations. In this chapter I use the term *perfect competition* to imply also *free trade*, or perfect competition in the spatial dimension.

2. An alternative concept often discussed in the literature is that of economies of scale. Economy of scale refers to the impact on per unit costs of production as the farm expands in size by increasing the level of all resources in fixed proportions. Few farms expand in this way. Most expand by increasing the proportion of capital to labor or by increasing the proportion of variable inputs to fixed capital. Thus, the concept of economies of size is more applicable in the real world.

3. For a recent assessment of the phenomenon of part-time farming in Pennsylvania and Massachusetts, see Hallberg, Findeis, and Lass (1987). This study points out not only that part-time farming is important to many farm families in these two states, but it is a long-term proposition and likely to remain an important source of income to many farm families well into the future.

Suggested Readings and References

Agricultural Policy Working Group. 1988. "Decoupling: A New Direction in Global Farm Policy." 517 C Street NE, Washington, D.C. February.

Cochrane, Willard W. 1958. *Farm Prices: Myth and Reality.* Minneapolis: University of Minnesota Press.

Drache, Hiram M. 1964. *The Day of the Bonanza.* North Dakota Institute for Regional Studies. Fargo: Lund Press.

Hallberg, M. C. 1988. "The U.S. Agricultural and Food System: A Postwar Historical Perspective." Northeast Regional Center for Rural Development, Pennsylvania State University. Publication no. 55. October.

_____. 1989. "Resource Adjustments in American Agriculture." Paper presented at Symposium on Surplus Capacity and Resource Adjustments in American Agriculture, St. Louis, January 23–25.

Hallberg, M. C., J. L. Findeis, and Daniel A. Lass. 1987. "Part-Time Farming in Pennsylvania and Massachusetts: Survey Results." Department of Agricultural Economics and Rural Sociology, Pennsylvania State University. A.E. & R.S. no. 194. September.

Hood, Kenneth. 1936. "An Economic Study of Part-Time Farming in the Elmira and Albany Areas of New York, 1932 and 1933."Cornell University Agricultural Experiment Station Bulletin no. 647.

Madden, J. Patrick. 1967. "Economies of Size in Farming." U.S. Department of Agriculture. Economic Research Service. AER-107. February.

Miller, Thomas A., Gordon E. Rodewald, and Robert G. McElroy. 1981. "Economies of Size in U.S. Field Crop Farming." U.S. Department of Agriculture.

Economics and Statistics Service. Agricultural Economic Report no. 472. July.

Office of Technology Assessment. 1986. *Technology, Public Policy, and the Changing Structure of American Agriculture.* Washington, D.C.: U.S. Government Printing Office.

Posnan, M. M., ed. 1966. *The Cambridge Economic History of Europe. The Agrarian Life of the Middle Ages.* Vol. 1. Cambridge: Cambridge University Press.

Schaller, Neill. 1988. "Alternative Agriculture Gains Attention." *Agricultural Outlook.* U.S. Department of Agriculture. Economic Research Service. AO-140. April, pp. 26–28.

U.S. Department of Agriculture. 1989. "Financial Characteristics of U.S. Farms, January 1, 1989." Economic Research Service. Agricultural Information Bulletin no. 579. December.

Advanced Readings

Baker, Derek, Milton Hallberg, and David Blandford. 1989. "U.S. Agriculture under Multilateral and Unilateral Trade Liberalization: What the Models Say." Department of Agricultural Economics and Rural Sociology, Pennsylvania State University. A.E. & R.S. 200. January.

Buxton, Boyd M., Tom McGuckin, Roger Selley, and Gayle Willet. 1985. "Milk Production: A Four-State Earnings Comparison." U.S. Department of Agriculture. Economic Research Service. Agricultural Economic Report no. 528.

Henderson, James M. and Richard E. Quandt. 1980. *Microeconomic Theory.* 3d ed. New York: McGraw-Hill.

Just, Richard E., Darrell L. Hueth, and Andrew Schmitz. 1982. *Applied Welfare Analysis and Public Policy.* Englewood Cliffs, N.J.: Prentice-Hall.

Kislev, Yoav and Willis Peterson. 1982. "Prices, Technology, and Farm Size." *Journal of Political Economy* 90 (3): 578–95.

Lee, John E. 1965. "Allocating Farm Resources between Farm and Non-Farm Uses." *Journal of Farm Economics* 47 (1): 83–92.

Ruttan, Vernon R. 1982. *Agricultural Research Policy.* Minneapolis: University of Minnesota Press.

Schmitz, Andrew, and David Seckler. 1970. "Mechanized Agriculture and Social Welfare: The Case of the Tomato Harvester." *American Journal of Agricultural Economics* 52 (4): 569–77.

U.S. Department of Agriculture. 1989. "Economic Indicators of the Farm Sector: National Financial Summary, 1988." Economic Research Service. ECIFS 8-1. September.

APPENDIX: Chronology of Legislation and Executive Orders Affecting U.S. Agriculture since 1862

PROVIDED HERE is a chronological listing of legislation having a major impact on the development, organization, and control of U.S. agriculture, beginning with the four major agricultural acts passed during President Lincoln's administration. A principal focus of this listing is on price and income policy for agriculture, which began, in large part, with the Agricultural Marketing Act of 1929. The major acts pertaining to this type of policy legislation (commonly referred to as farm bills) are set apart in the listing below in boldface type. I have included only the more recent and more significant legislation dealing with product grades and standards, product safety, and use of farm inputs. In general, legislation dealing with taxation is not included; however, the Tax Reform Act of 1986 appears to deserve a special place in this listing since it promises to have a significant impact on the future development and structure of agriculture. Also not included are the several legislative bills that have been debated and that have had a major impact on the final legislation adopted. In this category are the McNary-Haugen bills of the 1920s and the so-called Brannan plan proposed in 1949. The interested reader should consult Cochrane 1979, Tweeten 1979, and Christenson 1959.

The listing developed here is based on many sources. For additional details on any act, the reader is referred to Bowers, Rasmussen, and Baker 1984, Smith 1979, Cochrane and Ryan 1976, Cochrane 1979, Tweeten 1979, and Benedict 1953.

A full appreciation of the legislation listed here (or of a subset of it) would require a great deal of effort to uncover the major issues of the day leading up to that legislation and the political posturing of the administration, the Congress, and the agricultural lobbying groups which helped shape that legislation. For example, an examination of the issues leading up to the protectionist McKinley Tariff Act of 1890 and the Hawley-Smoot Tariff Act of

1930 would be most interesting and enlightening. Nonpoliticians have also played an important role in the formation of agricultural legislation through their writings, speeches, and debates. Three such examples are Upton Sinclair's *The Jungle,* published in 1906, John Steinbeck's *The Grapes of Wrath,* published in 1939, and Rachel Carson's *Silent Spring,* published in 1962. Another fruitful venture for the prospective student of policy would be to examine the "pure politics" of agricultural policy formation. James Giglio (1987), for example, has very ably studied the process of agricultural policy formation during the Kennedy years.

Homestead Act of 1862
- Gave fee simple title for a 160-acre tract of land to a U.S. citizen who (1) had farmed the claim for five consecutive years after filing on it and (2) had paid the requisite fees ($10 in the early years of this act).

Morrill Land-Grant College Act of 1862
- Authorized endowment, support, and maintenance of at least one college in each state for the main purpose of teaching agricultural and mechanical arts subjects.
- Each state was to receive 30,000 acres of public land or land script for each member of its congressional delegation.

Organic Act of the Department of Agriculture, 1862
- Signed by President Lincoln on May 15, 1862.
- "There shall be at the seat of Government a Department of Agriculture, the general design and duties of which shall be to acquire and to diffuse among the people of the United States useful information on subjects connected with agriculture and rural development, in the most general and comprehensive sense of, and to procure, propagate, and distribute among the people new and valuable seeds and plants."
- The department is now composed of the following agencies:

> Agricultural Cooperative Service (ACS)
> Agricultural Marketing Service (AMS)
> Agricultural Research Service (ARS)
> Agricultural Stabilization and Conservation Service (ASCS)
> Animal and Plant Health Inspection Service (APHIS)
> Commodity Credit Corporation (CCC)
> Cooperative State Research Service (CSRS)
> Economic Research Service (ERS)
> Extension Service (ES)
> Farm Credit Administration (FCA)
> Farm Credit System (FCS)
> Farmers Home Administration (FmHA)
> Federal Crop Insurance Corporation (FCIC)
> Federal Grain Inspection Service (FGIS)
> Food and Nutrition Service (FNS)
> Food Safety and Inspection Service (FSIS)
> Foreign Agricultural Service (FAS)
> Forest Service (FS)
> Human Nutrition Information Service (HNIS)
> National Agricultural Library (NAL)

National Agricultural Statistics Service (NASS)
Office of the Consumer Advisor (OCA)
Office of the General Counsel (OGC)
Office of Governmental and Public Affairs (OGPA)
Office of International Cooperation and Development (OICD)
Packers and Stockyards Administration (P&SA)
Rural Development Administration (RDA)
Rural Electrification Administration (REA)
Soil Conservation Service (SCS)
World Agricultural Outlook Board (WAOB)

Pacific Railway Act of 1862
- Authorized construction of a transcontinental rail line from Omaha to Sacramento.
- Federal subsidies were given western railroads in the form of (1) government credit varying from $16,000 to $48,000 per railroad mile and (2) grants of public domain land to a distance of 20 miles on each side of the rail line.

Tariff Act of 1864
- Continued the protectionist policies of previous tariff acts and increased the average level of tariffs on imports to 47 percent.

Desert Land Act of 1877
- Gave fee simple title to a 640-acre tract of land in the arid West on condition that the land be reclaimed by "conducting water on it."

Meat Inspection Act of 1884
- Authorized the organization of a Bureau of Animal Industry to conduct research on the control of animal diseases, to regulate exports of livestock and poultry, and to prevent the spread of animal diseases from state to state.

Hatch Experiment Station Act of 1887
- Provided federal grants (from the sale of public lands) to the states for support of agricultural experimentation and research.

Interstate Commerce Act of 1887
- Provided the authority for controlling passenger and freight rates across the nation.

Morrill Land-Grant College Act of 1890
- Authorized separate land-grant colleges for African Americans.

Sherman Antitrust Act of 1890
- Prohibited monopolistic-type business combinations such as "pools," "trusts," and "holding corporations," which acted as conspiracies in restraint of trade.

McKinley Tariff Act of 1890
- Provided for the highest protection yet afforded by any tariff laws of the United States—an average tariff rate on imports of about 50 percent.
- Provided a complete schedule of protective duties on farm products including meats, barley, wheat, eggs, corn, potatoes, and butter.
- Gave the president authority to impose by proclamation duties on sugar, molasses, tea, coffee, and hides if he considered that any country exporting these commodities to the United States "imposes duties or other exactions on the agricultural or other products of the United States, which, in view of the free introduction of sugar, molasses, tea, coffee and hides into the United States, he may deem to be

unjust or unreasonable."
- Provided for a bounty of 2 cents per pound to domestic sugar producers to offset the admission of raw sugar duty free. To protect American sugar refiners, a duty of one-half cent per pound on refined sugar was levied.

Carey Act of 1894
- Granted 1 million acres to each state containing arid land on condition that the state provide for the necessary reclamation.
- Usually the states contracted with construction companies to reclaim specified areas and to sell water rights to farmers as a means of recovering their investment.
- The land was sold for nominal prices but only to those who had contracted to purchase the water rights.

Land Reclamation Act of 1902
- Authorized funds derived from the sale of public lands for use in construction of storage and power dams and canal systems needed for irrigable lands in the West.
- Settlers on these lands would receive land free as under the Homestead Act but were to repay the cost of the structures (less interest) built over ten years.

Federal Food and Drug Act of 1906
- Prohibited the manufacture, sale, or transportation in interstate commerce of adulterated, misbranded, poisonous, or deleterious foods, drugs, medicines, or liquors.

Federal Meat Inspection Act of 1907
- Provided for mandatory postmortem examination of the carcasses of all cattle, sheep, swine, and goats prepared for human consumption and for transportation in interstate commerce. This act was really designed to protect the consumer but met with the general approval of farmers as well.

Federal Insecticide Act of 1910
- Prevented the manufacture, sale, or transportation of adulterated or misbranded insecticides and fungicides.
- Authorized state regulation of the sale of insecticides and fungicides.

Clayton Act of 1914
- Prohibited price discrimination when such discrimination lessened competition or created a monopoly. Prohibited "tie-in" arrangements. Prohibited the acquisition of stock of another company if such acquisition would lessen competition. Prohibited interlocking directorates. Declared labor unions and farmer organizations not to be conspiracies in restraint of trade.

Federal Trade Commission Act of 1914
- Provided for an investigative agency to enforce the Sherman and Clayton Act provisions.

Smith-Lever Act of 1914
- Formalized cooperative agricultural extension work.

Federal Farm Loan Act of 1916
- Provided for twelve farmland banks (National Farm Loan Associations, now known as Federal Land Bank Associations) for the purpose of making long-term loans to farmers with the policy goal "to provide capital for agricultural development, to create standard forms of investment based on farm mortgages and to equalize rates of interest on farm loans."

U.S. Grain Standards Act of 1916
- Promoted uniform and accurate grading of grain under the official grain standards of the United States.

U.S. Warehouse Act of 1916
- Provided for the licensing and supervision of warehouses and their operations.

Smith-Hughes Vocational Education Act of 1917
- Provided for vocational educational training in high schools.

Packers and Stockyards Act of 1921
- Authorized the regulation of meatpackers and livestock trading practices and charges at terminal livestock markets.

Commodity Exchange Act of 1922
- Intended to ensure fair practices and honest dealing on the futures exchanges.
- Limited the amount of price changes permitted in any given period of trading.
- Provided the authority for creation of the Commodity Exchange Authority to carry out the provisions of the act.

Capper-Volstead Act of 1922
- Gave producers the right to jointly market their products through the formation of a common marketing agency which would not (by its existence) be considered in restraint of trade in violation of the federal antitrust statutes.
- Prohibited restraint of trade and undue price enhancement.

Agricultural Credit Act of 1923
- Set up a Federal Intermediate Credit Bank in each of twelve federal districts for making intermediate length loans to farmers.

Agricultural Marketing Act of 1929
- Created the Federal Farm Board with a revolving fund of $500 million, which had authority to make loans to cooperative associations, to make advances to cooperative members, and to make loans to stabilization corporations for the purpose of controlling any surplus through purchase operations. The board created three national cooperative selling agencies:

    1. Farmers National Grain Corporation, through which a farmer could have his grain stored in bonded warehouses subject to sale at any time on orders from him or enter his grain in a pool, receiving an advance against it and being paid at the end of the pool period the average price received by the cooperative for pooled grain of the type and quality supplied by him
    2. National Wool Marketing Corporation
    3. American Cotton Cooperative Association

- The board established the Cotton Stabilization Corporation in June of 1930 as a means of absorbing excess cotton and supporting the cotton price structure.

Hawley-Smoot Tariff Act of 1930
- Raised tariffs on imports of agricultural and nonagricultural products to an average of over 52 percent.

Perishable Agricultural Commodities Act of 1930
- Provided for the licensing of commission men, dealers, and brokers handling fresh fruit and vegetables.

Agricultural Adjustment Act of 1933
- Aimed to restore purchasing power of agricultural commodities to the 1909–14 level—a goal later (1938) to become known as "parity."
- "Parity" was to be achieved by
  1. Voluntary acreage reductions in basic commodities and direct payments for participation (the so-called voluntary domestic allotment plan)
  2. Funds with which to implement the voluntary domestic allotment plan provided by an excise tax collected from processors
  3. Regulating marketing through voluntary agreements with processors, cooperative associations, and other handlers
- Basic commodities were defined to be wheat, cotton, corn, hogs, rice, tobacco, and milk.
- Authorized use of marketing orders and agreements.
- Provided for quantity and acreage controls for tobacco.
- Provided for price-fixing marketing agreements for tobacco by which buyers of tobacco were required to pay minimum prices.

Emergency Farm Mortgage Act of 1933 (Title II of Agricultural Adjustment Act of 1933)
- Provided for refinancing farm mortgages held by mortgagees other than Federal Land Banks, reduction of interest rates on all Land Bank loans, extension of time on obligations of Land Bank borrowers in need of such aid, a $200 million fund to be used for making "rescue" loans, more liberal loans to owners of orchard lands, and direct loans to farmers where local farm loan associations were not available.

Farm Credit Act of 1933
- Established a central bank and twelve regional banks for making loans to cooperatives and a system of twelve production credit corporations to organize, supervise and finance local Production Credit Associations so as to enable cooperating groups of farmers to borrow through them from the Intermediate Credit Banks. The twelve Land Banks for making mortgage loans were retained.

Executive Order 6084, March 1933
- Established the Farm Credit Administration, which was to administer the farm credit laws.

Executive Order 6340, October 16, 1933
- Established the Commodity Credit Corporation to grant nonrecourse loans to cotton, corn, and wheat growers enabling them to hold their crop until the price could advance as a result of the government's production control program.

Federal Surplus Relief Corporation, October 4, 1933
- Was operating agency for carrying out food purchase and distribution programs to remove surpluses of basic agricultural commodities.
- Renamed Federal Surplus Commodities Corporation in November 1935.

Tennessee Valley Authority Act of 1933 (TVA)
- Implemented a program of unified resource development in the Tennessee River drainage basin and adjoining territory within seven states.

International Wheat Agreement of 1933
- Was an agreement among twenty-two wheat exporting and importing countries.

- Provided for export quotas to be assigned to each of the exporting countries during two succeeding years, and for the importing countries to discontinue efforts to expand production.
- All countries were to reduce tariffs and other restrictions as prices improved.

Civilian Conservation Corps Act of 1933
- Was an emergency measure authorizing the recruitment of unemployed and unmarried men between the ages of 18 and 25 for work on improving national forests and parks, building roads and firebreaks, and checking erosion. These individuals were housed in "CCC Camps" and given food and lodging as well as a small monthly wage.

Jones-Connally Act of 1934
- Added rye, flax, barley, grain sorghum, peanuts, and cattle to the list of basic commodities established by the Agricultural Adjustment Act of 1933.

Jones-Costigan Sugar Act of 1934
- Added sugarcane and sugar beets to list of basic commodities.
- Gave the secretary of agriculture power to make rental payments or other benefit payments for acreage or marketing restrictions on sugar production.
- Imposed a processing tax on sugar.
- Established a system of import quotas for the amount of sugar that could be sold in the United States by foreign producers.
- Authorized U.S. sugar growers to be given an allotment for their share of the total U.S. acreage allotment.
- Required minimum wage payments to sugar field-workers and banned the use of child labor in beet fields.

Bankhead Cotton Control Act of 1934
- Implemented mandatory production controls via marketing quotas when approved in a referendum of two thirds of producers.
- Levied a high tax on all noncompliers.

Kerr-Smith Tobacco Control Act of 1934
- Implemented mandatory production controls via marketing quotas when approved in a referendum of three fourths of producers.
- Levied a high tax on all noncompliers.

Taylor Grazing Act of 1934
- Designed to (1) stop injury to public grazing lands by preventing overgrazing and soil deterioration, (2) provide for the improvement or development of public grazing lands, and (3) stabilize the livestock industry dependent upon the public range.

Reciprocal Trade Agreements Act of 1934
- Gave the president power to enter into trade agreements with foreign countries. Such agreements were to remain in force for three years and to continue in force thereafter unless terminated by either of the governments concerned.
- Gave the president power to revise tariff duties by not more than 50 percent up or down. Thus the Hawley-Smoot Tariff rates could, under this act, be substantially reduced in return for reciprocal tariff reductions by foreign countries.
- Preserved special concessions to Cuba so long as existing preferential duties were not modified by more than 50 percent.

- Represented a major change in U.S. trade policy away from protectionism.

Soil Conservation Act of 1935
- Established the Soil Conservation Service as a bureau in the Department of Agriculture.
- Was the parent legislation for subsequent soil conservation programs.
- Authorized (1) soil conservation surveys, research on soil conservation issues, and demonstration conservation projects, (2) the design and implementation of erosion prevention measures, and (3) the granting of aid to agencies and farmers for the implementation of soil conservation programs.

Warren Potato Act of 1935
- Added potatoes to list of basic commodities.

DeRouen Rice Act of 1935
- Provided for voluntary acreage restrictions for rice with payments financed by a processing tax.

Executive Order No. 7027, May 1, 1935
- Established the Resettlement Administration, which was to (1) aid in the resettlement and housing of destitute or low-income families, both rural and urban; (2) carry out certain land conservation projects; and (3) help farm families on relief to become independent by providing financial and technical assistance.

Agricultural Adjustment Act amendments, 1935
- Section 32 set aside 30 percent of customs receipts for encouragement of exports and domestic consumption of surplus agricultural commodities.
- Section 22 gave the President the authority to impose import quotas on agricultural commodities whenever he believed imports interfered with the agricultural adjustment programs of the United States.

Hoosac-Mills Decision of Supreme Court, 1936
- Declared illegal those features of the Agricultural Adjustment Act of 1933 dealing with government contracts with individual farmers and with processing taxes.

Flood Control Act of 1936
- Authorized the Department of Agriculture to institute investigations of watersheds and measures for runoff and water-flow regulation and for the prevention of soil erosion on watersheds.
- Provided that the upstream phase of flood control be administered by the Department of Agriculture and the downstream phase including dams and other structures, be administered by the Army Corps of Engineers.

Soil Conservation and Domestic Allotment Act of 1936
- Changed emphasis from "price" supports to "income" supports.
- Introduced an "income parity" goal.
- Provided funds for income support through Congress rather than from processing taxes.
- Replaced payments for acreage adjustment with payments for shifting acreage from soil-depleting to soil-conserving crops.

Rural Electrification Administration Act of 1936
- Provided long-term, low-interest loans to local rural electric administration cooperatives for the extension of electric lines to farms and rural communities.
- Provided up to 100 percent of the cost of constructing electric generation,

transmission, and distribution facilities.

Robinson-Patman Act of 1936

- Attempted to prevent large concerns from using their market power to secure unfair price advantage over competitors.

Pope-Jones Water Facilities Act of 1937

- Authorized the secretary of agriculture to provide facilities for water storage and utilization in arid and semiarid areas.

Sugar Act of 1937

- Maintained provisions of Jones-Costigan Sugar Act.
- Processing tax and government contracts with individual sugar producers were later repealed.

International Sugar Agreement of 1937

- Was agreement between twenty-one countries representing 85–90 percent of world's sugar supply.
- Importing countries agreed to limit production and exporting countries to observe the marketing quotas established by the agreement.

Agricultural Marketing Agreement Act of 1937

- Clarified legal status of and authorized marketing orders and agreements.

Bankhead-Jones Farm Tenancy Act of 1937

- Clarified legal status of activities previously carried out by the Resettlement Administration.
- Permitted secretary of agriculture to change name of Resettlement Administration to Farm Security Administration.
- Provided for loans to tenant farmers, share-croppers, and farm laborers for purchase of farms and for rehabilitation, and provided for the retirement of submarginal lands.

Fair Labor Standards Act of 1938

- Established minimum wages, maximum hours allowable without overtime pay, and child labor standards.
- Established criteria for claiming exemptions under the law and granted specific exemptions to agriculture.

Agricultural Adjustment Act of 1938

- Is one of the four legislative acts commonly considered part of "permanent" legislation for price and income policy for agriculture.
- Reenacted provisions of the Soil Conservation and Domestic Allotment Act of 1936.
- Was first legislation to explicitly use the term *parity*.
- Limited conservation payments to $10,000.
- Created the Federal Crop Insurance Corporation with a capital stock of $100 million, which initially was to provide crop insurance for wheat farmers. An insurance program for cotton was passed in 1941.
- Initiated mandatory nonrecourse loans for corn, wheat, and cotton producers and provided for marketing quotas if approved in a referendum of the producers voting if supplies reached certain levels. Cotton, wheat, and corn loan rates could be set between 52 percent and 75 percent of parity.
- Introduced the concept of an "ever-normal granary."

Wheeler-Case Act of 1939
- Authorized federal assistance for the purpose of developing small irrigation projects needed for the resettlement of farm families displaced by droughts.

Food Stamp Program and School Lunch Program of 1939
- Established the Federal Surplus Commodities Corporation, which was designed primarily as a means of disposing of agricultural surpluses.

Inter-American Coffee Agreement of 1940
- Established export quotas between fourteen producing nations and the United States.

Lend-Lease Act of 1941
- Was a program under which essential goods and services, including agricultural commodities, were provided to the Allies before and after U.S. entry into the war.
- Authorized higher price supports to encourage increased production of pork, dairy products, eggs, canned vegetables, garden seeds, flax, hemp, castor oil, tobacco, rice, beans, peanuts, soybeans, and peas.

Steagall Amendment (to Lend-Lease Act) of 1941
- Provided that for any agricultural commodities for which increased production was sought to satisfy war effort needs, prices would be supported at 85 percent of parity. These commodities were known as "Steagall" commodities to distinguish them from the then "basic" commodities.

Stabilization Act of 1942
- Prohibited price ceilings lower than either the parity price for the commodity or the highest price paid between January 1, 1942, and September 15, 1942.
- Provided that for cotton, corn, wheat, rice, tobacco and peanuts, loans would be made by Commodity Credit Corporation at 90 percent of parity for two years after the war ended for farmers who participated in the acreage or marketing quotas announced.
- Raised the support level on Steagall commodities from 85 to 90 percent of parity.

Farmers Home Administration Act of 1946
- Abolished the Farm Security Administration. It maintained the more conservative credit features but prohibited ventures into collective or cooperative farming and eliminated the farm labor camp programs.

Agricultural Research and Marketing Act of 1946
- Authorized generous support for study of the economic aspects of agriculture, particularly in respect to marketing.

National School Lunch Act of 1946
- Authorized regular federal appropriations for cash grants to the states to support lunch programs for public and private schools.

Bureau of Land Management Act of 1946
- Was created by executive reorganization.
- Is responsible for administering lands in federal grazing districts and other public lands not so assigned according to the provisions of the Taylor Grazing Act of 1934.

General Agreement on Tariffs and Trade Act of 1947
- A general agreement for working out conditions for freer trade among several nations (see Glossary).

Federal Insecticide, Fungicide, and Rodenticide Act of 1947
- Replaced the Federal Insecticide Act of 1910.
- Required registration of chemical pesticides before their sale or shipment in interstate or foreign commerce.
- Required prominent display of poison warnings on labels of highly toxic materials.
- Required coloring of insecticides to prevent their being mistaken for foodstuffs.
- Required instructions for use.
- Required furnishing of information to the administrator of the act concerning delivery, movement, and holding of pesticides.

Sugar Act of 1948
- Reenacted the quota system instituted under the Sugar Act of 1937.
- Stated quotas for Philippine suppliers in tons, and for other nations as a percentage of U.S. requirements above the domestic and Philippine quotas.
- Authorized that Philippine sugar enter duty free, that Cuban sugar enter at a partial duty rate, and all other sugar at the full duty.

Foreign Assistance Act of 1948
- The Marshall Plan act for assisting with European recovery from the war.
- Authorized the use of section 32 funds to subsidize exports of surplus agricultural commodities to participating nations or to occupied areas.

Commodity Credit Corporation Charter Act of 1948
- Is one of the four legislative acts commonly considered part of "permanent" legislation for price and income policy for agriculture.
- Provided a federal charter for the Commodity Credit Corporation.
- Specified the purposes and powers of the Commodity Credit Corporation.

Agricultural Act of 1948
- Authorized, through 1950, price supports for cotton, wheat, corn, tobacco, rice, peanuts, potatoes, milk, hogs, chickens, and eggs at 90 percent of parity. Steagall commodities were to be supported at not less than 60 percent of parity nor more than the 1948 support level.
- Maintained loan and price support procedures authorized in the Agricultural Adjustment Act of 1938.
- Authorized after 1950, a new parity formula based on 1910–14 and a more flexible program of price supports to replace the fixed percentages of parity in place during and following the war.

International Wheat Agreement of 1949
- Renewal of an agreement between the four major wheat exporters (Australia, Canada, France, and the United States) and thirty-seven importers.

Agricultural Act of 1949
- Is one of four legislative acts commonly considered part of "permanent" legislation for price and income policy for agriculture.
- Repealed previous price support provisions.
- Froze price supports at 90 percent of parity for 1950 if acreage allotments or quotas were in effect.
- Pegged minimum price support levels for 1951 at 80 percent of parity, and in 1952 at 75 percent of parity, except for tobacco which was to be maintained at 90 percent of parity. If there were no quotas in effect, the minimum price support

level was to be 50 percent of parity.
- Mandatory supports for dairy, tung nuts, honey, potatoes, and wool were to be between 60 and 90 percent of parity.
- Authorized donations of surplus commodities to needy at home and abroad under Section 416.
- It was the last major agricultural act without an expiration date.

Defense Production Act amendments of 1952
- Set price supports for six basic crops (wheat, corn, cotton, peanuts, rice, and tobacco) at 90 percent of parity through 1954.

Agricultural Act of 1954
- Established flexible price supports beginning in 1954.
- Authorized a Commodity Credit Corporation reserve for foreign and domestic relief.
- Under Title VII (the National Wool Act of 1954) provided for the support of wool and mohair prices by means of loans, purchases, payments, or other operations.
- Established the Special Milk Program to increase children's fluid milk consumption.

Agricultural Trade Development and Assistance Act of 1954
- Is one of four legislative acts commonly considered part of "permanent" legislation for price and income policy for agriculture.
- Is the basic act concerning selling and bartering surplus commodities overseas for relief.
- Is more commonly known as Public Law 480.
- Under Title I provided for "sales for foreign currency." Foreign currencies acquired could be used to develop new markets for U.S. farm commodities, for financing the purchase of goods or services for other friendly nations, for promoting economic development and trade, etc.
- Under Title II provided for "famine relief and other assistance" to any nation experiencing an emergency.
- Under Title III amended section 416 of the Agricultural Act of 1949 to expand donations of Commodity Credit Corporation surplus commodities for domestic and foreign relief and amended section 4 of the Commodity Credit Corporation Charter Act to broaden its authority dealing with barter of surplus commodities.

Amendment to Food and Drug Act, 1954
- Authorized administrator of the Federal Insecticide, Fungicide, and Rodenticide Act to set tolerance limits for the residues of pesticides in foods.
- Required the pretesting of a chemical pesticide before it could be used on food crops.
- Required manufacturers to provide detailed data demonstrating the usefulness of a chemical to agriculture and to provide scientific data on the toxicity of a chemical on animals.

Agricultural Act of 1956
- Authorized the Soil Bank Program for short- and long-term removal of land from production with annual rental payments to cooperators. Two programs authorized were (1) an acreage reserve program and (2) a conservation reserve program for long-term conversion to conservation uses.

- Raised price supports for oats, rye, barley, and sorghum to 70–76 percent of parity.
- Authorized a two-tiered system for rice whereby rice exports would move at a lower price than would rice for domestic consumption.

Great Plains Conservation Program of 1956
- Special land conservation program for the ten Great Plains states.

Agricultural Act of 1958
- Revised cotton supports to be no lower than 80 percent of parity in 1959 and no lower than 75 percent of parity in 1960 for farmers staying within their allotments.
- Set corn support level at 75–90 percent of parity if an acreage allotment program was in effect.
- Set rice support level at 75–90 percent of parity in 1959–60, no lower than 70 percent of parity in 1961, and no lower than 65 percent of parity in 1962 and thereafter.
- Terminated corn acreage allotments.
- Terminated acreage reserve program of Soil Bank Program.
- Extended the National Wool Act.

International Wheat Agreement of 1959
- Continued the International Wheat Agreement of 1949 for three years.

Tobacco Price Support Act of 1960
- Substituted "cost-of-production" for "parity" as the basis for price supports for tobacco.

Emergency Feed Grain Program of 1961
- Initiated a voluntary acreage reduction program for feed grains with payment-in-kind provisions.
- Authorized an acreage diversion program.

Trade Expansion Act of 1962
- Gave the president authority to enter into trade agreements with foreign nations.
- Gave the president authority to reduce duties then existing by not more than 50 percent.

Food and Agricultural Act of 1962
- Continued the feed grain acreage reduction program.
- Provided a two-tiered feed grain support program with price support payments in addition to nonrecourse loans.
- Permitted a mandatory production quota program for wheat if approved in a referendum of three-fourths of the producers.

International Wheat Agreement of 1962
- Continued the International Wheat Agreement of 1959 for three years.

Feed Grain Act of 1963
- Authorized price supports for corn at 65–90 percent of parity if a feed grain diversion program was in effect.
- Extended the existing feed grain program.

Agricultural Act of 1964
- Established a wheat certificate program in addition to diversion payments. The domestic marketing certificate was valued at 70 cents/bu. and the export certificate was valued at 25 cents/bu. The cost of certificates was passed on to millers.
- Initiated a payment-in-kind program for cotton millers (a subsidy enabling millers

to buy U.S. cotton at the same price as foreign buyers).
- Initiated voluntary programs for reducing cotton production.
- Implemented a paid land diversion program for wheat.

Food Stamp Act of 1964
- Made the Food Stamp Program a part of permanent agricultural legislation.

Meat Import Quota Act of 1964
- Gave the secretary of agriculture authority, subject to approval of the president, to impose quotas on imports of beef, veal, mutton, and goat meat whenever imports were expected to exceed a specified percentage of domestic supply. Quotas were to be allocated among importing nations according to their market shares in the 1959–63 base period.

Amendment to Food and Drug Act, 1964
- Prevented the use of a pesticide while it was under protest.
- Authorized each pesticide to carry a license number identification.
- Expedited procedures for suspending a previously registered pesticide found to be unsafe.

Food and Agricultural Act of 1965
- Was first in a series of comprehensive farm bills having a five-year life.
- Extended the feed grain program with price supports at 60–90 percent of parity.
- Extended voluntary acreage controls to wheat and cotton.
- Extended the wheat certificate program.
- Required cotton to be supported at 90 percent of the world price.
- Authorized a five- to ten-year cropland diversion program.
- Extended the National Wool Act.

Water Quality Act of 1965
- Required states to draw up minimum quality standards for all bodies of water within their boundaries.
- Required states to enforce these standards.
- States were required to allocate discharge permits to polluting firms in such a way that the quality standards were met.

Child Nutrition Act of 1966
- Extended the special milk program and provided for a program to offer breakfasts in schools.

Food for Peace Act of 1966
- Extended and amended the Agricultural Trade Development and Assistance Act (Public Law 480) of 1954.
- Shifted the focus of Public Law 480 from an agricultural surplus disposal program to an economic development program.
- Required developing countries to provide more of their own food requirements before qualifying for U.S. food aid.
- Required U.S. food aid shipments to be made up of "available commodities" rather than solely "surplus commodities."
- Required that countries now buying U.S. farm products with local currencies had to shift to buying for dollars or dollar credit by 1972.
- Authorized the use of foreign currencies from export sales to support family-planning programs when requested by the recipient country.

Agricultural Fair Practices Act of 1967
- Prohibited processors from discriminating against producers because they are members of an agricultural cooperative.

International Grains Agreement of 1967
- Extended the International Wheat Agreement of 1962 for three years.

National Environmental Policy Act of 1969
- Requires an environmental impact report.
- Authorizes the review and approval of projects with an environmental impact.

Agricultural Act of 1970
- Provided a more flexible approach to supply control through set-asides that called for reduced plantings but did not specify which nonquota crop had to be cut back.
- Retained the two-tiered support system.
- Limited government payments to $55,000 per crop per producer per year.
- Required support prices for corn to be no higher than 70 percent of parity and the loan rate to be no higher than 90 percent of parity.
- Authorized the wheat loan rate to be as high as 100 percent of parity.
- Made wheat certificates available to those participating in the set-aside program that were valued at the difference between the parity price and market price.
- Suspended cotton quotas; retained the loan rate for cotton at previous levels; pegged set-aside payments for cotton at the difference between 65 percent of parity and the market price.
- Extended the National Wool Act.

Environmental Protection Agency, 1970
- Was created by presidential reorganization to coordinate those activities relating to the environment that were previously the responsibility of the Departments of Agriculture, Interior, and/or Health, Education, and Welfare.

Occupational Safety and Health Act (OSHA) of 1970
- Set occupational safety and health standards and authorized enforcement of these standards. Standards affecting agricultural employers and employees cover temporary labor camps, field sanitation, storage and handling of anhydrous ammonia, pulpwood logging operations, and operation of slow-moving vehicles, farmstead equipment, and cotton gins.

Rural Environmental Assistance Program of 1971
- Is new name of the Agricultural Conservation Program.
- Redesigned the conservation program goals and projects to emphasize pollution prevention and environmental improvement.

Pesticide Control Act of 1972
- Replaced the Federal Insecticide, Fungicide, and Rodenticide Act of 1947.
- Authorized the regulation of pesticide use to protect humans and the environment covering all pesticide manufacture, sale, and use within states, as well as among states.
- Prohibited the sale or use of any pesticide in a manner inconsistent with its labeling.
- Required all pesticides to be classified for general or restricted use. Restricted-use pesticides may only be used by or applied under the supervision of certified applicators.

- Authorized states to implement the federal laws.

Rural Development Act of 1972
- Authorized grants and loans for promoting the establishment of business and industry in rural areas, for attracting industry to rural communities, and for improving rural life.
- Provided funds for research and extension programs in rural development.

Water Pollution Control Act of 1972
- Amended the Water Quality Act of 1965.
- Established the goal of eliminating *all* discharges of pollutants by 1985.
- Charged the Environmental Protection Agency (EPA) with setting quality standards rather than the states.
- Required EPA to issue discharge permits if the states cannot do so in a way that meets EPA standards.
- Authorized federal subsidies for the construction of municipal treatment plants.

Agriculture and Consumer Protection Act of 1973
- Initiated target prices and deficiency payments as a replacement for price support payments.
- Lowered the limit on deficiency payments to $20,000 per crop per year.
- Based target prices on cost of production.
- Maintained loan rates on wheat, feed grains, and cotton at or below market prices.
- Required the support price for milk to be no lower than 75 percent or higher than 90 percent of parity.
- Authorized disaster payments.
- Authorized a special food aid program for women and children (Women, Infants, and Children Program).
- Extended the National Wool Act.

Trade Act of 1974
- An act to foster economic growth and strengthen relations between the United States and foreign countries, reduce barriers to trade, establish fairness in international trading relations, provide safeguards against unfair competition, provide developing countries access to U.S. goods, and provide the U.S. access to goods of developed countries.
- Granted the president broad authority to make tariff reductions, to negotiate agreements on reducing or eliminating nontariff barriers to trade, grant preferential trade treatment to developing countries, liberalize relief and adjustment provisions of the Trade Expansion Act of 1962, and retaliate against countries engaging in unfair trade practices.

Rice Production Act of 1975
- Authorized a set-aside program and an acreage diversion program.
- Limited payments to $55,000 per farmer per year.
- Initiated a target price/deficiency payment program.
- Authorized disaster payments.
- Eliminated the quota program.

Farmer-to-Consumer Direct Marketing Act of 1976
- Encouraged direct marketing of agricultural commodities to consumers.

Toxic Substances Control Act of 1976

- Regulates industrial chemicals and products used in pesticides, drugs, and food.
- Requires the manufacturer or importer to submit information on the product to the Environmental Protection Agency.
- Places the burden of demonstrating safety on the chemical manufacturer.

Soil and Water Resources Conservation Act of 1977
- Requires the secretary of agriculture to develop and periodically update a National Program for Conservation of Soil and Water that provides guidance for USDA's soil and water conservation activities.

Food and Agriculture Act of 1977
- Raised price and income supports.
- Continued flexible production controls and target prices.
- Established a farmer-owned reserve program for grains.
- Established a two-tiered support program for peanuts.
- Kept loan rates at or near market prices.
- Raised payment limits for wheat, feed grains, and cotton but lowered them for rice.
- Eliminated the purchase requirement for food stamps beginning in 1977.
- Extended the Rice Production Act of 1975.
- Extended the National Wool Act.

Emergency Assistance Act of 1978
- Gave the secretary of agriculture authority to raise target prices for feed grains, wheat, and cotton.
- Authorized an emergency loan program and established a moratorium on Farmers Home Administration foreclosures.

Meat Import Act of 1979
- Provides for the imposition of import controls on certain fish, chilled and frozen beef, veal, mutton, and goat meat products. A target level of 1,204.6 million pounds was set in the legislation to be adjusted annually by a production adjustment factor and a countercyclical factor.

Federal Crop Insurance Act of 1980
- Expanded the experimental crop insurance program to cover all crops provided yield histories for the applicable area were available.
- Designed to replace disaster assistance.
- Allowed participants to choose among three yield-guarantee levels and three crop price-guarantee levels.
- Allowed for a premium subsidy of up to 30 percent.
- Relied on the private sector as much as possible for insurance delivery.

Staggers Rail Act of 1980
- Gave railroads freedom in rate making with little interference from the Interstate Commerce Commission.
- Allowed railroads to drop unprofitable branch lines.
- Allowed railroads to enter into long-term contracts with shippers.

Agriculture and Food Act of 1981
- Set specific target prices for each year of the bill.
- Eliminated rice allotments and marketing quotas.
- Lowered dairy support levels.

- Eliminated peanut acreage allotments but retained quotas.

Omnibus Budget Reconciliation Act of 1982
- Froze dairy price supports.
- Authorized two deductions of 50 cents/cwt. each from farmers' milk checks if dairy purchases exceeded specified levels.
- Increased the land diversion acreage.

No-Net-Cost Tobacco Program Act of 1982
- Established a producer-supported fund to repay the government for program costs.
- Required disposal of some nonfarm allotment holdings.

Extra-Long Staple Cotton Act of 1983
- Eliminated cotton marketing quotas and allotments.
- Tied extra-long staple cotton support to upland cotton support.

Payment-in-Kind Program of 1983
- Provided voluntary, massive acreage reduction by adding payments in kind to regular acreage reduction payments for grain, upland cotton, and rice.

Dairy and Tobacco Adjustment Act of 1983
- Froze tobacco price supports.
- Repealed the second 50 cent/cwt. deduction for dairy farmers.
- Initiated a voluntary dairy diversion program.
- Authorized a nonrefundable 15 cent/cwt. assessment on milk marketed by producers to finance dairy product research and promotion.

Migrant and Seasonal Agricultural Worker Protection Act of 1983
- Designed to give migrant and seasonal farmworkers assurances about pay, working conditions, and work related conditions.
- Required farm labor contractors to (1) obtain a certificate of registration before initiating labor contract activities, (2) disclose to migrant and seasonal agricultural workers information about wages, hours, other working conditions, and housing, (3) provide workers written statements of earnings and deductions, (4) ensure that transportation vehicles are safe and insured, and (5) ensure that provided housing meets safety and health standards.

Agricultural Program Adjustment Act of 1984
- Froze target price increases at levels provided for by the 1981 act.
- Authorized paid diversions for feed grains, upland cotton, and rice.
- Provided for a payment-in-kind program for wheat.

Food Security Act of 1985
- Sought to reduce loan rates to market-clearing levels and adjust them in future years on the basis of a moving average of market prices.
- Required marketing loans and certificates for cotton and rice. Authorized the same for other program commodities.
- Froze target prices at 1985 levels through 1987 and reduces them thereafter.
- Authorized acreage reduction, set-aside, and paid land diversion programs for wheat, feed grains, cotton, and rice.
- Initiated a "sodbuster" and a "swampbuster" program.
- Authorized a 40–45 million acre conservation reserve program.
- Permitted partial payment in kind for program participation.
- Authorized a 50/92 underplanting provision.

- Created the Export Enhancement Program (EEP) and the Targeted Export Assistance Program (TEAP).
- Required a milk-production termination program for 1986–87. Authorizes the same for subsequent years.
- Continued the two-tiered peanut price support program.
- Continued the soybean price support program through 1990.
- Continued the sugar price support program through 1990.
- Limited program payments to $50,000 per person per year.
- Limited disaster payments to $100,000 per person per year.
- Extended the National Wool Act.

Balanced Budget and Emergency Deficit Control Act of 1985
- Aimed at eliminating the federal budget deficit by October 1990.
- Mandated annual reductions in federal outlays (including agricultural outlays) to meet the target date set.

Farm Credit Restructuring and Regulatory Reform Act of 1985
- Authorized interest rate subsidy on farm loans.
- Restructured the Farm Credit Administration.

Tax Reform Act of 1986
- Reduced the tax burden on individuals and increased the tax burden on corporations.
- Eliminated many tax loopholes and tax shelters.
- Reduced the incentives for tax-motivated investments in agriculture in preference to future-income–motivated investments.

Food Security Improvements Act of 1986
- Modified the 50/92 underplanting provision.
- Reduced the funding for the Export Enhancement Program and the Targeted Export Assistance Program.

Immigration Reform and Control Act of 1986
- Designed to control unauthorized immigration to the United States with employer sanctions, increased appropriations for enforcement, and amnesty provisions.
- Required employers to verify the eligibility of each employee to work in the United States.

Omnibus Budget Reconciliation Act of 1987
- Reduced target prices for the 1988 and 1989 crop years.
- Limited loan rate reductions for the 1988 and 1989 crops.
- Established a 0/92 provision for wheat and feed grains.

Agricultural Credit Act of 1987
- Provided for credit assistance to farmers, strengthened the Farm Credit System, and facilitated the establishment of secondary markets for agricultural loans.

Disaster Assistance Act of 1988
- Provided for assistance to farmers hurt by the drought or other natural disasters in 1988.
- Made crop producers enrolled in the 1988 farm program with losses greater than 35 percent or more above normal eligible for special financial assistance.
- Made feed assistance available to livestock producers whose crop yields were significantly reduced by drought.

- Required recipients of disaster payments to purchase crop insurance.

Omnibus Trade and Competitiveness Act of 1988

- Gives the secretary of agriculture discretionary authority to trigger marketing loans for wheat, feed grains, and soybeans if it is determined that unfair trade practices exist.
- Extends the authority of the Export Enhancement Program.

United States–Canada Free Trade Agreement Implementation Act of 1988

- Implements a bilateral trade agreement between the United States and Canada whereby most agricultural tariffs will be phased out over a ten-year period and other trade rules will be relaxed.

Disaster Assistance Act of 1989

- Provided limited disaster payments to program and nonprogram crop producers.
- Allowed producers to plant certain alternative crops on program acres.

Omnibus Budget Reconciliation Act of 1989

- Reduced deficiency payments for the 1990 crop year.
- Mandated planting flexibility options for oilseeds on up to 25 percent of program acres.
- Reduced funding levels for the Export Enhancement and Targeted Export Assistance programs.
- Made the previously mandatory milk price support reduction for 1990 discretionary.

Omnibus Budget Reconciliation Act of 1990

- Mandates a 15 percent triple-base acreage reduction plan under which farmers (1) may plant any program commodity, any oilseed crop, or any other crop except fruits and vegetables without loss in program base acres, and (2) must forfeit deficiency payments on the triple-base acres. Farmers are permitted to optionally "flex" an additional 10 percent of their base acres under the same restrictions.
- Requires deficiency payments on feed grains, wheat, and rice to be based on the prior twelve-month period rather than five-month period, and reduces the 1991 acreage reduction percentages for feed grains and wheat.
- Implements a system of farmer assessments consisting of 1 percent of the loan value or support price on sugar, peanuts, honey, and tobacco; 1 percent of the incentive payments on wool and mohair; 5 cents/cwt. on milk in 1991 and 11.25 cents/cwt. subsequently; and 2 percent of the loan rate on oilseeds.

Food, Agriculture, Conservation, and Trade Act of 1990

- Freezes target prices at 1990 levels, and sets the loan rates for wheat, feed grains, cotton, and rice at 85 percent of a five-year moving average of market prices.
- Establishes a marketing loan program for oilseeds, prohibits cross- and offsetting-compliance as a condition for program eligibility.
- Establishes a $10.10/cwt. floor on the support price for milk.
- Reduces payment limits.
- Provides for the establishment of acreage reduction program requirements on the basis of stocks-to-use ratios rather than on the basis of carryover quantity.
- Establishes a Water Quality Incentive Program, a Wetlands Conservation Reserve Program, and an Integrated Farm Management Program.
- Requires certified applicators to maintain records on the use of restricted

pesticides.
- Authorizes changes in domestic price and income supports if an agreement is not reached in the Uruguay Round of GATT negotiations.
- Extends the National Wool Act of 1954.

GLOSSARY

Defined terms are in **boldface** type.

Absolute Advantage—The advantage a country is said to possess when it can produce a commodity at a lower cost per unit than can any other country.

Acre—One acre equals 0.405 **hectares.**

Acreage Allotment—An individual farm's share, based on its production history, of the **national program acreage** needed to produce sufficient supplies of a particular crop. Acreage allotments apply only to crops and under conditions specified by law. Acreage allotments are currently used only for flue-cured (not burley) tobacco.

Acreage Reduction Program (ARP)—A voluntary land retirement program in which farmers reduce their planted acreage from their **base acreage.** This is generally an unpaid reduction and is generally required for participation in other agricultural programs such as the **nonrecourse loan** and **deficiency payment** programs.

Ad Valorem Duty—A duty (tax) on imports (exports) expressed as a fixed percentage of the import (export) value.

Adjusted Base Price—In **parity** calculations, the average price received by farmers in the most recent ten years, divided by the index (1910-14 = 100) of average prices received by farmers for all farm products in the same ten years.

African, Caribbean, and Pacific States (ACP States)—A group of sixty countries to which special economic relations were extended by the **European Economic Community** as a result of the **Lome Convention of 1975.** Manufactured goods and some agricultural products exported from these countries enjoy duty-free entry into the European Economic Community.

African Development Bank (AFDB)—An association of independent African nations to assist in the economic and social development of these nations and to promote economic cooperation among them.

Agency for International Development (AID or USAID)—The agency of the U.S. government responsible for assistance programs in underdeveloped countries friendly to the United States and for execution of all **Public Law 480** programs overseas.

Agricultural Conservation Program (ACP)—A program to encourage farmers and ranchers to carry out resource-conserving practices. This program is designed to restore and improve soil fertility, minimize erosion caused by wind and water, and conserve resources and wildlife. Cost-sharing is offered only for farm

325

conservation measures considered necessary to meet the most urgently needed conservation problems and which would not otherwise be carried out to the extent needed to meet the public interest. To be eligible, the farmer must in general request cost-sharing before beginning the practice. In lieu of cash reimbursement, cost-sharing assistance may be in the form of partial payment by the government for the purchase price of materials and services needed by the farmer for performing approved practices. The farmer bears the balance of the cost, which amounts to about 50 percent, and, in addition, supplies the necessary labor and management. Materials and services are obtained through private sources where practical. This program is administered by the **Agricultural Stabilization and Conservation Service** and the **Soil Conservation Service**.

Agricultural Marketing Service (AMS)—An agency of the U.S. Department of Agriculture that establishes standards for grades of agricultural products, operates grading services, and administers federal **marketing orders**.

Agricultural Protection Program—This program was established by the Food and Agricultural Act of 1977 to protect farmers in the event of a U.S. imposed trade embargo. The program applied to wheat, corn, sorghum, barley, rye, oats, cotton, rice, soybeans, and flaxseed. If the president or any other member of the executive branch of the federal government, because of supply shortages, suspended commercial export sales of any of the above commodities to any country or area with which other commercial trade was continued, the secretary of agriculture was required under this act to set the **loan rate** for that commodity, if a loan program was in effect, at 90 percent of its **parity price**. Loan rates at 90 percent of parity would become effective on the day of suspension, would be determined by the commodity's parity price on the day of suspension, and would remain in effect for the duration of the export suspension.

Agricultural Research Service (ARS)—An agency of the U.S. Department of Agriculture that conducts basic, applied, and developmental research on regional, national, and international concerns in the fields of livestock; plants; soil, water, and air quality; energy; food safety and quality; food processing, storage, and distribution; nonfood products; and international development.

Agricultural Stabilization and Conservation Service (ASCS)—An agency of the U.S. Department of Agriculture charged with administering farm price and income support programs as well as some conservation and forestry cost-sharing programs. Local ASCS offices are maintained in nearly all farming counties of the country.

All-Risk Crop Insurance—*See* **Federal All-Risk Crop Insurance**.

Andean Common Market (ANCOM)—An association of Bolivia, Colombia, Ecuador, Peru, and Venezuela formed in 1969 to promote regional economic integration, elimination of trade barriers among members, and to establish a common external tariff. Chile was an original member but withdrew in 1976.

Animal and Plant Health Inspection Service (APHIS)—An agency of the U.S. Department of Agriculture that conducts regulatory and control programs to protect animal and plant health.

Arbitrage—Transactions in commodity markets, financial markets, or foreign exchange markets by which the arbitrageur profits from price differences in the seller's and buyer's markets with little or no new investment in physical facilities. The price disparity in the seller's and buyer's markets cannot persist in the long run since the actions of one or more arbitrageurs will drive the prices together, squeezing out any excess profits.

Asian Development Bank (ADB)—An organization of Asian countries formed in 1966 to foster economic growth and cooperation in Asia and to contribute to the acceleration of the region's economic development.

Association of Southeast Asian Nations (ASEAN)—An association of Indonesia, Malaysia, the Philippines, Singapore, Thailand, and Brunei formed in 1967 to promote political, economic, and social cooperation among members.

Bank for Cooperatives—A bank providing credit to all agricultural and aquatic cooperatives. There is one Central Bank for Cooperatives in Washington, D.C., and twelve district Banks for Cooperatives. *See* **Farm Credit Administration.**

Base Acreage—Base acreage is used to compute allowable planting and **diverted acreage** in qualifying for benefits under government commodity programs. The apportionment of a farm's base acreage is determined from historical planting practices.

Base Price—The average price for an item in a specified period—such as 1910–14, 1935–39, 1957–59—used, for example, in **parity** calculations.

Basic Commodities—Corn, cotton, peanuts, rice, tobacco, and wheat were designated as basic in the Agricultural Act of 1949 and require **price support.**

Blended Credit—A form of export subsidy which combines government export credit and credit guarantees with commercial credit to reduce the effective interest rate to the recipient country. It is available only when appropriations are provided by Congress. Target countries are selected based on the magnitude of surpluses and competitive need, as well as on diplomatic and domestic political considerations.

Bushel—One bushel of wheat, soybeans, or potatoes weighs 60 **pounds,** or 27.216 **kilograms,** or 0.0272155 **metric tons;** one bushel of corn, grain sorghum, or rye weighs 56 pounds, or 25.4016 kilograms, or 0.0254 metric tons; one bushel of barley, buckwheat, or apples weighs 48 pounds, or 15.4224 kilograms, or 0.021772 metric tons; one bushel of oats weighs 32 pounds or 14.5152 kilograms, or 0.014515 metric tons.

Cairns Group—A group of thirteen countries—Argentina, Australia, Brazil, Canada, Chile, Colombia, Hungary, Indonesia, Malaysia, New Zealand, the Philippines, Thailand, and Uruguay—seeking global liberalization of agricultural trade, including a phasing out of domestic farm subsidies and an immediate termination of export subsidies. The Cairns Group proposal for trade liberalization consists of (1) an initial freeze in subsidies, nontariff barriers, and other barriers to trade in the short term, (2) a reduction of government supports to agreed-upon targets over the medium term, and (3) a strengthening of GATT rules covering market access and agricultural subsidies over the longer term.

Call Price—A (market) price trigger above which the secretary of agriculture was to call for farmers to repay the nonrecourse **farmer-owned reserve** loan plus any accumulated interest. The secretary of agriculture was required to notify

Congress fourteen days in advance of issuing a call. For wheat the call was automatic when the market price reached 175 percent of the **loan rate**. For corn the call price was to be set at "an appropriate level." The 1981 Agricultural and Food Act abandoned the notion of a call price altogether and replaced it with other incentives to encourage loan redemption (e.g., higher interest rates).

Cargo Preference—A provision of farm legislation that requires exports of government-owned commodities or products shipped under government-financed arrangements to be carried on United States flag vessels.

Caribbean Basin Initiative—A U.S. initiative signed in 1983 that gave twenty-seven Caribbean states duty-free access to U.S. markets for most of their exports in return for certain changes in tax and general economic policy. Textiles, clothing, footwear, tuna, and petroleum are excluded. Sugar and beef are subject to special treatment.

Caribbean Community (CARICOM)—An organization formed in 1973 by enlargement of its predecessor, the Caribbean Free Trade Association. Current members of CARICOM are Antigua, the Bahamas, Barbados, Belize, Dominica, Grenada, Guyana, Jamaica, Montierrat, St. Christopher-Nevis, St. Lucia, St. Vincent, Trinidad, and Tobago. The Leeward and Windward Islands form the East Caribbean Community subregional group. CARICOM is designed to promote increased agricultural production in the region and to foster intraregional trade.

Cartel—An organization of independent producers or producing countries formed to regulate the production, pricing, or marketing practices of its members in order to limit competition and maximize the combined market power of members.

Centner—A unit of weight equal to 50 **kilograms** in West Germany and Scandinavia and 100 kilograms in the Soviet Union.

Central American Common Market (CACM)—An organization of Costa Rica, El Salvador, Guatemala, Honduras, and Nicaragua formed in 1960 to promote economic development in its member states through a **customs union** and through industrial integration.

Child Nutrition Program—A program enacted in 1966 to strengthen and expand food service programs for needy children in schools, day care centers, and summer camps.

Coarse Grains—Corn, barley, oats, grain sorghum, rye, and millet.

Collective Good—A good for which the consumption by one consumer does not reduce or prevent consumption by another consumer. Examples include national defense, a view of a mountain, clean air, and clean water. Collective goods must be produced by public action because the private sector cannot secure the benefits for themselves in sales of the product. It is impossible to force an individual to pay for a collective good according to the benefit this individual derives from the good.

Commodity Credit Corporation (CCC)—An agency of the U.S. Department of Agriculture empowered to make **nonrecourse loans** to farmers and store commodities and in this way carry out the **price support** provisions of the agricultural legislation. It was established in 1933. Its charter gives it authority to "buy, sell, make loans, store, transfer, export and otherwise engage in operations" necessary to carry out the aims of the commodity provisions of the legislation.

Commodity Distribution Programs—Programs to provide primarily staple food products direct to needy households. **Section 32** commodities are distributed to households meeting specific eligibility standards—normally participation in some welfare program or unemployed.

Common Agricultural Policy (CAP)—A set of regulations by which member states of the **European Economic Community** seek to merge their individual agricultural programs into a unified effort to promote regional agricultural development, fair and rising standards of living for the farm population, stable agricultural markets, increased agricultural productivity, and methods of ensuring food supply security in the member states. The principal elements of the CAP are **variable levies, export restitutions, intervention prices, threshold prices, target prices, and sluice gate prices.**

Common Market—A regional grouping of countries which levies common external duties on imports from nonmember countries but which eliminates tariffs, quotas, and other restrictions on trade among member countries. Also referred to as a **customs union** or a tariff union. The best known example is the **European Economic Community**, but others include the Belgium-Luxembourg Economic Union, ANCOM, ASEAN, CACM, CAEU, ECCAS, ECOWAS, EFTA, LAIA, OAS, and OAU (see entries in this Glossary).

Comparative Advantage—A concept based on the notion that a country will (in the absence of trade distortions and under perfect competition) export that commodity for which its relative (opportunity) costs of production are the lowest of any country capable of producing this commodity.

Complier and Noncomplier or Cooperator and Noncooperator—These terms refer to compliance or noncompliance with the provisions of a government program. Compliers (or cooperators) comply with the provisions and hence qualify for program benefits. Noncompliers (or noncooperators) are not eligible for program benefits. Moreover, when **marketing quotas** apply, noncompliers are subject to penalties.

Conservation Compliance—A provision of the Food Security Act of 1985 requiring farmers with highly erodible cropland (an **erodibility index** of 8 or more) to implement a conservation plan by 1990 which must be approved by the local conservation district and completely implemented by 1995 to maintain eligibility for commodity program benefits.

Conservation Reserve Program (CRP)—A program authorized by the Food Security Act of 1985 under which farmers voluntarily contract (via a bid system) to take cropland out of production for a period of ten to fifteen years and devote it to soil-**conserving uses.** The farmer receives an annual rental payment for the contract period and may receive assistance in cash or in kind for carrying out approved conserving practices on this acreage.

Conserving Base—A conserving base is a specified amount or share of cropland historically devoted to **conserving uses** or the amount of conservation land recommended by good farming practices.

Conserving Use—A conserving use is a cultural practice or use approved under requirements of commodity programs or of general cropland retirement programs. These uses include permanent or rotation cover of grasses and legumes; summer or winter cover crops consisting principally of small grains, annual legumes, or annual grasses; small grain cover crops when used for any purpose other than grain; idle cropland, including clean tillage and summer fallowed cropland; and volunteer cover. The specific uses that qualify in this category differ somewhat among programs and from year to year. In general, more uses qualify for **conserving base** acreage than for **diverted acreage**.

Consultative Group on International Agricultural Research (CGIAR)—An organization that serves to coordinate and fund a majority of the international agricultural research centers such as the International Rice Research Institute (IRRI) in the Philippines, Centro Internacional de Mejoramiento de Maiz y Trigo (CIMMYT) in Mexico, and International Crops Research Institute for the Semi-Arid Tropics (ICRISAT) in India. CGIAR is sponsored by the **World Bank** and the United Nations.

Consumer Subsidy Equivalent—An estimate of the amount of subsidy consumers would need to maintain their economic well-being if all agricultural programs were removed. It is negative when the net effect of all programs affecting that commodity is to increase the price consumers pay for food, and positive when consumers pay less for food than they would in the absence of such programs.

Cost, Insurance, and Freight (c.i.f.)—Terms of sale frequently used in international trade whereby the seller's price includes the costs of the goods being sold and all transportation charges and insurance expenses to the named point or destination. The seller's liability for risk ends after the goods have been shipped and a bill of lading proving such shipment has been obtained by the buyer. The buyer's liability begins when the goods have arrived at the destination point.

Council for Mutual Economic Assistance (CMEA or COMECON)—An organization of Bulgaria, Cuba, Czechoslovakia, the German Democratic Republic, Hungary, Mongolia, Poland, Romania, the USSR, and Vietnam formed in 1949 to further economic cooperation among Eastern European nations and the Soviet Union. Mongolia joined in 1962 and Cuba in 1972. Yugoslavia is not a full participant. Finland, Iraq, and Mexico have joined as "cooperants."

Council of Arab Economic Unity (CAEU)—An organization of the twelve members of the League of Arab States formed in 1957 to promote an Arab **common market** and joint commercial ventures. The member states are Egypt (suspended), Mauritania, Somalia, Sudan, Iraq, Jordan, Kuwait, Palestine, Syrian Arab Republic, United Arab Emirates, People's Democratic Republic of Yemen, and Yemen Arab Republic.

Countervailing Duty—A levy, permitted under the GATT, article VI, imposed by an importing country on imported goods to offset subsidies provided by the government of an exporting country. Under U.S. law, such duties can only be imposed after the U.S. International Trade Commission has determined that the imports are causing or threatening to cause material injury to a U.S. industry.

Cross-Compliance—This term refers to commodity program requirements that restrict program benefits to those who comply with all commodity programs applicable to a given farm as a condition of program eligibility for any single commodity. In recent years cross-compliance has been used more frequently to refer to soil **conservation compliance** as a condition for commodity program eligibility. The Food, Agriculture, Conservation and Trade Act of 1990 prohibits cross-compliance as a condition for program eligibility.

Crop Year—*See* **marketing year.**

Customs Union—A **common market** that does not permit free movement of all factors of production among member states. The **European Economic Community** is in fact a customs union although one aim of its new initiative (EUROPE 1992) is to remove all barriers to the free movement of most factors of production.

Deadweight Loss—*See* **social costs.**

Decoupling—A term used to describe policy aimed at ensuring that government programs do not interfere with farmers' current or future production decisions. In a decoupled world, farmers would make production decisions based solely on market prices.

Deficiency Payment—A method of supporting prices of farm products via direct payments. Deficiency payments equal the difference between actual domestic market price (or **nonrecourse loan** rate if higher than market price) for a commodity and a higher fixed or guaranteed **target price.** Deficiency payments are calculated as (target price − market price) × **(normal yield)** × **(farm program acreage).** Deficiency payments are currently available for producers of wheat, rice, corn, oats, barley, sorghum, and upland cotton.

Devaluation—A reduction in the value of a country's currency relative to the value of the currency of a foreign country. A devaluation of the U.S. dollar, for example, reduces the "price" of U.S. goods in terms of the currency of importers and thus serves to increase the demand (i.e., cause the importers' aggregate demand curve to shift to the right) for U.S. goods on foreign markets.

Disaster Payments—Payments to producers of feed grains, wheat, rice, upland cotton, peanuts, soybeans, sugarcane, and sugar beets whose crops are subjected to natural disasters such as drought or flood.

Diversion Payments—Payments made to farmers for not planting a specified portion of their **acreage base** and diverting this land to **conserving uses.**

Diverted Acreage—Acreage withdrawn from crop production and devoted to approved **conserving uses** under production adjustment programs. In some years planting of nonsurplus crops was allowed on diverted acreage.

Dumping—The sale of a commodity in a foreign market at less than fair value. Generally recognized as an unfair trade practice because it can disrupt markets and injure producers of competitive products in an importing country. When that happens, article VI of the GATT permits the imposition of special antidumping duties equal to the difference between the price sought in the importing country and the normal value of the product in the exporting country. The Hawley-Smoot Tariff Act of 1930 permits the U.S. Department of Commerce to levy antidumping duties on imports equivalent to the dumping margin when it is determined that products are being dumped onto U.S. markets.

Economic Community of Central African States (ECCAS)—An association of ten central African nations established in 1981 to promote free trade among member states as well as the harmonious economic and cultural development of the member states. Member states are Burundi, Cameroon, Central African Republic, Chad, Congo, Equatorial Guinea, Gabon, Rwanda, Sao Tome and Principe, and Zaire.

Economic Community of West African States (ECOWAS)—An association of sixteen western African nations established in 1975 to abolish tariff and nontariff barriers to trade among the members, to establish a common external tariff, and to establish a West African **common market** within fifteen years. The member states are Benin, Burkino Faso, Cape Verde, Côte d'Ivoire, Gambia, Ghana, Guinea, Guinea-Bissau, Liberia, Mali, Mauritania, Niger, Nigeria, Senegal, Sierra Leone, and Togo.

Effective Rate of Protection—A measure of protection that incorporates the level of price protection afforded the commodity under consideration as well as the level of subsidy applied to inputs used in producing that commodity. It is calculated as the difference between the value-added per unit of output in domestic prices and the value-added per unit of output in world prices expressed as a percentage of the value-added per unit of output in world prices.

Eligible Farm—One with an effective allotment or feed grain base established under a farm commodity program. For the various farm commodity programs, specific eligibility requirements are specified for program participation.

Embargo—*See* **Agricultural Protection Program.**

Emergency Feed Assistance Program—A program providing for the sale of **Commodity Credit Corporation**-owned grain at 75 percent of the basic **loan rate** to livestock producers whose feed harvest was reduced by natural disaster. This program is only available if the secretary of agriculture has declared the producer's county a natural disaster county.

Emergency Feed Program—A federal cost-sharing program to assist livestock owners who have been severely affected by a natural disaster. This program is only available if the secretary of agriculture has declared the owner's county a natural disaster county.

Erodibility Index—A measure of the inherent potential of a soil to erode considering length and steepness of slope, rainfall intensity and duration, and soil characteristics. Cropland containing primarily soil with an erodibility index of 8 or greater may be enrolled in the **conservation reserve program,** and is subject to the **conservation compliance** provisions of the current legislation.

Eurodollar—U.S. dollar–denominated deposits in banks and other financial institutions outside the United States. Originating from, but not limited to, the large quantity of U.S. dollar deposits held in Western Europe.

European Currency Unit (ECU)—A weighted average of all **European Economic Community** currencies (except for those of Spain and Portugal) which fluctuates against third-country currencies and is used for internal European

Economic Community accounting purposes. For agriculture, the ECU is the unit of account in which common farm prices, subsidies, and import levies in the member states are established.

European Economic Community (EEC)—A **common market** created in 1958 providing for the gradual elimination of customs duties and other intraregional trade barriers, a common external tariff against other countries, and gradual adoption of other integrating measures, including a **common agricultural policy** and guarantees of free movement of labor and capital. The original six members were Belgium, France, West Germany, Italy, Luxembourg, and the Netherlands. Denmark, Ireland, and the United Kingdom became members in 1973, Greece in 1981, and Spain and Portugal in 1986.

European Free Trade Association (EFTA)—A regional free-trade area established in 1958 and including Austria, Finland, Iceland, Norway, Sweden, and Switzerland. EFTA is concerned with the elimination of tariffs on manufactured goods originating in and traded among the member nations. Agricultural products, for the most part, are not included in the schedule for tariff reductions.

European Monetary System (EMS)—A monetary system intended to move Europe toward closer economic integration and avoid the disruptions in trade that can result from fluctuations in currency exchange rates. The EMS member countries deposit gold and dollar reserves with the European Monetary Cooperation Fund in exchange for the issuance of ECUs. All **European Economic Community** member countries participate in the exchange except Greece and the United Kingdom.

Ever-Normal Granary—The principle of using stored grain when it is needed, such as during times of drought, and rebuilding grain stocks to a safe level during good production years. Its purpose is to ensure adequacy of supply, as well as to aid in stabilizing farm prices and incomes.

Export Credit Guarantee Programs (GSM-102 and GSM-103)—Under these programs, the **Commodity Credit Corporation** guarantees for a fee payments due U.S. exporters under deferred-payment sales contracts when the foreign buyer defaults on the contract. GSM-102 was authorized in 1980 and provides for Commodity Credit Corporation credit guarantees for up to three years. The Intermediate Export Credit Guarantee Program (GSM-103) was authorized by the Food Security Act of 1985 and provides for Commodity Credit Corporation credit guarantees for up to ten years.

Export Credit Sales Program (GSM-5)—Under this program, the **Commodity Credit Corporation** is authorized to finance commercial export sales of agricultural commodities by purchasing the exporter's accounts receivable.

Export Enhancement Program (EEP)—A program authorized by the Food Security Act of 1985 which permits the U.S. Department of Agriculture to use **Commodity Credit Corporation**–owned commodities as export bonuses to exporters to make U.S. agricultural commodities competitive in the world market and to offset adverse effects of unfair trade practices or subsidies on the part of competing nations.

Export-Import Bank—An institution of the U.S. government that administers programs to assist private exporting firms in their trade with foreign countries by providing various forms of assistance, including direct lending and the issuance of guarantees or insurance to minimize risk.

Export License—A government document authorizing a firm to export specific goods in specific quantities over a specified time period to a particular country. Some countries require such a license for a limited number of commodities, while others require a license for most exports.

Export Quota—Control applied to exports by an exporting country to limit the amount of goods leaving that country. Such controls are usually applied in time of war or during some other emergency requiring conservation of domestic supplies.

Export Restitution—A subsidy used by the **European Economic Community** to promote exports of farm goods that cannot be sold within the European Economic Community at target prices.

Export Subsidy—A government grant in cash or in kind, made to a private enterprise, for the purpose of facilitating or expanding exports. Other terms may be used synonymously, such as export *payment* or *differential* or *assistance* or *aid*. In the European Economic Community the term *restitution payment* is used.

Externality—A situation in which the action of one economic agent affects the utility or production possibility of another economic agent in a way that is not accounted for by any usual market activity. The situation may involve only consumers (e.g., the effect of smoking by one individual on a nonsmoker), only producers (e.g., in honey and clover production, insecticide applications on one farm that kill bees on another farm, or nitrogen applications on a farm that eventually kill fish in a downstream body of water), or producers and consumers (e.g., industrial or agricultural pollution that contaminates drinking-water supplies).

Fannie Mae—*See* **Federal Agricultural Mortgage Corporation.**

Farm—Beginning in 1978 the Bureau of the Census defines a *farm* as any place that has $1,000 or more in gross sales of farm products per year. The previous definition (used for the 1959, 1964, 1969, and 1974 censuses) counted as a farm any place with less than 10 acres from which $250 or more of agricultural products were sold or normally would have been sold during the census year, or any place of 10 acres or more from which $50 or more of agricultural products were sold or normally would have been sold during the census year.

Farm Bill—The popular, generic, term given to current federal legislation for agriculture.

Farm Credit Administration—A farm credit system established to provide a dependable source of both long- and short-term credit on a sound basis for farmers and cattle raisers and their cooperative associations. These credit services are supplied in each of the twelve farm credit districts by a **Land Bank**, an **Intermediate Credit Bank**, and a **Bank for Cooperatives**. In addition, there is the Central Bank for Cooperatives in Washington, D.C. Farmers obtain individual farm real estate mortgage loans from local Land Bank Associations and loans for production expenses and operating capital from a local **Produc-**

tion Credit Association. These associations are farmer owned but obtain their financing through the Land Banks and Intermediate Credit Banks, which sell debentures in the open market.

Farm Program Acreage—The individual farm program acreage for **deficiency payment** purposes under the Food and Agricultural Act of 1977 determined by multiplying the **program allocation factor** by the number of acres planted for harvest on the individual farm. The secretary of agriculture is required to make deficiency payments, when price conditions mandate them, on no less than 80 percent of the wheat and feed grain acreage planted for harvest on the farm during the current year. This provision of the 1977 act replaces historical farm **acreage allotments** authorized by previous legislation.

Farmer Mac—*See* **Federal Agricultural Mortgage Corporation.**

Farmer-Owned Reserve (FOR) Program—The 1977 legislation required the secretary of agriculture to administer a producer-held storage program for wheat and, at his discretion, a similar program for feed grains. This was to be accomplished through an extended price support loan program of three to five years' duration. Producers were to receive storage payments and the secretary of agriculture could adjust or waive interest charges on farmer-held reserves. This program was designed to protect against production shortfalls and to provide a buffer against unusually sharp price movements.

Under current law, the quantity of wheat held is not to be less than 300 million **bushels** nor more than 450 million bushels. The maximum amount may be adjusted upward, however, to meet any commitments assumed by the United States in an international agreement on grain reserves. For feed grains, reserve quantities are not to be less than 600 million bushels or more than 900 million bushels.

When the market price is greater than 105 percent of the current **target price** for the crop, the secretary of agriculture may charge an interest fee to encourage producers to redeem loans and market their grain. The secretary must cease making storage payments whenever the market price is 95 percent or more of the current target price. If a producer redeems his loan before the market price reaches the **release price,** he will be subject to a penalty. The secretary of agriculture is required to recover storage payments and to assess penalty interest or other charges.

Under the Food and Agricultural Act of 1977, the loan was to be called when the wheat market price reached 175 percent of the current **loan rate**—the **call price.** Again, the secretary of agriculture was to determine the actual level at which the loan would be called. The Agricultural and Food Act of 1981 rescinded the call price but gave the secretary of agriculture authority to increase the interest rate to encourage redemption of the loan when market price exceeds the release price.

Farmers Home Administration (FmHA)—An agency of the U.S. Department of Agriculture providing credit to those in rural America unable to get credit from other sources at reasonable rates.

Fast-Track—A legislative procedure authorized by the Omnibus Trade Act of 1988 by which a trade agreement can be negotiated by the administration and submitted to congress for a yes or no vote without the possibility of an amendment.

Federal Agricultural Mortgage Corporation—An institution authorized by the Agricultural Credit Act of 1987 which provides a secondary or resale market for agricultural mortgage loans made by government credit institutions, banks, or insurance companies, enabling lenders to obtain cash for further lending. Mortgages from lenders are pooled into securities and sold on the bond market. Such bonds are referred to as **Farmer Mac** bonds. This instrument is patterned after similar secondary financial markets such as the Federal National Mortgage Association (**Fannie Mae** bonds), the Government National Mortgage Association (**Ginnie Mae** bonds), and the Federal Home Loan Mortgage Corporation (**Freddie Mac** bonds).

Federal All-Risk Crop Insurance—A voluntary risk management tool available to all farmers that protects them from the economic effects of unavoidable adverse natural events. Administrative costs are appropriated by Congress and 30 percent of the insurance costs are federally subsidized. This program is administered by the Federal Crop Insurance Corporation (FCIC)—a wholly owned federal corporation.

Federal Grain Inspection Service (FGIS)—An agency of the U.S. Department of Agriculture that establishes official U.S. standards for grains and administers a nationwide inspection system to certify those standards.

Feed Grains—Corn, grain sorghum, barley, rye, and oats. For price support operations, the Agriculture and Consumer Protection Act of 1973 specified corn and grain sorghum as feed grains, and barley if designated by the secretary of agriculture, up till the 1977 crop year.

Findley Payment—A direct payment made to producers to compensate for a reduction in the **nonrecourse loan** level from the basic or statutory **loan rate** by up to 20 percent to make the commodity more competitive on the world market. The secretary of agriculture may pay producers the difference between the basic loan rate and the higher of the twelve-month average market price or the announced loan rate.

Flexible Acres—A portion of the **permitted acres** not eligible for **deficiency payments** as specified in the **Triple-Base Plan** and which could be planted to a crop other than the program crop without reducing the program crop **base acreage**.

Food and Drug Administration (FDA)—An agency of the U.S. Department of Health and Human Services with primary responsibility for attesting to the efficacy, safety, and wholesomeness of food products, drugs for humans and animals, cosmetics, and medical devices.

Food Security Wheat Reserve—A 4-million **metric ton** reserve program providing wheat to be used for emergency food donations.

Food Grains—Cereals used for human food; primarily wheat and rice.

Food Stamp Program—A U.S. Department of Agriculture program designed to help low-income households achieve an adequate and nutritious diet. The program began as a pilot operation in 1961 and was made part of permanent legislation in the Food Stamp Act of 1964. Recipients are provided with food stamps having an equivalent cash value. Eligibility is determined on the basis of income levels in relation to established poverty guidelines. Level of assistance is based on a "Thrifty Food Plan" covering the cost of commodities needed to achieve a balanced diet. Higher levels of assistance are provided for lower income families and larger family sizes.

Freddie Mac—*See* **Federal Agricultural Mortgage Corporation.**

Free on Board (f.o.b.)—A price quotation indicating that the seller assumes all responsibility and costs for delivering the goods and loading them on a stated carrier at a specified location.

General Agreement on Tariffs and Trade (GATT)—An agreement negotiated in 1947 among twenty-three countries, including the United States, to increase (liberalize) international trade by reducing tariffs and other trade barriers and thereby contribute to global economic growth and development. This agreement provides a code of conduct for international commerce and a framework for periodic multilateral negotiations on trade liberalization and expansion. Eight negotiation sessions (Rounds) have been held, including the first Geneva Round (1947), the Annecy Round (1949), the Torquay Round (1950–51), the second Geneva Round (1955–56), the Dillon Round (1960–61), the Kennedy Round (1963–67), the Tokyo Round (1973–79), and the Uruguay Round (initiated in 1987 and scheduled to end in 1991). The first seven Rounds focused primarily on tariff cuts on manufactured products. Attempts to bring agricultural trade into the negotiations in a significant way were unsuccessful. The Uruguay Round was the first to focus on agriculture. The aim of this Round was to bring discipline to agricultural trade by reducing distortions caused by import barriers and subsidies.

The agreement is currently subscribed to by ninety-six signatory governments, which together account for more than 80 percent of world trade, and is informally adhered to by an additional thirty or so (mostly developing) nations.

Under GATT, quantitative trade restrictions and nonagricultural export subsidies are generally prohibited, and export **dumping** is subject to prescribed legal action. Domestic quotas are permitted when a country is attempting to curtail production, and export restrictions are allowed during periods of severe domestic shortages. Several policies such as **variable levies** (as implemented by the **European Economic Community**), minimum import prices, and **voluntary export restraints** not in use when GATT was drafted have not come under its jurisdiction.

Generalized System of Preferences (GSP)—A framework under which developed countries give preferential tariff treatment to (principally) manufactured goods imported from certain developing countries. It is one element in a coordinated effort of the industrial world to bring developing countries more fully into the international trading system. These schemes institute a system of nonreciprocal tariff preferences for the developing countries. GSP was authorized in the United States by the U.S. Trade Act of 1974 as amended in 1984. Under this arrangement, 136 developing countries were extended temporary duty-free treatment for some 3,000 products. In February 1988, the U.S. government removed GSP privileges from Taiwan, South Korea, Singapore, and Hong Kong because these countries had strengthened their economies to the point where they no longer need GSP treatment.

Generic Certificate—The Food Security Act of 1985 authorizes the **Commodity Credit Corporation** to issue certificates in lieu of cash to farmers for participation in **acreage reduction, paid acreage diversion, conservation reserve, marketing loan,** and **disaster** programs, and to grain merchants for the **Export Enhancement Program** and the **Targeted Export Assistance Program**. These certificates have a fixed dollar value and an eight-month life. They can be used to reacquire commodities pledged as collateral under the various loan programs, they can be sold or transferred to others, or they can be returned to the **Commodity Credit Corporation** for cash at 100 percent of face value during the sixth through eighth month of their life, at 85 percent of face value during the first six months after expiration, and at 50 percent of face value during the next twelve months.

Ginnie Mae—*See* **Federal Agricultural Mortgage Corporation**.

Government Payments—Payments to producers authorized under various federal government programs. They include commodity **price support** payments, payments for diverting land from crop production to **conserving uses** or to nonsurplus crops, and payments to implement conservation practices. Government payments have been variously called direct payments, compensatory payments, income payments, diversion payments, set-aside payments, and **deficiency payments**. *See* **payment limitation** for current payment limits.

Gramm-Rudman-Hollings—The three senators who sponsored the Balanced Budget and Emergency Deficit Control Act of 1985, which requires automatic cuts in federal expenditures if necessary to reduce the budget deficit to zero over the period 1986–90. This act was amended in 1987 to extend the period to 1993. If Congress and the president cannot agree on a targeted reduction for any given year, automatic cuts affecting almost all federal programs take place. Programs exempted from the cuts are Social Security, interest on federal debt, veterans' compensation and pensions, Medicaid, Aid to Families with Dependent Children, the **Women, Infants, and Children Program**, Supplement Security Income, the **Food Stamp Program**, and **child nutrition programs**.

Green Rate of Exchange—An administratively determined exchange rate used by the **European Economic Community** to convert agricultural prices from the **European currency unit** to the currencies of member countries. It was established to ensure that farm prices remain uniform throughout the European Economic Community despite the devaluation or revaluation of individual member currencies.

Gross Domestic Product (GDP)—A measure of the market value of goods and services produced by a nation. GDP excludes receipts from that nation's business operations in foreign countries as well as the share of reinvested earnings in foreign affiliates of domestic corporations.

Gross National Product (GNP)—A measure of the market value of goods and services produced by a nation. GNP includes receipts from that nation's business operations in foreign countries as well as the share of reinvested earnings in foreign affiliates of domestic corporations.

Hard Currency—Currency which may be exchanged for that of another nation without restriction—for example, the U.S. dollar, the British pound, or the West German deutsche mark. *Compare* **soft currency.**

Hectare—One hectare equals 2.471 **acres.**

Horizontal Integration—A form of market control under which a single organization controls, via ownership or contractual arrangement, two or more firms each performing similar activities at the same level of the production or marketing process.

Import Duty—A tariff levied by a country on imports of goods from selected countries. This tariff may or may not be **ad valorem.**

Import License—A government document authorizing imports of specific goods in specific quantities over a specified time period into the country. Some countries require such a license for a limited number of commodities, while others require a license for most imports.

Import Quota—The maximum quantity or value of a commodity allowed to enter a country during a specified period of time.

Incentive Payment—A **deficiency payment** that increases with the *quality* of the commodity. The incentive payment is used only in the case of wool. It is designed to encourage an increase in the production of high-quality wool—a commodity for which the United States is a deficit producer.

Indemnity Payment—Payments to beekeepers, for example, who through no fault of their own suffer losses of honey bees as a result of the utilization of economic poisons near or adjacent to property on which beehives are located.

Inter-American Development Bank (IDB)—A financial institution established in 1959 to further the economic and social development of its twenty-seven Latin American member countries.

Intermediate Credit Bank—A bank providing funds to **Production Credit Associations.** There are twelve such banks. *See* **Farm Credit Administration).**

International Bank for Reconstruction and Development (IBRD)—An organization established in 1944 to help countries reconstruct their economies after World War II. It lends money to governmental agencies, or guarantees private loans for such projects as agricultural modernization or infrastructure development. The IBRD is also known as the **World Bank.**

International Cocoa Agreement (ICOA)—An agreement among all major exporters and importers of cocoa, except the United States, for stabilizing prices and marketings of cocoa. The initial agreement was signed in 1973.

International Coffee Agreement (ICA)—An agreement among sixty-seven countries, including the United States, concerning the pricing and market sharing of coffee. The initial agreement was signed in 1962.

International Dairy Arrangement (IDA)—An agreement negotiated under GATT auspices in 1979, signed by eighteen major importing and exporting countries, and aimed at expanding and liberalizing trade in dairy products. The United States withdrew in 1985 because the **European Economic Community** violated a major purpose of the agreement by selling butter and other dairy products at prices below the negotiated minimum export price.

International Monetary Fund (IMF)—Established in 1946 to act as a lender of last resort for countries experiencing foreign exchange deficiencies and to monitor currency exchange relationships among nations.

International Olive Oil Agreement (IOOA)—An agreement signed by seventeen olive oil exporters and importers (not including the United States) designed to ensure fair competition, the delivery of commodity in accordance with contract specifications, and the stabilization and development of olive oil markets.

International Sugar Agreement (ISA)—An agreement among the major sugar producers and the United States as a major consuming country which seeks to stabilize sugar trade at equitable prices and to aid those countries whose economies are largely dependent on the production and export of sugar. The initial agreement was signed in 1959.

International Wheat Agreement (IWA)—An agreement currently containing two conventions: the Wheat Trade Convention signed by sixty countries, including the United States, which provides a forum for the periodic exchange of information among member countries on the world grain situation, and the Food Aid Convention signed by eleven countries, excluding the United States, which commits the signatories to minimum annual food aid contributions of edible grains. The initial agreement was signed in 1949.

Intervention Price—A price fixed under the **Common Agricultural Policy** at which **European Economic Community** intervention (buying) agencies are obliged to make open market purchases, thus setting a floor to domestic wholesale prices. The intervention price is set slightly below the **target price.**

Inventory Reduction Payment—A **loan deficiency** payment made in-kind to farmers who agree to forgo obtaining a loan or receiving a **deficiency payment** and who do not plant more than their crop acreage base less one-half the **acreage reduction program** requirement.

Kilogram—2.2046 pounds.

Kilometer—0.62137 miles.

Land Bank—A borrower-owned bank providing long-term farm and rural real estate loans of from five to forty years. There are twelve such banks. *See* **Farm Credit Administration.**

Land Retirement Program—A multi-year voluntary land retirement program. The land idled must be planted to soil-conserving cover crops or trees. The government generally pays the landowner an annual rental fee plus the cost of establishing a cover crop or trees. *See also* **Soil Bank** and **Conservation Reserve Program.**

Latin American Economic System (LAES)—An organization of the sovereign states of Latin America, including Cuba, formed in 1975 seeking to promote coordination of economic development efforts in the region and to offer a permanent consultative system for adoption of joint positions on international economic and social issues.

Latin American Free Trade Association (LAFTA)—An organization formed in 1960 by Argentina, Bolivia, Brazil, Chile, Colombia, Ecuador, Mexico, Paraguay, Peru, Uruguay, and Venezuela to liberalize trade among member nations and eventually create a Latin American common market.

Latin American Integration Association (LAIA)—A treaty signed in 1980 replacing the **Latin American Free Trade Association.** Its principal tool of integration is the Regional Tariff Preference implemented in 1984 and designed to lower tariffs on products imported from member countries below tariffs on similar products imported from nonmember countries.

Liter—0.264179 gallons, or 1.056716 **quarts.**

Loan Deficiency Payment—A payment made to farmers who are eligible to put their crops under loan but who forgo this opportunity. The loan deficiency payment was authorized by the Food Security Act of 1985 on wheat, feed grains, upland cotton, and rice at any time a **marketing loan** is in effect.

Loan Rate—Price at which the **Commodity Credit Corporation** will purchase the farmer's commodity previously placed under loan to the Commodity Credit Corporation. The Agricultural and Food Act of 1981 established minimum loan rates for wheat, feed grains, and rice, and set soybean and cotton rates by a formula reflecting an average of previous years' market prices. Statutory loan rates for most crops are currently set at between 75 and 85 percent of a five-year moving average of past market prices. *See* **price support.**

Lome Convention of 1975—An agreement between the **European Economic Community** and the former African, Caribbean, and Pacific (ACP) colonies of the European Economic Community member states. The agreement covers some aid provisions as well as trade and tariff preferences for the ACP countries when shipping to the European Economic Community. The agreement also includes the **STABEX** scheme for stabilizing export receipts.

Low-Input Sustainable Agriculture (LISA)—A method of farming that emphasizes minimizing the application of chemical inputs (fertilizer, pesticides and herbicides) that are potentially hazardous to the environment, and/or encourages using rotation and cultivation practices that might reduce environmental conflicts while maintaining farm profitability and productivity. Practices alternative to the more common, high-input type include crop rotations and mechanical cultivations to control weeds and pests, integrated pest-management strategies (IPM), planting of legumes, animal manures and compost applications, overseeding of legumes into maturing grain crops, and minimum tillage practices.

Market Basket of Farm Foods—The average quantities of domestic farm-produced foods purchased annually per household for consumption at home by urban wage earners, clerical worker families, and single persons living alone.

Marketing Agreement—A marketing agreement may contain more diversified provisions than a **marketing order,** but it is enforceable only on those producers or handlers who voluntarily enter into the agreement with the secretary of agriculture.

Marketing Board—A central government authority which directs the marketing of a commodity. Export management is the most frequently performed function. Producers give up their right to the commodity at harvest; all storage and marketing functions are managed by the government authority. Producers receive an advance on commodities delivered or stored on the farm, with subsequent payments being made as marketing is completed. All producers receive the same price adjusted for location and quality differences. Marketing boards have traditionally been used in the United Kingdom, Canada, and Australia, but never in the United States.

Marketing Certificate—*See* **generic certificate** and **payment-in-kind.**

Marketing Loan—A **nonrecourse loan** with a repayment rate which may be less than the announced **loan rate.** The difference between the loan rate and the repayment rate, then, constitutes an income support payment to producers. The repayment rate is generally some percentage of the loan rate or world market price.

Marketing Order—A marketing order is a legal instrument which sets the limits within which an agricultural industry can operate a program of self-regulation. It defines the terms of trade for handler and producer, the commodity to be regulated, and the area to be covered. It provides for an advisory committee to administer the order, the number of members, and terms of office. It lists the economic tools to be used and has procedures for financing provisions of the order.

The basic purpose is to improve returns to producers through orderly marketing. An order may establish and maintain minimum quality standards and provide for an orderly flow of products to market to avoid unreasonable fluctuations in supplies and prices. A marketing order may contain only certain specified types of provisions. However, once approved by a required number of producers (usually two thirds of those voting) of the regulated commodity, the order is binding on all handlers of the commodity in the area of regulation.

Marketing Quota—A means of regulating the marketing, when supplies become excessive, of a commodity to which the quota is applicable under law. A national marketing quota is the quantity of the crop that, in general, will provide adequate and normal market supplies. Each producer is given a portion of the national quota based on past production. When marketing quotas are in effect (usually after approval by two thirds or more of the eligible producers voting in a referendum) growers who produce the commodity on acreage in excess of their farm **acreage allotment** are subject to marketing penalties on the "excess" production and may be ineligible for **nonrecourse loans.** Marketing quotas once approved in a producer referendum are mandatory on all producers. Quota provisions have been suspended for wheat, feed grains, cotton, and rice. Quotas are still used for domestically consumed peanuts but not for exported peanuts. For certain tobaccos, a poundage limitation is applicable, as well as acreage allotments, when approved in a grower referendum.

Marketing Year—Dates that signify the beginning and ending of the year for the various commodities as follows: corn, grain sorghum, soybean meal and oil, burley tobacco, and milk, October 1–September 30; peanuts, rice and cotton, August 1–July 31; flue-cured tobacco, July 1–June 30; wheat, oats, and barley, June 1–May 31; soybeans, September 1–August 31.

Meter—3.2808 feet.

Metric Ton—*See* **ton (metric).**

Monetary Compensatory Amounts—Border taxes and subsidies for member countries in the **European Economic Community** intended to equalize support prices between countries as exchange rates vary.

Most Favored Nation (MFN) Treatment—A commitment that a country will extend to another the lowest tariff rates it applies to any third country. A country is under no obligation to extend MFN treatment to another country unless they are both members of GATT or unless MFN treatment is specified in a bilateral agreement.

Multifiber Arrangement (MFA)—An agreement between seventeen developed and thirty-four developing countries designed to facilitate regulation of world trade in textiles and clothing under the aegis of GATT. The agreement was initiated in 1974, extended three times since, and is due to expire in 1991 unless extended. It provides the legal framework under which bilateral agreements are negotiated that are designed to restrain disruptive imports and dumping, and to increase trade flows. The United States has bilateral agreements limiting textile imports with forty-eight nations—forty-five under the MFA and three under section 204 of the Agricultural Act of 1956.

Multiple-Peril Crop Insurance—A term sometimes used to refer to the **Federal All-Risk Crop Insurance.**

National Program Acreage—The national program acreage as specified by the Food and Agricultural Act of 1977 is the number of harvested acres (based on the weighted national average farm program payment yields) required to meet estimated domestic and export needs (less imports) plus any desired increase or decrease in carryout stocks. This provision was removed by the Food, Agriculture, Conservation, and Trade Act of 1990.

Nominal Rate of Protection—A measure of protection that considers only price protection of the commodity under consideration but not of any inputs used to produce this commodity. It is calculated as the ratio of domestic price to world price at the border of the country under consideration. *Compare* **effective rate of protection.**

Nonbasic Commodities—All commodities not classified as basic by the Agricultural Act of 1949. For price support operations, mandatory price support programs on nonbasic commodities have included those for tung nuts, honey, milk, butterfat, the products of milk and butterfat, wool, mohair, barley, grain sorghum, oats, and rye.

Nonpoint Source Pollution—Pollution that enters the environment from broad areas via water runoff from a field or feedlot, such as areas in which fertilizers or other chemicals have been applied or animal manure is deposited, rather than from concentrated discharge points.

Nonrecourse Loan—A commodity loan made by the government to farmers or farmer cooperative marketing associations as a means of implementing a program to provide a floor below which market prices will not be permitted to fall. The loan is secured by a commodity stored in approved facilities, on or off the farm. It is **nonrecourse** in the sense that the government accepts the commodity as full satisfaction for the loan at the farmer's discretion. If the farmer chooses to repay the loan, he or she may do so at any time before maturity, for most commodities. If the loan is to be redeemed, interest and service charges are added to the face value of the loan. This nonrecourse feature allows farmers to gain from any price rise with no risk of loss.

When supply conditions warrant, the **Commodity Credit Corporation** offers a loan extension, or a "reseal" privilege. This permits withholding of surplus commodities for longer periods of time. Storage costs accruing during reseal periods are paid by the government. This enables farmers who stored their commodities under loan to earn income for storage.

Nonrecourse loans serve several functions: (1) they provide farmers a cash return for the commodity at the support level; (2) they strengthen market prices of the commodity through withdrawal of supplies from the market, especially at harvest (this was the original and principal purpose of the nonrecourse loan program); and (3) they tend to even out marketings through the year because farmers who obtain loans on their crops at harvest time can market the crops throughout the season. Frequently, nonrecourse loans are the first step in the government acquisition of commodities under price support operations.

For most commodity programs, in many years, producers had to comply with planting restrictions to obtain nonrecourse loans. Yet, insofar as loan operations maintained market prices above equilibrium levels, **noncompliers** benefitted because they sold their production at the supported market prices. Market prices could drop below the **loan rate** when program participation is low and thus supplies withheld from the market are small.

Nontariff Barrier to Trade—Any non-monetary restriction to trade, including quotas, embargoes, licensing, exchange controls, state trading, bilateral agreements, and regulations on health, safety, sanitation, packaging, and labeling.

Normal Crop Acreage—The acreage on a farm normally devoted to a group of designated crops. When a set-aside program is in effect, a farm's total planted acreage of such designated crops plus **set-aside acreage** cannot exceed the normal crop acreage if the farmer wants to participate in the commodity loan program or receive **deficiency payments.**

Normal Yield—A term designating the average historical yield established for a particular farm or area. Normal production would be the normal acreage planted to a commodity multiplied by the normal yield.

Off-Setting Compliance—A provision that requires a producer participating in an **acreage diversion** or **acreage reduction** program not to offset that reduction by planting more than his acreage base for that crop on another farm under his management. The Food, Agriculture, Conservation, and Trade Act of 1990 prohibits off-setting compliance as a condition for commodity program eligibility.

Oilseeds—Soybeans, peanuts, cottonseed, sunflower seeds, flaxseed, rapeseed, safflower, castor beans, mustard seeds, and sesame seeds used for the production of edible and inedible oils as well as for the production of high-protein animal meals.

Organization for Economic Cooperation and Development (OECD)—An organization consisting of the United States, Canada, the **European Economic Community** member countries, the **European Free Trade Association** member countries, Australia, New Zealand, Japan, and Turkey established in 1960 to study and discuss trade and related matters.

Organization of African Unity (OAU)—An organization consisting of all independent African countries except the Republic of South Africa, established in 1963 to promote self-government, respect for territorial boundaries, and social progress.

Organization of American States (OAS)—An organization consisting of the United States, Mexico, and most Central American, South American, and Caribbean countries established in 1948 to determine political, defense, economic, and social policies for the Inter-American system.

Paid Acreage Diversion—A voluntary land retirement system in which farmers are paid for forgone production on their **base acreage**. This policy tool has not been used since the mid 1980s.

Pareto Optimum—A term used to refer to an economic state or position from which it is impossible to make one person better off without making another person worse off. Such a state is attained by a competitive equilibrium—hence the attractiveness of the perfectly competitive model for use as a norm in analyzing policy choices. It is important to note that a Pareto optimal state does not necessarily imply a *just* or *equitable* state. It may well be, for example, that the initial distribution of wealth upon which a Pareto optimal state depends is, in the eyes of many, *unjust* or *inequitable*. Or it may be that the growth path of the economy consistent with this Pareto optimal state is not consistent with society's desires.

Parity—When used alone, the term is synonymous with equality. In agriculture, this is the price which will give agricultural commodities the same purchasing power in terms of goods and services farmers buy that the commodities had in a specified base period. This concept was first introduced in the Agricultural Adjustment Act of 1933 and was first explicitly incorporated in the Agricultural Adjustment Act of 1938. The base period now used for parity calculations is 1910–14, a particularly good period for farmers.

Parity Index—The index of prices paid by farmers for commodities and services, interest, taxes, and farm wage rates used for producing farm products and in farm family living on a 1910–14 base.

Parity Price—A price for an individual commodity such that it would purchase, on a given date, a quantity of a standard list of goods equal to those that could have been bought in the base period (1910-14) at the prices then prevailing. The parity price for a commodity is computed as the product of the **adjusted base price** and the **parity index.**

Parity Ratio—This measures the general or overall relationship between the price level of the commodities farmers sell and the prices of items farmers buy. More specifically, it is 100 times the ratio of (1) the index of prices received by farmers to (2) the index of prices paid by farmers, including an allowance for interest, taxes, and wages paid hired labor. When this ratio is 100 (i.e., the two indexes are equal), agricultural prices in general are considered at **parity**. In general, parity ratio is a traditional, although questionable, measure of whether farmers are getting "fair" prices in relation to their costs.

Payment-In-Kind (PIK)—PIK refers to payment in commodities in lieu of currency. Starting in 1956 a PIK program has been used by the **Commodity Credit Corporation** for export and domestic commodity programs. PIK certificates are issued to producers, buyers, and exporters. They state a dollar value, redeemable for specified commodities and products from Commodity Credit Corporation stocks or in face value cash equivalent.

In 1961 and again in 1983 a PIK program was used to reduce Commodity Credit Corporation stocks and to limit further accumulation of stocks of wheat, feed grains, cotton, and rice. In this case producers were paid in kind to idle acreage of the in-kind commodity.

Payment Limitation—A limit set by law on the amount of money any one individual farmer may receive in farm program payments, such as **deficiency** and **disaster payments**, in any one year under the commodity programs. Under the Agricultural and Consumer Protection Act of 1973, the limit on government payments to wheat, feed grain, and upland cotton producers was $20,000, and to rice producers was $55,000. Under the Food and Agriculture Act of 1977, the limit on government payments to wheat, feed grain, and upland cotton producers was $40,000 for the 1978 crop and $45,000 for the 1979 crop. Under the Food Security Act of 1985 the limits were $50,000 on deficiency payments, paid land diversion payments, and **conservation reserve program** payments; $250,000 on **Findley payments, loan deficiency payments,** gains on **marketing loan** repayments, and disaster payments; $250,000 on gains from honey marketing loan repayments; and $500,000 maximum to any one person for all payments. Under the Food, Agriculture, Conservation, and Trade Act of 1990 the limits are $75,000 on Findley payments, loan deficiency payments, gains on marketing loan repayments, and disaster payments; $200,000 in 1991 and down to $125,000 by 1994 on payments to wool, mohair, and honey producers; and $250,000 maximum to any one person for all payments.

Permanent Legislation—The statutory legislation upon which most agricultural programs are based. For the major commodities this is principally the Agricultural Adjustment Act of 1938, the Agricultural Act of 1949, and the Commodity Credit Corporation Charter Act of 1948 for price and income provisions, and the Agricultural Trade and Development and Assistance Act of 1954. Although these laws are frequently amended (now every five years) for a given number of years, they would once again become law if current amendments, such as the 1990 farm bill, were to lapse or new legislation was not enacted at the termination of the amendment.

Permitted Acres—The maximum acreage of a crop which may be planted for harvest. It is determined by multiplying the crop **base acreage** by the acreage reduction requirement (announced by the **Commodity Credit Corporation**) less any diversion requirement (e.g., **acreage reduction program**) applicable.

Point-Source Pollution—Pollution originating from a distinct source, such as the outflow from a pipe or concentrated animal production facilities.

Pound—453.59 grams.

Price Support—The Agricultural Act of 1949 directed the secretary of agriculture to support prices of some commodities within a specific range or at a specific level of parity. Support was mandatory for the **basic commodities** and for several other commodities. In subsequent legislation the secretary of agriculture was given discretion in setting the level of price support for commodities. Support is authorized by law but is or is not undertaken at the discretion of the secretary. In determining whether support is to be undertaken and at what level, the secretary must consider the following factors: (1) supply of the commodity in relation to demand; (2) levels at which prices of other commodities are being supported; (3) the availability of funds; (4) the perishability of the commodity; (5) the importance of the commodity to agriculture and the national economy; (6) the ability to dispose of stocks which would be acquired through price support operations; (7) the need for offsetting temporary losses from export markets; and (8) the ability and willingness of producers to keep supplies in line with demand.

Price Support Loan—*See* **nonrecourse loan.**

Prices Paid Index—*See* **parity index.**

Prices Received Index—An index of average prices received by farmers for fifty-five of the most important products sold by farmers (1910–14 = 100).

Producer Subsidy Equivalent—An estimate of the total monetary value of government assistance to producers from agricultural policies affecting both output and input markets. Included here are all policy measures affecting producer and consumer prices, responsible for direct transfers to producers without raising commodity prices to consumers, lowering input costs, resulting in research and development expenditures, and/or providing tax concessions to farmers.

Production Credit Associations (PCAs)—Over 400 local associations providing short-term credit directly to farmers and farm-related businesses from funds provided by the **Intermediate Credit Banks.** *See* **Farm Credit Administration.**

Program Allocation Factor—The program allocation factor is determined by dividing the **national program acreage** by the number of acres that the secretary of agriculture estimates will be harvested in the current year. The allocation factor is used for determining the **farm program acreage.** The Food, Agriculture, Conservation and Trade Act of 1990 removes this provision.

Program Yield—The crop yield of record for an individual farm that determines the level of production eligible for direct payments on that farm. Program yields are currently established at the average yield for the 1981–85 crops exclusive of the high and low years.

Protectionism—Use of tariff or nontariff barriers to trade or subsidies to shield domestic producers from foreign competition thus hampering the operation of normal competitive processes and fostering inefficient domestic production.

Public Law 480—Enacted in 1954 to expand foreign markets for U.S. agricultural products, combat hunger, and encourage economic development in developing countries. This law makes U.S. agricultural commodities available through low-interest, long-term credit under Title I of the act, and as donations for famine or other emergency relief under Title II of the act. Under Title I, the recipient country agrees to undertake agricultural development projects to improve its own food production or distribution. Title III authorizes "food for development" projects.

Public Policy—A course of action consciously chosen from among the available alternatives and designed to achieve a specified need or set of needs of the body politic, including specification of the means by which this course of action is to be implemented.

Purchase Agreement—Purchases of commodities by the government are means of supporting prices by reducing the supply in the market. When purchase agreement programs apply, the **Commodity Credit Corporation** is required to buy offered commodities from eligible producers. The transaction price is the **loan rate.** Purchase procedures supply price protection similar to loans except that the farmer does not receive payments under purchase programs until the commodity is delivered to the government. Purchases provide producers who do not have immediate need for cash or who cannot meet loan storage requirements a less complex form of price protection than the loan program. Purchase agreements require an advance formal agreement between the farmer and the Commodity Credit Corporation; other purchase programs do not require any advance arrangement.

For some commodities, notably dairy products, the government initiates purchase operations to ensure that market prices are maintained at the support level. Purchases of dairy products are made from processors rather than from producers. Purchase programs of this type are called direct purchase programs. They are not continuous programs but are instituted by the government when markets are depressed.

Quart—Dry equals 1.101 **liters.** Liquid equals 0.946 liters. An "imperial" quart equals 1.136 liters for both dry and liquid.

Quintal—One tenth of a **metric ton.**

Reciprocity—A principle of the **General Agreement on Tariffs and Trade (GATT)** that seeks an equality of trade-related concessions or benefits among all participants in a negotiation.

Release Price—A (market) price trigger at which a farmer may sell his **farmer-owned reserve** grain and repay the loan without penalty. Under the Agriculture and Food Act of 1981 the release price for wheat was to be established by the secretary of agriculture within a range of 140–160 percent of the support price. For corn the secretary was to establish the release price "as appropriate."

Reserves (Disaster)—The authority for the secretary of agriculture to maintain and dispose of disaster reserves was amended by the Food and Agriculture Act of 1977. If neither wheat, feed grains, nor soybeans are available through the **price support** program at a location where they may be economically used to lessen the impact of natural disasters, the secretary may purchase these commodities and other livestock forages through the **Commodity Credit Corporation.** Disaster reserves were initially authorized by the Agriculture and Consumer Protection Act of 1973.

Reserves (International)—The Food and Agriculture Act of 1977 encouraged the president to enter into negotiations with other nations to develop and maintain an international system for food reserves. These reserves were to be used to provide humanitarian food relief.

School Breakfast Program—A program authorized by the 1975 legislation providing financial and commodity assistance to schools that agree to serve nourishing breakfasts and to provide these meals free or at reduced or full prices depending on the income of the child's family.

School Lunch Program—A program to provide assistance to schoolchildren through direct commodity distribution, or cash subsidies. Over time this program has been expanded to encompass both breakfast and lunch. Meals are given to children from low-income households for free or at subsidized levels depending upon the income levels of their parents.

Section 22—Section of Public Law 320 (which amends the Agricultural Adjustment Act of 1933) passed in 1935 that authorizes the imposition of quotas or fees on imports of price-supported commodities when these measures are necessary to prevent imports from interfering with the operation of U.S. support programs on the products involved. The president is authorized to impose tariffs of up to 50 percent of the value of the imported products or quotas of up to 50 percent of the quantity of imports in a representative period. This authorization is currently in use to prevent imports of dairy products, peanuts, sugar, and tobacco.

Section 32—Section of Public Law 320 (which amends the Agricultural Adjustment Act of 1933) passed in 1935 which allows 30 percent of all customs receipts to be used for (1) encouraging exports of agricultural commodities, (2) encouraging domestic consumption of surplus agricultural commodities by the poor and by schoolchildren, and (3) reestablishing farmers' bargaining power. The funds may be used as export subsidies and payments to farmers and to cover costs of distribution of goods to charitable institutions and schools and direct to the needy.

Section 201—A provision of the U.S. Trade Act of 1974 as amended which empowers the president to raise import duties or impose import restrictions on goods entering the United States which injure or threaten to injure domestic industries producing like goods.

Section 301—A provision of the U.S. Trade Act of 1974 as amended which empowers the president to take all appropriate action, including retaliation, to obtain the removal of any act, policy, or practice of a foreign government which violates an international agreement or is unjustified, unreasonable, or discriminatory, and which burdens or restricts U.S. commerce.

Section 416—Section of the Agricultural Act of 1949 which authorizes overseas donations of surplus agricultural commodities acquired by the **Commodity Credit Corporation** as a result of its price-support operations. Prior to 1985, donations under this provision were limited to dairy products, wheat, and rice. The Food Security Act of 1985 expanded the list of commodities to include any food product acquired by the Commodity Credit Corporation. The Food Security Act of 1985 also mandated that certain minimum quantities of uncommitted stocks of grain, oilseeds, and dairy products be distributed as long as surpluses persist.

Sequester—Across-the-board **Gramm-Rudman-Hollings** spending cuts.

Set-Aside Acreage—Acreage withdrawn from crop production and devoted to approved **conserving uses.** The program is voluntary, but producers must comply when a set-aside program is in effect in order to be eligible for **nonrecourse loans** or **deficiency payments.**

Sheet and Rill Erosion—Sheet erosion is of a type that leads to a generally uniform removal of topsoil over all of a field as a result of strong rains. If the rainwater moves fast enough, the land will be scoured and soil removed such that small channels, or rills, remain. This is rill erosion.

Slippage—A term used to refer to the case of farmers thwarting the government's attempt to reduce output through voluntary cropland diversion programs. Participating farmers do so through diversion of inferior cropland and by substituting nonland inputs for land. In addition, nonparticipating farmers may bring new land into production in anticipation of higher prices as a result of the **acreage reduction program.** Slippage is calculated as the proportion of acreage set aside or put into a reserve program for which there is no corresponding reduction in production of the crops for which output reduction is sought.

Sluice Gate Price—A price fixed by the **Common Agricultural Policy** to restrict imports. Imports priced at less than this level are charged a tax equal to the difference between this price and the import price. Quantitative restrictions may also be imposed on such imports.

Social Costs—A term used to refer to the losses society suffers when a **Pareto optimal** state or, what is the equivalent in economics, a competitive equilibrium cannot be or is prevented from being achieved. Most frequently used in connection with an evaluation of one or more government policies that prevent achievement of a competitive equilibrium. Equivalent terms include *social loss, net social loss, welfare loss, net welfare loss,* and *deadweight loss.* All refer to welfare (income and/or utility) losses to some group or groups recovered by no other individual or group. Social costs are calculated by summing (1) net income losses to producers, (2) losses to consumers in terms of utility forgone, and (3) taxpayer costs incurred to prevent attainment of a competitive equilibrium.

Sodbuster Legislation—Legislation provided by the Food Security Act of 1985 designed to curb or discourage the plowing up of land now planted to grasses to grow erosive crops. This legislation denies all program benefits to farmers who grow supported crops on highly erodible new cropland.

Soft Currency—A currency that may not be exchanged for the currency of another country freely or without restrictions. *Compare* **hard currency.**

Soil Bank—A long-term **land retirement program** authorized by the Agricultural Act of 1956, reestablished by the Cropland Adjustment Program of 1965, and again by the Food Security Act of 1985. *See also* **conservation reserve program.**

Soil Conservation Service (SCS)—An agency of the U.S. Department of Agriculture responsible for developing and administering national soil and water programs in cooperation with landowners, operators, and others.

Southern Cone Common Market—An agreement signed in 1991 by Argentina, Brazil, Paraguay and Uruguay providing for a common external tariff and free movement of goods and services between member nations by 1994.

Special Drawing Rights (SDR)—International reserve assets created by the

International Monetary Fund and allocated to individual member nations. Within conditions set by the International Monetary Fund, SDRs can be used by a nation with a deficit in its balance of international payments to settle debts with another nation or with the International Monetary Fund.

Special Milk Program—A program to encourage children to drink milk by reimbursing participating schools and institutions for part of the cost of milk served. Children are offered this milk free or at reduced or full prices depending on the income of the child's family.

STABEX—A compensatory finance scheme established under the **Lome Convention of 1975** aimed at stabilizing the export earnings of the ACP states. It is a commodity-specific arrangement providing compensation to ACP states associated with the **European Economic Community** for shortfalls in individual agricultural commodity export earnings. Export earning shortfalls are calculated for each commodity separately, so excess export earnings of one commodity do not offset shortfalls in export earnings of another commodity.

Stone—14 pounds.

Summer Food Service Program—A program providing meal service to children from areas where poor economic conditions exist. Sponsors must qualify for assistance under this program by operating programs in areas where at least one-third of the children would qualify under the **School Lunch** or **School Breakfast** programs, or they must provide meals as part of an organized program for children enrolled in camps.

Super 301—A provision of the Omnibus Trade Act of 1988 which requires the U.S. Trade Representative to name specific countries engaging in unfair trade practices, allows 18 months of negotiations during which the named countries and the United States must work to eliminate the unfair practices, and if negotiations fail, the president is obligated to institute some form of retaliatory trade action against the named country or countries.

Swampbuster Legislation—Same as for **sodbuster legislation** but for wetlands.

Target Price—The price established by law which serves as a trigger point for **deficiency payments** to U.S. farmers if the five-month-average market price for the commodity falls below that level. In the **European Economic Community,** this is the price at which the authorities aim to manage the market.

Targeted Export Assistance Program (TEAP)—A program authorized by the Food Security Act of 1985 whereby priority assistance may be provided producers of commodities who have been found to have suffered from unfair trade practices under **section 301** of the U.S. Trade Act of 1974 or who have suffered from retaliatory actions of various nations.

Tariffication—The process of conversion of **non-tariff barriers to trade** to their tariff equivalent. This was one of the concessions sought by the United States in the Uruguay Round of the GATT negotiations.

Temporary Emergency Food Assistance Program (TEFAP)—A program authorized in December 1981 to allow donation of **Commodity Credit Corporation** owned commodities to states in amounts determined by the number of unemployed and needy persons. This program was originally established as a means of disposing of the growing government stocks of dairy products. The food distributed under this program is given free by charitable organizations to eligible recipients.

Threshold Price—A price fixed under the **Common Agricultural Policy** at a level that will bring the selling price of imported products up to the level of the **target price** in the region of the **European Economic Community** with the least supplies. It is equal to the target price less transportation costs.

Ton (Long)—One long ton weighs 2,240 **pounds**, or 1.016 **metric tons**.

Ton (Metric)—One **metric ton** weighs 2,204.622 **pounds**, or 10 **quintals**.

Ton (Short)—One short ton (or 1 ton) weighs 2,000 **pounds**, or 0.907 **metric tons**.

Triple-Base Plan—A program mandated by the Omnibus Budget Reconciliation Act of 1990 under which a producer's **base acreage** would be divided into three portions: (1) program acres, (2) **flexible acres**, and (3) **conserving-use** acres. The conserving-use acres are subject to the restrictions that normally apply to **acreage reduction program** acres. The program acres must be planted to a program crop for which **deficiency payments** will be paid. The flexible acres can be planted to any program crop, oilseed crop, or nonprogram crop other than fruits and vegetables. Production on the flexible acres is not eligible for deficiency payments, **nonrecourse loans**, or **marketing loans**.

Under the 1990 act, a 15 percent triple-base plan was mandated—that is, 15 percent of the farmer's base acres must be "flexed." Optionally, a farmer may flex an additional 10 percent of his base acres without losing any base acres in subsequent years.

Two-Price or **Two-Tier Plan**—A pricing scheme that differentiates between the domestic and export markets by establishing prices for a commodity such that domestic consumers pay a higher price for the commodity than do foreign consumers. As with any price discrimination scheme, this is of advantage to domestic producers so long as domestic demand is less elastic than is foreign demand, and so long as foreign buyers are prevented from reselling the commodity in the domestic market.

Underplanting (50/92 and 0/92) Provision—A provision of the Food Security Act of 1985 (the Findley amendment) under which farmers planting between 50 and 92 percent (later changed to between 0 and 92 percent) of their **base acreage** to the program commodity and devoting the remaining base acres to a **conserving use** are eligible to receive **deficiency payments** on 92 percent of the base acreage. This provision is available to cotton and rice producers and during 1986–88 was available to wheat and feed grain producers.

United Nations Conference on Trade and Development (UNCTAD)—A group within the United Nations which focuses special attention on international economic relations and on measures that might be taken by developed countries to accelerate economic development in developing countries.

Value-Added Tax (VAT)—A tax levied on processors and merchants on the amount by which they increase the value of items they purchase and sell. It is the most common form of taxation in the **European Economic Community**.

Variable Levy—An import tax which varies as frequently as import prices vary imposed to ensure that the import price plus the levy will equal a predetermined minimum import or **sluice gate price**. The variable levy is used extensively by the **Common Agricultural Policy** of the **European Economic Community.**

Vertical Integration—A form of market control under which a single organization controls, via ownership or contractual arrangement, two or more adjacent stages in the production and/or marketing of a commodity.

Voluntary Export Restraint—An agreement between an exporter and one or more importers to voluntarily limit exports to an agreed-upon quantity.

Water Quality Protection Program—A program authorized by the Food, Agriculture, Conservation, and Trade Act of 1990 to provide incentive payments to farmers of up to $3,500 per year to adopt water quality improvement practices on cropland acres near wellheads, in areas inhabited by threatened or endangered species, or where agricultural production poses a threat to underground or surface water quality. Cost sharing funds are also available, with priority given to producers who improve wildlife habitat.

Weights and Measures—*See* **acre, bushel, centner, hectare, kilogram, kilometer, liter, meter, pound, quart, quintal, stone,** and **ton.**

Wetlands Reserve Program—A program authorized by the Food, Agriculture, Conservation, and Trade Act of 1990 under which farmers voluntarily contract (via a bid system) to take wetlands out of production for a period of thirty years or more. Priority will be given to enrolling wetlands that enhance wildlife habitat.

Women, Infants, and Children (WIC) Program—This program combines direct commodity distribution with nutrition education. Targeted recipients are single mothers with very low incomes and preschool children.

World Bank—*See* **International Bank for Reconstruction and Development.**

World Price—The price at which a fungible commodity moves in international markets. Often interpreted as the c.i.f. **(cost, insurance, and freight)** price of a commodity with specific characteristics at the main port of entry of a major importing country.

BIBLIOGRAPHY

Agricultural Commodity Policy

Bergland, Bob S. 1981. *A Time to Choose: Summary Report on the Structure of Agriculture*. Washington, D.C.: U.S. Department of Agriculture.

Binswanger, Hans P., and Vernon W. Ruttan, eds. 1978. *Induced Innovation, Technology, Institutions, and Development*. Baltimore: Johns Hopkins University Press.

Brandow, G. E. 1961 *Interrelations among Demands for Farm Products and Implications for Control of Market Supply*. Agricultural Experiment Station Bulletin 680. University Park: Pennsylvania State University.

_____. 1977. "Policy for Commercial Agriculture." *A Survey of Agricultural Economics Literature*. vol. 1, edited by Lee R. Martin. Minneapolis: University of Minnesota Press.

Chisholm, A. H., and R. Tyers, eds. 1982. *Food Security: Theory, Policy, and Perspectives from Asia and the Pacific Rim*. Lexington, Mass.: Lexington Books.

Cochrane, Willard W. 1958. *Farm Prices: Myth and Reality*. Minneapolis: University of Minnesota Press.

_____. 1969. *The World Food Problem*. New York: Thomas Y. Crowell Co.

Cochrane, Willard W., and M. E. Ryan. 1976. *American Farm Policy, 1948–73*. Minneapolis: University of Minnesota Press.

Edwin, Ed. 1974. *Feast or Famine: Food, Farming, and Farm Politics in America*. New York: Charterhouse.

Fishel, Walter L. ed. 1971. *Resource Allocation in Agricultural Research*. Minneapolis: University of Minnesota Press.

Gardner, Bruce L. 1979. *Optimal Stockpiling of Grains*. Lexington, Mass.: Lexington Books.

_____. 1981. *The Governing of Agriculture*. Lawrence: Regents Press of Kansas.

_____. 1981. *The Economics of Agricultural Policies*. New York: Macmillan Co.

Hadwiger, Don F., and Ross B. Talbot. 1982. *Food Policy and Farm Programs*. Proceedings of the Academy of Political Science. vol. 34, no. 3. Montpelier, Vt.: Capital City Press.

Halcrow, H. 1977. *Food Policy for America*. New York: McGraw-Hill Book Co.

_____. 1984. *Agricultural Policy Analysis*. New York: McGraw-Hill Book Co.

Hallett, Graham. 1981. *The Economics of Agricultural Policy*. 2d ed. New York: John Wiley and Sons.

Hathaway, Dale E. 1963. *Government and Agriculture*. New York: Macmillan Co.
_____. 1964. *Problems of Progress in the Agricultural Economy*. Chicago: Scott, Foresman and Co.

Heady, E. O. 1962. *Agricultural Policy under Economic Development*. Ames: Iowa State University Press.

Heady, Earl O., Howard G. Diesslin, Harald R. Jensen, and Glenn L. Johnson, eds. 1958. *Agricultural Adjustment Problems in a Growing Economy*. Ames: Iowa State College Press.

Heady, Earl O., and Larry R. Whiting. 1975. *Externalities in the Transformation of Agriculture: Distribution of Benefits and Costs from Development*. Ames: Iowa State University Press.

Hildreth, R. J., ed. 1968. *Readings in Agricultural Policy*. Lincoln: University of Nebraska Press.

Hillman, J. S., ed. 1984. *United States Agricultural Policy for 1985 and Beyond*. Tucson: University of Arizona Press.

Hoos, Sidney, ed. 1979. *Agricultural Marketing Boards—An International Perspective*. Cambridge, Mass.: Ballinger Publishing Co.

Iowa State University Center for Agricultural Adjustment. 1959. *Problems and Policies of American Agriculture*. Ames: Iowa State University Press.

Iowa State University Center for Agricultural and Economic Development. 1961a. *Adjustments in U.S. Agriculture: A National Basebook*. Ames: Iowa State University Press.

_____. 1961b. *Goals and Values in Agricultural Policy*. Ames: Iowa State University Press.

_____. 1963. *Farm Goals in Conflict*. Ames: Iowa State University Press.

_____. 1969. *Food Goals, Future Structural Changes, and Agricultural Policy*. Ames: Iowa State University Press.

Johnson, D. Gale. 1947. *Forward Prices for Agriculture*. Chicago: University of Chicago Press.

_____. 1973. *World Agriculture in Disarray*. New York: Macmillan.

_____. ed. 1981. *Food and Agricultural Policy for the 1980s*. Washington, D.C.: American Enterprise Institute.

Johnson, D. Gale, and G. Edward Schuh, eds. 1983. *The Role of Markets in the World Food Economy*. Boulder: Westview Press.

Johnson, G. L., and C. L. Quance, eds. 1972. *The Overproduction Trap in U.S. Agriculture*. Baltimore: Johns Hopkins University Press.

Johnson, Paul R. 1984. *The Economics of the Tobacco Industry*. New York: Praeger.

Knutson, Ronald D., J. B. Penn, and William T. Boehm. 1983. *Agricultural and Food Policy*. Englewood Cliffs, N.J.: Prentice-Hall.

Lowenberg-DeBoer, James. 1986. *The Microeconomic Roots of the Farm Crisis*. New York: Praeger.

Manchester, Alden C. 1983. *The Public Role in the Dairy Economy*. Boulder: Westview Press.

O'Rourke, A. Desmond. 1978. *The Changing Dimensions of U.S. Agricultural Policy*. Englewood Cliffs, N.J.: Prentice-Hall.

Paarlberg, Don, ed. 1977. *Food and Agricultural Policy*. Washington, D.C.: American

Enterprise Institute.

Pasour, E. C., Jr. 1990. *Agriculture and the State: Market Processes and Bureaucracy.* New York: Holmes & Meier Publishers.

Rausser, G., and K. R. Farrell, eds. 1984. *Alternative Agricultural and Food Policies and the 1985 Farm Bill.* Berkeley and Los Angeles: University of California Press.

Ritson, Christopher. 1982. *Agricultural Economics: Principles and Policy.* Boulder: Westview Press.

Roberts, Ivan, Graham Love, Heather Field, and Nico Klijn. 1989. *U.S. Grain Policies and the World Market.* Australian Bureau of Agricultural and Resource Economics. Policy Monograph no. 4. Canberra: Australian Government Publishing Service.

Robinson, Kenneth L. 1989. *Farm and Food Policies and Their Consequences.* Englewood Cliffs, N.J.: Prentice-Hall.

Runge, C. Ford, ed. 1985. *The Future of the North American Granary: Politics, Economics, and Resource Constraints in North American Agriculture.* Ames: Iowa State University Press.

Ruttan, Vernon W. 1982. *Agricultural Research Policy.* Minneapolis: University of Minnesota Press.

Ruttan, Vernon W., Arley D. Waldo, and James P. Houck. 1969. *Agricultural Policy in an Affluent Society.* New York: Norton.

Schertz, Lyle P., et al. 1979. *Another Revolution in U.S. Farming?* Washington, D.C.: U.S. Department of Agriculture.

Schickele, Rainer. 1954. *Agricultural Policy.* New York: McGraw-Hill Book Co.

Schmitz, Andrew, Alex F. McCalla, Donald O. Mitchell, and Colin A. Carter. 1981. *Grain Export Cartels.* Cambridge, Mass.: Ballinger Publishing Co.

Strange, Marty. 1988. *Family Farming: A New Economic Vision.* Lincoln: University of Nebraska Press.

Sumner, Daniel A., ed. 1988. *Agricultural Stability and Farm Programs: Concepts, Evidence, and Implications.* Boulder: Westview Press.

Sumner, Daniel A. and Julian M. Alston. 1985. *Removal of Price Supports and Supply Controls for U.S. Tobacco.* National Planning Association, Food and Agricultural Committee. NPA Report no. 220. Washington, D.C.

Timmer, C. Peter, Walter P. Falcon, and Scott R. Person. 1983. *Food Policy Analysis.* Baltimore: Johns Hopkins University Press.

Tomek, William G., and Kenneth L. Robinson. 1972. *Agricultural Product Prices.* Ithaca: Cornell University Press.

Tweeten, Luther. 1979. *Foundations of Farm Policy.* 2d ed., rev. Lincoln: University of Nebraska Press.

World Bank. 1986. *World Development Report 1986.* International Bank for Reconstruction and Development. New York: Oxford University Press.

Agricultural Development

Clark, Colin. 1957. *The Conditions of Economic Progress.* New York: St. Martin's

Press.

Clark, Colin and M. R. Haswell. 1964. *The Economics of Subsistence Agriculture.* New York: St. Martin's Press.

Meier, Gerald M. 1984. *Leading Issues in Economic Development.* 4th ed. New York: Oxford University Press.

Schultz, T. W. 1964. *Transforming Traditional Agriculture.* New Haven: Yale University Press.

Todaro, Michael P. 1981. *Economic Development in the Third World.* 2d ed. New York: Longman.

Union of International Associations. 1989. *Yearbook of International Organizations, 1988/89.* 25th ed. Munich: K. G. Saur.

Valdes, Alberto. 1981. *Food Security for Developing Countries.* Boulder: Westview Press.

Agricultural History in the United States

Atack, Jeremy, and Fred Bateman. 1987. *To Their Own Soil: Agriculture in the Antebellum North.* Ames: Iowa State University Press.

Baker, Gladys L., Wayne D. Rasmussen, Vivian Wiser, and Jane M. Porter. 1963. *Century of Service: The First 100 Years of the United States Department of Agriculture.* Washington, D.C.: U.S. Department of Agriculture. Economic Research Service, Agricultural History Branch.

Benedict, M. R., and O. C. Stine. 1956. *The Agricultural Commodity Programs.* New York: Twentieth Century Fund.

Benedict, Murray. 1953. *Farm Policies of the United States: 1790–1950.* New York: Twentieth Century Fund.

Berger, Michael. 1979. *The Devil Wagon in God's Country.* Hamden, Conn.: Archon Books.

Berry, Wendell. 1977. *The Unsettling of America: Culture and Agriculture.* San Francisco: Sierra Club Books.

Bidwell, Percy Wells, and John I. Falconer. 1925. *History of Agriculture in the Northern United States, 1620-1860.* Washington, D.C.: Carnegie Institution of Washington.

Boorstein, Daniel J. 1973. *The Americans: The Democratic Experience.* New York: Vintage.

Bowers, Douglas E., Wayne D. Rasmussen, and Gladys L. Baker. 1984. "History of Agricultural Price-Support and Adjustment Programs, 1933–84." U.S. Department of Agriculture. Economic Research Service. Agriculture Information Bulletin no. 485. December.

Campbell, Christiana McFadyen. 1982. *The Farm Bureau and the New Deal.* Ames: Iowa State University Press.

Carson, Rachael. 1962. *Silent Spring.* Boston: Houghton Mifflin Co.

Christenson, Reo M. 1959. *The Brannan Plan.* Ann Arbor: University of Michigan Press.

Cochrane, Willard W. 1979. *The Development of American Agriculture: A Historical*

Analysis. Minneapolis: University of Minnesota Press.

Davis, Chester C. 1950. "The Development of Agricultural Policy since the End of the World War." In U.S. Department of Agriculture, *Farmers in a Changing World. Yearbook of Agriculture, 1950*. Washington, D.C.

Drache, Hiram M. 1964. *The Day of the Bonanza*. North Dakota Institute for Regional Studies. Fargo: Lund Press.

_____. 1976. *Beyond the Furrow*. Danville, Ill.: Interstate Printers and Publishers.

Edwards, Everett E. 1940. "American Agriculture—The First 300 Years." In U. S. Department of Agriculture, *Farmers in a Changing World: Yearbook of Agriculture, 1940*. Washington, D.C.

Fite, Gilbert C. 1964. *American Agriculture and Farm Policy since 1900*. New York: Macmillan Co.

_____. 1981. *American Farmers, The New Minority*. Bloomington: Indiana University Press.

Fornari, Harry. 1973. *Bread upon the Waters*. Nashville: Aurora Publishers.

Gates, Paul W. 1960. *The Farmer's Age: Agriculture 1815–1860*. New York: Holt, Rinehart and Winston.

Giglio, James N. 1987. "The New Frontier Agricultural Policy: The Commodity Side, 1961–1963." *Agricultural History* 61 (3): 53–70.

Gray, L. C., and Esther K. Thompson. 1933. *History of Agriculture in the Southern United States to 1860*. vols 1 and 2. Washington, D.C.: Carnegie Institution.

Griswold, A. Whitney. 1948. *Farming and Democracy*. New York: Harcourt, Brace and Co.

Hadwiger, Don F. and Ross B. Talbot. 1965. *Pressures and Protests: The Kennedy Farm Programs and Wheat Referendum of 1963*. San Francisco: Chandler Publishing.

Hightower, Jim. 1972. *Hard Tomatoes, Hard Times: The Failure of the Land Grant College Complex*. Washington, D.C.: Agribusiness Accountability Project.

Kirkendall, Richard S. 1966. *Social Scientists and Farm Politics in the Age of Roosevelt*. Columbia: University of Missouri Press.

Kirschner, Don. 1970. *City and Country: Rural Responses to Urbanization in the 1920s*. Westport, Conn.: Greenwood Publishing Co.

Matusow, Allen J. 1967. *Farm Policies and Politics in the Truman Years*. Cambridge: Harvard University Press.

Morgan, Dan. 1979. *Merchants of Grain*. New York: Viking Press.

Opie, John. 1987. *The Law of the Land*. Lincoln: University of Nebraska Press.

Peterson, Trudy Huskamp. 1979. *Agricultural Exports, Farm Income, and the Eisenhower Administration*. Lincoln: University of Nebraska Press.

Rasmussen, Wayne D., and Gladys L. Baker. 1972. *The Department of Agriculture*. New York: Praeger.

Robbins, Roy M. 1962. *Our Landed Heritage: The Public Domain, 1776–1936*. Lincoln: University of Nebraska Press.

Saloutos, Theodore. 1982. *The American Farmer and The New Deal*. Ames: Iowa State University Press.

Schlebecker, John T. 1975. *Whereby We Thrive—A History of American Farming, 1607–1972*. Ames: Iowa State University Press.

Schultz, T. W. 1945. *Agriculture in an Unstable Economy*. New York: McGraw-Hill Book Co.

Shannon, Fred A. 1945. *The Farmer's Last Frontier: Agriculture, 1860–1897*. New York: Farrar and Rinehart.

Shover, John. 1976. *First Majority—Last Minority: The Transforming of Rural Life in America*. DeKalb: Northern Illinois University Press.

Sinclair, Upton. 1905. *The Jungle*. New York: Grosset and Dunlap.

Smith, Maryanna S. 1979. "Chronological Landmarks in American Agriculture." U.S. Department of Agriculture. Economics, Statistics, and Cooperatives Service. Agriculture Information Bulletin no. 425. May.

Soth, Lauren. 1957. *Farm Trouble*. Princeton: Princeton University Press.

Steinbeck, John. 1939. *The Grapes of Wrath*. New York: Viking Press.

Taylor, George Rogers. 1968. *The Transportation Revolution, 1815–1860*. New York: Harper and Row.

Trager, James. 1975. *The Great Grain Robbery*. New York: Ballantine.

True, A. C. 1928. "A History of Agricultural Extension Work in the United States, 1785–1923." U.S. Department of Agriculture. Miscellaneous Publication no. 15.

_____. 1929. "A History of Agricultural Education in the United States 1785–1925." U.S. Department of Agriculture. Miscellaneous Publication no. 36.

_____. 1937. "A History of Agricultural Experimentation and Research in the United States, 1607–1925." U.S. Department of Agriculture. Miscellaneous Publication no. 251.

Wellford, Harrison. 1973. *Sowing the Wind*. New York: Grossman Publishers.

Wik, Reynold. 1972. *Henry Ford and Grass-Roots America*. Ann Arbor: University of Michigan Press.

Agricultural Organizations

Dyson, Lowell K. 1986. *Farmers' Organizations*. New York: Greenwood Press.

Guither, Harold. 1980. *The Food Lobbyists*. Lexington, Mass.: Lexington Books.

Shannon, Fred A. 1957. *American Farmers Movements*. Princeton: D. Van Nostrand.

Agricultural Trade

Anderson, Kym, and Yujiro Hayami. 1986. *The Political Economy of Agricultural Protection*. Sydney: George Allen and Unwin. Sydney.

Goldin, Ian, and Odin Knudsen. 1990. *Agricultural Trade Liberalization: Implications for Developing Countries*. Organization for Economic Cooperation and Development. Paris: OECD.

Houck, James P. 1986. *Elements of Agricultural Trade Policies*. New York: Macmillan Company.

Johnson, D. Gale, Kenzo Hemini, and Pierre Lardinois. 1985. *Agricultural Policy and Trade: Adjusting Domestic Programs in an International Framework*. New

York: New York University Press.

Krugman, Paul R., and Maurice Obstfeld. 1988. *International Economics*. Glenview, Ill.: Scott, Foresman and Co.

McCalla, Alex F., and Timothy E. Josling. 1985. *Agricultural Policies and World Markets*. New York: Macmillan Co.

Miller, Geoff. 1986. *The Political Economy of International Agricultural Policy Reform*. Department of Primary Industry. Canberra: Australian Government Publishing Service.

Organization for Economic Cooperation and Development. 1987. *National Policies and Agricultural Trade*. Paris: OECD.

_____. 1990. *Modelling the Effects of Agricultural Policies*. Economic Studies no. 13. Paris: OECD.

Paarlberg, Philip L., and Jerry A. Sharples. 1984. "Japanese and European Community Agricultural Trade Policies: Some U.S. Strategies." ERS/USED. Foreign Agricultural Economic Report. no. 204. August.

Parikh, Kirit S., Gunther Fischer, Klaus Frohberg, and Odd Gulbrandsen. 1988. *Towards Free Trade in Agriculture*. Dordrecht, the Netherlands: Martinus Nijhoff Publishers.

Sorenson, V. L. 1975. *International Trade Policy: Agriculture and Development*. East Lansing: Michigan State University Press.

Tarditi, Secondo, Kenneth J. Thomson, Pierpaolo Pierani, and Elisabetta Croci-Angelini. 1989. *Agricultural Trade Liberalization and the European Community*. Oxford: Clarendon Press.

Valdes, Alberto, and Zietz, Joachim. 1980. "Agricultural Protection in OECD Countries: Its Cost to Less Developed Countries." Research Report 21. Washington, D.C.: International Food Policy Research Institute.

Vousden, Neil. 1990. *The Economics of Trade Protection*. New York: Cambridge University Press.

Environment and Agriculture

Anderson, Terry L. 1983. *Water Crisis: Ending the Policy Drought*. Baltimore: Johns Hopkins University Press.

Batie, S., and Robert G. Healy. 1980. *The Future of American Agriculture as a Strategic Resource*. Washington, D.C.: Conservation Foundation.

Headley, J. C., and J. N. Lewis. 1967. *The Pesticide Problem: An Economic Approach to Public Policy*. Baltimore: Johns Hopkins University Press.

Phipps, Tim T., Pierre R. Crosson, and Kent A. Price, eds. 1986. *Agriculture and the Environment*. Washington, D.C.: Resources For The Future.

Whittlesey, Norman K., ed. 1986. *Energy and Water Management in Western Irrigated Agriculture*. Boulder: Westview Press.

Western European Agriculture

Buckwell, Allen E., David R. Harvey, Kenneth J. Thomson, and Kevin A. Parton. 1982. *The Costs of the Common Agricultural Policy*. London: Croom Helm.

Bureau of Agricultural Economics. 1985. *Agricultural Policies in The European Community: Their Origins, Nature and Effects on Production and Trade*. Canberra: Australian Government Publishing Service.

Clout, Hugh. 1984. *A Rural Policy for the EEC*. New York: Methuen.

Commission of the European Community. 1983. *The Agricultural Situation in the Community: 1982 Report*. Brussels: European Economic Community.

Fennell, Rosemary. 1979. *The Common Agricultural Policy of the European Community*. London: Crosby Lockwood Staples.

Harris, S., A. Swinback, and G. Wilkinson. 1983. *The Food and Farm Policies of the European Community*. New York: John Wiley and Sons.

Hill, Brian E., and K. A. Ingersent. 1982. *An Economic Analysis of Agriculture*. 2d ed. London: Heinemann Educational Books.

Lodge, Juliet. 1983. *The European Community*. Phoenix: Oryx Press.

Marsh, John S., and Pamela J. Swanney. 1980. *Agriculture and the European Community*. Studies on Contemporary Europe no. 2. Boston: George Allen and Unwin.

Nicholson, Frances, and Roger East. 1986. *From The Six to The Twelve*. Chicago: St. James Press.

Organization for Economic Cooperation and Development. 1987. National Policies and Agricultural Trade. Paris: OECD.

Slicher Von Bath, B. H. 1963. *The Agrarian History of Western Europe, A.D. 500-1850*. London: Arnold.

Tracy, Michael. 1982. *Agriculture in Western Europe: Challenge and Response 1880-1980*. 2d ed. London: Granado.

Welfare Economics

Baumol, W. J. and W. E. Oats. 1975. *The Theory of Environmental Policy*. Englewood Cliffs, N.J.: Prentice-Hall.

Buchanan, James M., and Robert D. Tollison, eds. 1972. *Theory of Public Choice*. Ann Arbor: University of Michigan Press.

Buchanan, James M., Robert D. Tollison, and Gordon Tullock, eds. 1980. *Toward a Theory of the Rent-Seeking Society*. College Station: Texas A & M University Press.

Collard, David. 1978. *Altruism and Economy: A Study in Non-Selfish Economics*. New York: Oxford University Press.

Graaff, J. de V. 1957. *Theoretical Welfare Economics*. London: Cambridge University Press.

Henderson, James M., and Richard E. Quandt. 1980. *Microeconomic Theory*. 3d ed. New York: McGraw-Hill Book Co.

Just, Richard E., Darrell L. Hueth, and Andrew Schmitz. 1982. *Applied Welfare Analysis and Public Policy*. Englewood Cliffs, N.J.: Prentice-Hall.

Mishan, E. J. 1971. *Cost-Benefit Analysis*. Boston: George Allen and Unwin.

Pareto, Vilfredo. [1927] 1971. *Manual of Political Economy*. Translated by Ann S. Schwier. Edited by Ann S. Schwier and Alfred N. Page. New York: A. M. Kelly.

Sen, A. K. 1982. *Choice, Welfare, and Measurement*. Cambridge: MIT Press.

Tinbergen, Jan. 1955. *On the Theory of Economic Policy*. 2d ed. Amsterdam: North-Holland Publishing.

"Political" Economics

Anderson, James E. 1979. *Public Policy-Making*. 2d ed. New York: Holt, Rinehart and Winston.

Bartlett, Randall. 1973. *Economic Foundations of Political Power*. New York: Macmillan Co.

Beard, Charles A. 1934. *The Economic Basis of Politics*. New York: Alfred A. Knopf.

Breton, A. 1974. *The Economic Theory of Representative Government*. Chicago: Aldine.

Browne, William P. 1988. *Private Interests, Public Policy, and American Agriculture*. Lawrence: University Press of Kansas.

Downs, Anthony. 1957. *An Economic Theory of Democracy*. New York: Harper and Row.

Greevy, David U., Chadwick R. Gore, and Marvin I. Weinberger, eds. 1984. *The PAC Directory: Book II The Federal Committees*. PAC Research, Inc. Cambridge, Mass: Ballinger Publishing Co.

Hadwiger, Don and William P. Browne. 1978. *The New Politics of Food*. Lexington, Mass.: Lexington Books.

Hardin, Garrett and John Baden, eds. 1977. *Managing the Commons*. San Francisco: W. H. Freeman.

Jones, Charles O. 1977. *An Introduction to the Study of Public Policy*. 2d ed. North Scituate, Mass.: Duxbury Press.

Lindblom, Charles E. 1968. *The Policy-Making Process*. Englewood Cliffs, N.J.: Prentice-Hall.

_____. 1977. *Politics and Markets*. New York: Basic Books.

Madden, J. P., and D. E. Brester, eds. 1970. *A Philosopher among Economists—Selected Works of John M. Brewster*. Philadelphia: J. T. Murphy Co.

Mishan, E. J. 1982. *What Political Economy Is All About*. London: Cambridge University Press.

Myrdal, Gunnar. 1944. *An American Dilemma*. New York: Harper and Row.

_____. 1958. *Value in Social Policy*. Edited by Paul Streeton. New York: Harper Bros.

_____. 1960. *Beyond the Welfare State*. New Haven: Yale University Press.

Niskanen, William. 1971. *Bureaucracy and Representative Government*. Chicago:

Aldin.

Rivlin, Alice M. 1971. *Systematic Thinking for Social Action*. Washington, D.C.: Brookings Institution.

Schmid, Allan A. 1978. *Property, Power and Public Choice*. New York: Praeger.

Sen, A. K. 1970. *Collective Choice and Social Welfare*. San Francisco: Holden-Day.

Talbot, Ross B., and Don F. Hadwiger. 1968. *The Policy Process in American Agriculture*. San Francisco: Chandler Publishing Co.

Whynes, David K., and Roger A. Bowles. 1981. *The Economic Theory of the State*. New York: St. Martin's Press.

INDEX